Applied Probability and Statistics

BAILEY · The Elements of Stochastic Processes with Applications to the Natural Sciences
BARTHOLOMEW · Stochastic Models for Social Processes, *Second Edition*
BENNETT and FRANKLIN · Statistical Analysis in Chemistry and the Chemical Industry
BHAT · Elements of Applied Stochastic Processes
BLOOMFIELD · Fourier Analysis of Time Series: An Introduction
BOX and DRAPER · Evolutionary Operation: A Statistical Method for Process Improvement
BROWNLEE · Statistical Theory and Methodology in Science and Engineering, *Second Edition*
BURY · Statistical Models in Applied Science
CHERNOFF and MOSES · Elementary Decision Theory
CHOW · Analysis and Control of Dynamic Economic Systems
CLELLAND, deCANI, BROWN, BURSK, and MURRAY · Basic Statistics with Business Applications, *Second Edition*
COCHRAN · Sampling Techniques, *Second Edition*
COCHRAN and COX · Experimental Designs, *Second Edition*
COX · Planning of Experiments
COX and MILLER · The Theory of Stochastic Processes, *Second Edition*
DANIEL · Application of Statistics to Industrial Experimentation
DANIEL and WOOD · Fitting Equations to Data
DAVID · Order Statistics
DEMING · Sample Design in Business Research
DODGE and ROMIG · Sampling Inspection Tables. *Second Edition*
DRAPER and SMITH · Applied Regression Analysis
DUNN and CLARK · Applied Statistics: Analysis of Variance and Regression
ELANDT-JOHNSON · Probability Models and Statistical Methods in Genetics
FLEISS · Statistical Methods for Rates and Proportions
GNANADESIKAN · Methods for Statistical Data Analysis of Multivariate Observations
GOLDBERGER · Econometric Theory
GROSS and CLARK · Survival Distributions
GROSS and HARRIS · Fundamentals of Queueing Theory
GUTTMAN, WILKS and HUNTER · Introductory Engineering Statistics, *Second Edition*
HAHN and SHAPIRO · Statistical Models in Engineering
HALD · Statistical Tables and Formulas
HALD · Statistical Theory with Engineering Applications
HARTIGAN · Clustering Algorithms
HILDEBRAND, LAING and ROSENTHAL · Prediction Analysis of Cross Classifications
HOEL · Elementary Statistics, *Third Edition*
HOLLANDER and WOLFE · Nonparametric Statistical Methods
HUANG · Regression and Econometric Methods
JAGERS · Branching Processes with Biological Applications

continued on back

Planning of Experiments

A WILEY PUBLICATION IN APPLIED STATISTICS

Planning of Experiments

D. R. COX
Reader in Statistics
Birkbeck College
University of London

John Wiley & Sons, Inc.
New York · London · Sydney

13 14 15 16 17 18 19 20

ISBN 0 471 18183 8

Library of Congress Catalog Card Number: 58–13457

PRINTED IN THE UNITED STATES OF AMERICA

Preface

THIS BOOK IS AN ACCOUNT OF THE IDEAS UNDERLYING MODERN work on the statistical aspects of experimental design. I have tried, so far as is possible, to avoid statistical and mathematical technicalities, and to concentrate on a treatment that will be intuitively acceptable to the experimental worker, for whom the book is primarily intended. The great majority of the book requires no specialised knowledge for its understanding, although some parts are quite closely argued and are not necessarily easy reading. In particular, I have tried to deal in a careful though elementary way with a number of topics that are not given much prominence in other books on experimental design. Examples of such subjects are the justification and practical difficulties of randomization, the relation of covariance to randomized blocks and to the calculation of adjustments, the different kinds of factors that can occur in factorial experiments, the choice of the size of experiment, and the different purposes for which observations may be made.

Chapters 1 to 9 describe general concepts and such key designs as randomized blocks and Latin squares. The remaining chapters deal briefly with more advanced topics such as incomplete block designs, fractional replication, etc. The object with the complex designs has been to indicate what types are available, and when they are likely to be useful, rather than to give a comprehensive study of their properties. The worker who feels that these designs are of possible use to him will probably need to consult a more advanced textbook or to obtain expert help.

Most of the discussion, particularly in the first half of the book, is in terms of examples, drawn from many fields of application. Obtaining suitable examples has not been easy. Real experiments are seldom sufficiently simple to illustrate just the single point being dealt with, and considerable oversimplification has often been necessary, both in order to bring out clearly the point at issue and in order to make the experiment as a whole understandable to the general reader. A reference to an account of the original experiment has been given wherever possible.

Detailed methods of statistical analysis, and in particular the analysis of variance, receive only incidental mention. There are two reasons for this. To have included anything like an adequate account of the statistical techniques would have greatly lengthened the book and would, moreover, have meant duplicating material that is in any good modern textbook on statistical methods. Further, it seems to me that although the subject-matter of this book is, of course, intimately connected with analysis of variance, it is better for the person who has not yet gained a thorough mastery of advanced statistical methods to use modern designs for their intuitive reasonableness, rather than to regard them as in some way essentially dependent on analysis of variance.

The book is meant in the first place for private reading and reference. I hope, however, that it will be found useful as supplementary reading for courses in statistics. It would also be possible to use the book, in conjunction with a textbook on statistical analysis, to give a comprehensive nonmathematical course on the design and analysis of experiments. If this were done it would be necessary to select material rather carefully, depending on the special interests of the students, since the practical importance of different parts of the subject varies greatly between different applied fields.

I wish to thank Messrs. S. L. Anderson, J. D. Biggers, W. G. Cochran, and D. V. Lindley for their very helpful criticism. Of course I am solely responsible for errors of fact and judgement that remain.

I am grateful to the authors and to the *Biometrika* trust for permission to use parts of Tables 1 and 12 of *Biometrika Tables for Statisticians* by E. S. Pearson and H. O. Hartley and to the authors and to University College, London, for permission to print an extract from *Tables of Random Sampling Numbers,* Tracts for Computers XXIV, by M. G. Kendall and B. Babington Smith. Thanks are due similarly to Prof. W. G. Cochran and Miss G. M. Cox, and to John Wiley & Sons, Inc., for permission to print a portion of the tables of random permutations from their book, *Experimental Designs.*

Parts of Chapters 4 and 5 were written at Princeton University. Support there from the Office of Naval Research is acknowledged with thanks.

D. R. Cox

July, 1958

Contents

CHAPTER 1

Preliminaries

1.1 COMPARATIVE EXPERIMENTS

This book is about the planning of experiments in which the effects under investigation tend to be masked by fluctuations outside the experimenter's control. Large uncontrolled variations are common in technological experiments and in many types of work in the biological sciences, and it is in these fields that the methods to be described are most used. It is likely, however, that acquaintance with the simpler methods is of some value in most branches of experimental science.

The following are some typical situations in which large erratic fluctuations occur.

Example 1.1. Most agricultural field trials have as their object the comparison of a number of varieties of some crop, or of a number of alternative manurial treatments, or of a number of systems of management, etc. The experimental area is divided into plots and the different varieties, or whatever is under comparison, are assigned one to each plot. The yield, or some other property, is then measured or estimated for each plot and from the observations a comparison of varieties is made. Experience shows that even if the same variety were to be sown on all plots there would still be substantial variations in yield from plot to plot, the main features of this variation being that

(*a*) neighboring plots tend to give yields more alike than distant plots;
(*b*) there may be systematic trends or locally periodic variations across a field;
(*c*) if the experiment is repeated in a different field or in a different year, there may be a substantial change in the mean yield.

It would be common for the yields on individual plots in a field to vary by as much as $\pm 30\%$ from their mean, and a systematic difference of 5% between varieties might be of considerable practical importance. We shall be concerned with methods for arranging the experiment so that we may with confidence and accuracy separate the varietal differences, which interest us, from the uncontrolled variations, which do not.

The aim of such experiments is the comparison of varieties rather than the absolute determination of the yield per acre likely from a given variety under given conditions. There are two reasons for this. First, the differences between varieties determine any practical recommendations that may be based on the experiment; i.e., a choice of which of the two varieties is to be preferred depends not on the absolute yields but on how much more one variety is likely to yield than another, and on differences between any other properties that are considered important. Second, it is common for the difference between varieties to remain relatively constant even when the substantial changes in mean yield mentioned in (*c*) occur. This implies that it is much more economical to make a direct comparison of varieties than to estimate, in separate experiments for each variety, the mean yield under representative conditions and then to compare the estimates.

To sum up the discussion of this example, we are concerned with an experiment in which

(*a*) the object is to compare a number of varieties (or treatments);
(*b*) in the absence of varietal differences there is a substantial variation in yield from plot to plot;
(*c*) differences between varieties are comparatively stable, even though the mean level of response may fluctuate.

It is convenient to introduce a standard terminology. We shall refer to the plots as *experimental units*, or more briefly as *units*, and to the varieties, fertilizers, etc. under comparison as *treatments*. The formal definition of an experimental unit is that it corresponds to the smallest division of the experimental material such that any two units may receive different treatments in the actual experiment. For example suppose that in order to estimate the yield from the plots, two sub-areas are taken on each plot and the crop on these harvested and weighed. These sub-areas are not the experimental units, because the two sub-areas on one plot always receive the same treatment.

Example 1.2. Many experiments in industrial technology have a similar form to Example 1.1. The object may be to compare a number of alternative methods of processing, or to assess the effect of a modification to a standard process. The experiment consists in dividing the raw material into batches and then processing one batch by one process in the first period (day, hour, etc.), another batch in the next period by, in general, a different process, and so on. Or there may be several sets of machinery in use simultaneously. An observation (mean strength, yield of product, etc.) is made for each batch. In the absence of process differences the observation will vary from batch to batch and in addition to apparently random variation there may be smooth trends following, for example, hour-to-hour and day-to-day variations in temperature and relative humidity, and also sudden discontinuities corresponding to the introduction of fresh consignments of raw material.

Example 1.3. When slates, on which are bases of *Balanus balanoides*, are exposed in sea-water, the setting of further barnacles of this type occurs rapidly. Knight-Jones (1953), in investigating the mechanism of setting, exposed untreated slates and slates that had been treated with a variety of chemical reagents. By finding which reagents produced a substantial decrease in the amount of setting he was able to infer something about the chemical processes involved.

This experiment has a feature additional to those of Examples 1.1 and 1.2 in that the comparison of treatments is of interest only insofar as it aids in revealing the nature of the phenomenon under investigation. The experiment is concerned with comparisons because it is advisable to include as a control a series of untreated slates. This is to ensure that any observed decrease in the rate of setting after treatment is not due to a change in the natural rate of setting, which is subject to erratic fluctuations.

The experimental units are slates, the observation is the number of barnacles setting in a three-day period, and the treatments are the control and the various chemical reagents.

Example 1.4. One method of determining the potency of a drug is by direct comparison with an agreed standard in the following way: The drug is applied at a constant rate to an experimental animal and the dose at which death, or some other recognizable event, occurs is noted. This critical dose is called the tolerance or threshold. This is repeated for a number of animals using the drug under analysis and the standard. The tolerances vary from animal to animal but by comparing the mean log tolerances (see § 2.2) for the drug and for the standard a measure of potency is obtained. Here each animal is an experimental unit receiving one of two possible treatments, the drug and the standard.

An alternative procedure would be to measure the potency directly by, say, the mean log tolerance, without using a standard. This is usually unsatisfactory because the tolerance varies appreciably from group to group of animals, so that results in different laboratories and at different times would be only very roughly comparable. Experience shows that differences of log tolerance between a drug and a suitable standard are often little affected by systematic differences between groups of animals, so that the introduction of a standard into the experiment leads to a measure of potency that can be reproduced to within close limits at different times and places.

This simple form of comparative bioassay is discussed fully by Finney (1952).

Example 1.5. The clinical investigation of the use of new medical treatments raises similar problems of experimental design. It is almost always advisable to include a control treatment in the investigation, as well as the new treatment, because the effect of the new treatment may, except in dramatic cases, be shown by a comparatively small change in the proportion of cures. There are several cogent reasons, which will be discussed in detail later, why the determination of the proportion of cures for the control treatment should be part of the experiment and not just based on past experience. In this application each patient is an experimental unit, receiving one of two or more possible treatments.

In the treatment of serious diseases there is the complication that it will be considered unethical to withhold a treatment that is suspected to give increased chance of survival. This makes it imperative to conclude the experiment as soon as there is reasonable evidence that a particular treatment is in fact superior (Armitage, 1954).

An essential difference between the above experiments and many experiments in physics and chemistry is that in the latter, once the experimental technique is mastered and the apparatus working correctly, closely reproducible results are obtained. More precisely the uncontrolled variations are small compared with the effects to be expected when a change is imposed on the system. Therefore, if the system is altered and the observation changes, the imposed alterations may safely be assumed to be the cause of the change in the observation. In such cases the methods described in this book are of little value, except as a safeguard against errors arising from defects in the apparatus. However, as soon as the effects under investigation become comparable with the uncontrolled variations, the problems we shall be concerned with become important.

Examples 1.1–1.5 are all of the same form. We have a number of experimental units and a number of alternative treatments. The experiment consists in applying one treatment to each unit and making one (or more) observations, the assignment of treatments to units being under the experimenter's control. When the object of such an experiment is the comparison of treatments rather than the determination of absolute values, the experiment will be called *comparative*.*

The main planned investigations that are not comparative experiments are concerned with determining the properties of defined sets of things, such as the mean fiber diameter of a consignment of wool, the number of species of beetle in a particular area, or the characteristics of children in an area who watch television (*a*) frequently, (*b*) infrequently.

It is especially important to distinguish between the type of comparison that would be made in the last example and the type that would be made in a comparative experiment. The crucial distinction is that in the experiment the choice of treatment for each unit is made by the experimenter, whereas in the planned survey the observer has no control at all over what makes a particular individual fall in one group rather than another. Interesting conclusions can be drawn from planned surveys, particularly if comparisons are made within similar groups of individuals, for instance within groups of children of the same age, educational background, social class, etc. Nevertheless, much more cogent conclusions about causal effects can be drawn from experiments than from planned surveys. From this point onwards we restrict attention almost entirely to comparative experiments.

The discussion of the planning of such experiments falls into two almost distinct parts, dealing with the principles that should govern

(*a*) the choice of treatments to be compared, of observations to be made, and of experimental units to be used;

(*b*) the method of assigning treatments to the experimental units and the decision about how many units should be used.

Most of this book is about (*b*), but there is some attempt to discuss the first set of questions in Chapter 9.

It is convenient to discuss first the requirements for a good experiment.

* All measurements, including counting, are in a sense comparative, but this does not affect the distinction between comparative and other experiments, since within the framework of a particular experiment, measurements can usually be regarded as absolute.

1.2 REQUIREMENTS FOR A GOOD EXPERIMENT

We shall assume in this section that the treatments, the experimental units, and the nature of the observations have been decided on. The requirements for a good experiment are then that the treatment comparisons should as far as possible be free from systematic error, that they should be made sufficiently precisely, that the conclusions should have a wide range of validity, that the experimental arrangement should be as simple as possible, and finally that the uncertainty in the conclusions should be assessable.

These requirements will now be discussed in turn.

(i) Absence of Systematic Error

This means that if an experiment of the given design were done using a large number of experimental units it would almost certainly give a correct estimate of each treatment comparison. Some examples should make the point clear.

Example 1.6. Consider an industrial experiment to compare two slightly different processes, A and B, on the same machinery, in which A is always used in the morning and B in the afternoon. No matter how many lots are processed it is impossible, from the results of the experiment alone, to separate the difference between the processes from any systematic change in the performance of the machinery or operatives from morning to afternoon, unconnected with the difference between A and B. Such systematic changes do sometimes exist. The difficulty is not met by a calculation of statistical significance; this may tell us that the apparent difference between A and B is unlikely to be a purely random one but cannot determine which of two or more possible explanations of the difference is the right one.

Of course it would be foolish to suggest that such an experiment is useless. Previous experimental work, or general knowledge of the process, or supplementary measurements on relevant variables (e.g., temperature, relative humidity) may suggest that any difference between conditions in the morning and afternoon is unimportant. Then, provided that it is clearly understood that the interpretation of the experiment rests on this extra assumption, no great harm may be done. But suppose that a surprising result is obtained, or a result that is in apparent contradiction with later work. Then unless the evidence for the absence of morning-afternoon differences is strong, the experiment may lose much of its cogency.

It is therefore a sound principle to plan an experiment so that such difficulties are as far as possible avoided, i.e., to ensure that experimental units receiving one treatment differ in no systematic way from those receiving another treatment.

Difficulties similar to those just discussed arise whenever the comparisons under test get completely mixed up with differences between batches of experimental material, between observers, between different experimental methods, and so on. They are also liable to occur when all the units receiving one treatment are collected together in single groups and not left to respond independently.

Example 1.7. In animal feeding trials one possible plan is to have all animals receiving one treatment together in a single pen. This to some extent simulates practical conditions and also is very convenient in organizing the experimental work. If, however, we have one large pen of animals receiving the experimental ration, it is impossible to separate ration differences from systematic differences between pens or say from the presence in one pen of some disease wholly unconnected with the experimental treatments.

For example Yates (1934) has described an experiment on pigs in which the animals were divided into small groups housed separately, so that each treatment was tested on several entirely independent sets of pigs. It was found that pigs receiving no green food fell sick. Yates remarked that had the pigs receiving no green food been in a single pen it would probably have been concluded that the sickness was due to extraneous causes, particularly since previous experiments had suggested that green food was unnecessary. The fact that several independent sets of pigs receiving no green food fell sick and that no other pigs did so was, however, strong evidence that the treatment was responsible.

Another way of putting the difficulty is that in the experiment with single pens the experimental units are, in accordance with the definition of § 1.1, pens of animals, not single animals. Hence this is an experiment without replication for which further assumptions are needed before valid conclusions can be drawn.

The decision about what method of design to use in such experiments is not easy and the example is quoted primarily to illustrate the logical point involved. There are further discussions of animal feeding trials by Lucas (1948) and by Homeyer (1954).

A common type of experiment, of which Example 1.3 is an instance, involves applying a treatment, noting a change in the observation as compared with that expected in the absence of the treatment, and concluding that the treatment has caused the change. For such an experiment to be convincing by itself, the treated units must be compared with a control series of units, receiving no treatment, but included in the experiment under the same conditions as the treated units, and not being systematically different from them. To say that a certain observation has been obtained in the past, and that the treated units now give a different observation, is not by itself necessarily cogent evidence of a treatment effect, since there may be systematic differences among the experimental units or a systematic change in the external conditions. If past experience has shown that the observations on untreated units vary in a stable way, it may be in order to dispense with special control units, particularly in preliminary work. However this procedure is the same as allowing possible systematic differences between units in an experiment, such as in Example 1.6, and is best avoided in the great majority of cases.

A classical example of an experiment that was largely vitiated by the absence of controls is the following.

Example 1.8. McDougall (1927), to examine a possible Lamarckian effect in rats, taught some rats to choose between a lighted and an unlighted exit. He then bred from them and for each generation measured the speed with which the above task was learned. A Lamarckian effect would be shown by a steady increase in speed with generation number and this was in fact found. Certain other explanations, such as selection, were ruled out but there were no control units, i.e., no rats bred under the same conditions, but from untrained parents. Therefore it was possible that the effect was due to systematic uncontrolled variations in the experimental conditions.

Crew (1936) repeated the experiment with controls and found no apparent Lamarckian effect. Agar et al. (1954), in an experiment continued over a period of 20 years, found an initial increase in speed similar to McDougall's, but the same for the control as for the "treated" rats. They concluded that the effect was due to secular changes in the health of the colony of rats.

We can sum up as follows: experimental units receiving one treatment should show only random differences from units receiving any other treatment, including the control, and should be allowed to respond independently of one another. When it is impossible or impracticable to achieve this, any assumption about the absence of systematic differences should be explicitly recognized and as far as possible checked by supplementary measurements or by previous experience.

We shall see later how it is possible to ensure the absence of the main sources of systematic error by means of a randomization procedure.

(ii) Precision

If the absence of systematic errors is achieved by randomization (Chapter 5), the estimate of a treatment contrast obtained from the experiment will differ from its true value* only by random errors. It should be noted that the term random will be used throughout in its technical statistical sense. Roughly speaking this means that it refers to variations showing no reproducible pattern. For example, the variations of yield in a field described briefly in Example 1.1 are not random, because of the trends, correlation between yields on adjacent plots, etc.

The probable magnitude of the random errors in the estimate of the treatment contrast can usually be measured by the *standard error*. The precise definition and method of calculation of this is described in textbooks on statistical methods, for example in that of Goulden (1952, pp. 17–20), but for the present purpose a sufficiently good idea of its meaning can be grasped as follows:

In about one case out of three the estimate will be in error by more than plus or minus the standard error.

* The true value is defined more precisely in Chapter 2.

In about one case out of twenty the estimate will be in error by more than plus or minus twice the standard error.

In about one case out of a hundred the estimate will be in error by more than plus or minus two and one-half times the standard error.

These statements require some qualification depending on the form of the distribution of the errors and on the accuracy of the standard error, which itself has to be estimated; these points need not concern us at the moment.

The value of the standard error, and hence the precision of any particular experiment, will depend on

(*a*) the intrinsic variability of the experimental material and the accuracy of the experimental work;

(*b*) the number of experimental units (and on the number of repeat observations per experimental unit);

(*c*) the design of the experiment (and on the method of analysis if this is not fully efficient).

In most of the experiments where statistical design is useful, only a very limited increase in precision can be achieved by modifying the experimental material or by increasing the precision of measuring devices. This is partly because there is often an intrinsic variability that is very difficult to remove and partly because experiments under very controlled conditions, e.g., in greenhouses, in small-scale industrial plants, etc., cease to be representative of practical conditions. The point will be discussed again in Chapter 9.

If there is one observation per experimental unit, then, other things being equal, the standard error of the estimate of the difference between two treatments is inversely proportional to the square root of the number of units for each treatment. In fact, the standard error is

$$\text{standard deviation} \times \sqrt{\left(\frac{2}{\text{no. of units per treatment}}\right)}, \qquad (1)$$

or if there are differing numbers of observations on the two treatments A, B, it is

$$\text{standard deviation} \times \sqrt{\left(\frac{1}{\text{no. of units for } A} + \frac{1}{\text{no. of units for } B}\right)}. \qquad (2)$$

Here the standard deviation is a statistical measure of the random dispersion of the observations on experimental units treated alike (Goulden, 1952, p. 17).*

* Note that the standard deviation refers to the variation of the observations on individual units, whereas the standard error refers to the random variation of an estimate from a whole experiment.

From equation (1), the standard error is halved by a fourfold increase in the number of experimental units, but a hundredfold increase in the number of units is necessary to divide the standard error by ten. Although in theory the standard error can be made arbitrarily small by increasing the number of units, this is an expensive method of increasing precision.

The gain due to taking repeat observations on the experimental units is less than or equal to the gain from a corresponding increase in the number of units. It can be assessed from formulas similar to, but a little more complicated than, (1) and (2).

The third method of increasing precision is by improved design and it is with this that we shall be most concerned. The general idea is that whatever knowledge is available about the experimental units should be used to reduce the effective standard deviation in (1) and (2). It is sometimes possible to obtain an increase in precision equivalent to a substantial increase in the number of experimental units.

Our requirement about precision is, roughly speaking, that the standard error should be sufficiently small for us to be able to draw cogent conclusions, but not too small. If the standard error is large the experiment is, by itself, almost useless, whereas an unnecessarily small standard error implies a waste of experimental material. In the majority of cases the object is the estimation of treatment differences, and in these cases formulas (1) and (2) enable us to predict, when we are designing the experiment, the precision to be obtained with any given number of units or, alternatively, the number of units necessary for a given precision. For this we must know something about the standard deviation, i.e., the variability of the units, but approximate information from previous similar experiments is often available. Occasionally the object is not the estimation of treatment differences but is to reach an irreversible decision on, say, which of a number of treatments is the best. In this case if one treatment is much better than the rest and the units are tested in sequence, the experiment can be ended after a small number of observations, even though the precision of estimation is still low. This raises special problems. The whole question of the choice of number of units will be discussed in detail in Chapter 8.

(iii) Range of Validity

When we estimate the difference between two treatments, we obtain conclusions referring to the particular set of units used in the experiment and to the conditions investigated in the experiment. If we wish to apply the conclusions to new conditions or units, some additional uncertainty is involved over and above the uncertainty measured by the standard error. The only exception to this statement is when the units in the experiment are chosen from a well-defined population of units by a proper statistical sampling procedure.

The wider the range of conditions investigated in the experiment, the greater is the confidence we have in the extrapolation of the conclusions. Therefore if we can arrange, without decreasing the accuracy of the experiment, to examine a wide range of conditions, this is desirable. This is particularly important in experiments to decide some practical course of action and rather less so where the object is purely to gain insight into some phenomenon.

Example 1.9. "Student" (1931) mentions some experiments done by the Irish Department of Agriculture in connection with the introduction of Spratt–Archer barley. This was almost everywhere a great success; yet in one district the farmers refused to grow it, alleging that their own native race of barley was superior. After some time the Department, to demonstrate Spratt–Archer's superiority, produced a single-line culture of the native barley and tested it against the Spratt–Archer in the district in question. "Student" reports that to the Department's surprise the farmers were perfectly right: the native barley gave the higher yield. At the same time, the reason became clear: the barley in question grew more quickly and was able to smother the weeds, which flourished in that area; Spratt–Archer, growing less strongly to begin with, was, however, the victim of the weeds. Thus the original experiments, carried out on well-farmed land, were definitely misleading when their conclusions were applied elsewhere.

Similar points arise with other types of experiment. A new experimental technique that works very well when special attention is devoted to it may be quite unsuited to routine use. A new industrial process that works well under special supervision during an experiment may not be successful in routine production. Or, to take a more specific example, a modification to a textile process tested on a homogeneous batch of raw material may in fact be quite critically dependent on the oil content of the raw material. The difference between varieties of wheat may be dependent on soil and weather conditions, and so on.

There are several consequences of these remarks. First it is important, even in purely technological experiments, to have not just empirical knowledge about what the treatment differences are, but also some understanding of the reasons for the differences. Such knowledge will indicate what extrapolation of the conclusions is reasonable. Secondly we should, in designing the experiment, artificially vary conditions if we can do so without inflating the error. For example in comparing two methods of drawing wool, it may sometimes be expected that the difference between the methods is unaffected by the oil content of the wool. It would often be advantageous to include both lightly and heavily oiled wool in the experiment with the hope of providing a direct check on the independence of the difference between methods and the oil content. The snag is, of course, that if several such supplementary factors are included the experiment may become difficult to organize, and also there is the possibility, if the system is a complicated one, that no clear-cut conclusions can be drawn, owing to no one set of conditions having been thoroughly investigated. This leads to the third point that it is important to recognize explicitly what are the restrictions on the conclusions of any particular experiment.

These considerations are rather less important in purely scientific work, where the best thing is usually to try to gain thorough insight into some very special situation rather than to obtain in one experiment a wide range of conclusions.

(iv) Simplicity

This is a very important matter which must be constantly borne in mind but about which it is difficult to make many general remarks. There are several considerations involved. If the experiment is to be done by relatively unskilled people, it may be difficult to ensure adherence to a complicated schedule of alterations. If an industrial experiment is to be run under production conditions, it will be important to disturb production as little as possible, i.e., to have a few long runs of the different processes rather than frequent changes. In scientific work, particularly in the preliminary stages of an investigation, it may be important to retain flexibility; the initial part of the experiment might suggest a much more promising line of enquiry, so that it will be a bad thing if a large experiment has to be completed before any worth-while results are obtained. Nevertheless there certainly are cases where a fairly complicated arrangement is advantageous and it is a matter of judgement and experience to decide how far it is safe to go in any particular application.

The above remarks apply to simplicity of design. It is also desirable to have simple methods of analysis. Fortunately the requirements of efficiency in design and simplicity in analysis are highly correlated and for nearly all the methods in this book, straightforward schemes of full statistical analysis are available, provided that certain assumptions to be described later are satisfied. If only estimates of the treatment differences are required, with no estimates of precision, few of the designs require more than simple averaging.

The use of electronic computers for the analysis of experimental results is an important recent development, particularly for those fields where either very large amounts of data are involved or where the time taken on the experimental work is comparable to or smaller than the time it would take to analyze the results by conventional methods. Once suitable programs have been written, the time taken to make a statistical analysis on an electronic computer is likely to be very small in all ordinary circumstances.

(v) The Calculation of Uncertainty

The previous requirements have not been statistical; this last one is. It is desirable that we should be able to calculate, if possible from the data themselves, the uncertainty in the estimates of the treatment differences. This usually means estimating the standard error of the differences, from which limits of error for the true differences can be calculated at any required level of probability, and from which the statistical significance of the differences between the treatments can be measured.

To be able to make this calculation rigorously we must have a set of experimental units responding independently to one treatment and differing only in a random way from the sets of units for the other treatments. A comparison, not necessarily straightforward, of the observations on units receiving the same treatment then gives a valid measure of error. The use of randomization, discussed in detail in Chapter 5, to eliminate systematic differences between units treated differently, automatically makes differences random and justifies the statistical analysis under weak assumptions. The distinction between such an analysis and that of Example 1.6 should be carefully noted.

In experiments with very small numbers of experimental units it may not be possible to obtain an effective estimate of the error standard deviation from the observations themselves. In such cases it will be necessary to use the results of previous experiments to estimate the standard deviation (see also § 8.3); the disadvantage of this is that we need to assume that the amount of random variation is unchanged.

As a general rule, methods of statistical analysis will not be described in this book. This is partly because there are a number of excellent accounts of such methods available, and partly because their inclusion would not only greatly increase the length of the book but would also tend to distract attention from considerations of design.

SUMMARY

We deal mostly with experiments of the following form: there are a number of alternative *treatments* one of which is applied to each *experimental unit*, an *observation* (or several observations) then being made on each unit. The object is to be able to separate out differences between the treatments from the uncontrolled variation that is assumed to be present;

this may of course be only the first step towards understanding the phenomena under investigation.

Once the treatments, the experimental units, and the nature of the observations have been fixed, the main requirements are that

(*a*) experimental units receiving different treatments should differ in no systematic way from one another, i.e., that assumptions that certain sources of variation are absent or negligible should, as far as practicable, be avoided;

(*b*) random errors of estimation should be suitably small, and this should be achieved with as few experimental units as possible;

(*c*) the conclusions should have a wide range of validity;

(*d*) the experiment should be simple in design and analysis;

(*e*) a proper statistical analysis of the results should be possible without making artificial assumptions.

REFERENCES*

Agar, W. E., F. H. Drummond, O. W. Tiegs, and M. M. Gunson. (1954). Fourth (final) report on a test of McDougall's Lamarckian experiment on the training of rats. *J. Exp. Biol.*, **31**, 307.

Armitage, P. (1954). Sequential tests in prophylactic and therapeutic trials. *Q. J. of Medicine*, **23**, 255.

Crew, F. A. E. (1936). A repetition of McDougall's Lamarckian experiment. *J. Genet.*, **33**, 61.

Finney, D. J. (1952). *Statistical method in biological assay.* London: Griffin.

Goulden, C. H. (1952). *Methods of statistical analysis.* 2nd ed. New York: Wiley.

Homeyer, P. G. (1954). Some problems of technique and design in animal feeding experiments. Chapter 31 of *Statistics and Mathematics in Biology*. Ames, Iowa: Iowa State College Press. Edited by O. Kempthorne et al.

Knight-Jones, E. W. (1953). Laboratory experiments on gregariousness during setting in *Balanus balanoides* and other barnacles. *J. Exp. Biol.*, **30**, 584.

Lucas, H. L. (1948). Designs in animal research. *Proc. Auburn Conference on Applied Statistics*, 77.

McDougall, W. (1927). An experiment for testing the hypothesis of Lamarck. *Brit. J. Psychol.*, **17**, 267.

"Student" (1931). Agricultural field experiments. *Nature*, **127**, 404. Reprinted in "*Student's*" *collected papers*. Cambridge, 1942.

Yates, F. (1934). A complex pig-feeding experiment. *J. Agric. Sci.*, **24**, 511.

* These are explicitly referred to in the text. There are some general references on p. 294.

CHAPTER 2

Some Key Assumptions

2.1 INTRODUCTION

In many experiments several types of observation are made on each experimental unit. For example in comparing varieties of sugar beet, yield of roots, yield of tops, yield of sugar, and possibly plant number would be measured, as well as perhaps observations on the incidence of disease, the frequency of bolting, and the chemical analysis of the sugar. In comparing methods of spinning wool yarn, it would be common to measure the yarn irregularity, the yarn strength, and the end-breakage rate in spinning, as well as possibly making tests on fabric woven from the yarns. In a preliminary account it is, however, convenient to suppose that only one observation is made on each experimental unit. This observation may be derived by calculation from a number of experimental readings. For example, measures of yarn irregularity are often obtained by computing a so-called coefficient of variation from a trace showing the changes in thickness along the length of the yarn. Again, in learning experiments in experimental psychology, one observation for analysis is usually a measure of the rate of learning. This is derived from the raw data which consist, for example, of a record of success or failure at each attempt at the experimental task.

The following assumption, or some simple modification of it, underlies the use of most of the designs described in this book. The observation obtained when a particular treatment is applied to a particular experimental unit is assumed to be

$$\begin{pmatrix}\text{a quantity depending} \\ \text{only on the} \\ \text{particular unit}\end{pmatrix} + \begin{pmatrix}\text{a quantity depending} \\ \text{on the treatment} \\ \text{used}\end{pmatrix} \qquad (1)$$

and to be unaffected by the particular assignment of treatments to the other units. This can be put more vividly as follows. Denote the alternative treatments by the letters $T_1, \ldots, T_t$; then it is assumed that the observation obtained on any unit when, say, T_1 is applied differs from the observation that would have been obtained had, say, T_2 been applied by a constant, $a_1 - a_2$. There are constants $a_1, \ldots, a_t$, one for each treatment, and the object of the experiment is to estimate differences such as $a_1 - a_2$; we call such differences the *true treatment effects.*

The essential points about this assumption are that

(*a*) the treatment term in (1) adds on to the unit term rather than, for example, multiplying;

(*b*) the treatment effects are constant;

(*c*) the observation on one unit is unaffected by the treatment applied to other units,

and these three points will be discussed separately in the subsequent sections.

The assumption is particularly important if a full statistical analysis is to be made of the observations. It is, however, still required even if the experiment is analyzed just by calculating simple averages, in the sense that a gross departure from the assumption will affect the whole qualitative interpretation of the results. It is usually possible to check the assumptions to a certain extent from the data, but never possible to avoid completely making some assumption or other. Too much attention should not be paid to the details of the following sections at a first reading.

2.2 ADDITIVITY

The first consequence of the additive law (1) is that the difference between two treatments, say T_1 and T_2, is usually* appropriately estimated by

$$\begin{pmatrix}\text{mean of all observations}\\ \text{on } T_1\end{pmatrix} - \begin{pmatrix}\text{mean of all observations}\\ \text{on } T_2\end{pmatrix}. \tag{2}$$

If the treatment effects and uncontrolled variations are relatively small any functional law for combining unit and treatment terms would be equivalent to the additive law (1) to a first approximation. In other cases, however, it may be worth considering whether some other form may not be more appropriate. The most important alternative form is multiplicative, replacing expression (1) by

$$\begin{pmatrix}\text{a quantity depending}\\ \text{only on the}\\ \text{particular unit}\end{pmatrix} \times \begin{pmatrix}\text{a quantity depending}\\ \text{on the treatment}\\ \text{used}\end{pmatrix}. \tag{3}$$

If this is the appropriate form, we work with the logarithms of the original observations. Since $\log (xy) = \log x + \log y$, equation (3) is thereby converted into form (1).

* The exceptions are incomplete block designs (Chapter 11) and certain types of confounded design (Chapter 12).

Example 2.1. In Example 1.4 we discussed briefly a simple comparative assay for measuring the potency of a drug by comparison with a standard. A natural working hypothesis is that any dose x of the experimental drug is equivalent in all relevant respects to a dose ρx of the standard, where ρ is the potency of the drug with respect to the standard and is constant. Or equivalently the tolerance of an animal (say in mg) for the experimental drug is $1/\rho$ times what it would have been with the standard. This is of the form (3) and is reduced to (1) by working with log tolerances rather than with tolerances.

Example 2.2. Consider a field trial comparing the effect of a number of alternative treatments on the incidence of a certain disease. The treatments are applied one to each plot and after a suitable time the disease is measured, say by counting the number of diseased plants out of one hundred on each plot. It is reasonable to expect that if the proportion diseased varies appreciably over the whole experiment, the difference between proportion diseased for two treatments will be rather greater when the level of disease is fairly high than when it is low. However if the level of disease is very high, it may be that all treatments are ineffective so that differences between treatments decrease again.

At any rate there seems to be no general reason for expecting a constant additive effect for one treatment as compared with another. There are several ways of proceeding. If the experiment is divided into sections within each of which the natural level of disease is fairly constant, it would be reasonable to estimate treatment differences separately for each section. Then by comparing the estimates with the overall level of disease for the section, the change, if any, of treatment effects with level of disease could be assessed. This is probably the best procedure, if it can be used; it amounts to allowing the data to determine the appropriate scale of measurement. Alternatively, if the proportion diseased varied, say from 5 to 50 per cent, it might be reasonable to assume constant proportional differences, and therefore to take logarithms. Or, occasionally, more complicated assumptions might seem justifiable, such as that the treatments have a constant effect on the probit of the proportion diseased. [The probit is a quantity derived by a particular mathematical transformation of a proportion (Goulden, 1952, p. 395).]

Example 2.3. A rather similar example concerns feeding or management trials on pigs. Suppose that two treatments A and B are under comparison and that at the end of the experiment the pigs are examined by a judge and a total score out of 100 assigned to each pig. Then, because of the upper limit to the scale, the following might happen: a pig which would have scored 50 with treatment A would score 70 with treatment B, but a very good pig, which would have scored 85 with treatment A, would score 90 if given B. That is, we are measuring on a scale on which the treatment effects are not additive. A conventional way of attempting to deal with this is to work not with the total score x but with $\log[(x + \frac{1}{2})/(100\frac{1}{2} - x)]$, which should often nullify the restriction at the upper and lower ends of the scale. For two values a certain distance apart and near the top, or bottom, of the scale differ much more after transformation than do two values initially the same distance apart but near the center of the scale.

In all these examples the comparison of treatments by the mean difference (2) is valid in the narrow sense that this will estimate the average treatment difference over the units used in the experiment, i.e., the mean observation that would have been obtained if T_1 had been applied to all units minus the corresponding mean for T_2. But if the assumption (1) does not hold, this difference is rather an artificial quantity. Thus in Example 2.1 the difference in mean tolerances depends on the particular animals, and if these vary appreciably in tolerance from laboratory to laboratory a comparison of mean tolerances would not be independent of laboratories. Moreover the mean difference, even if it was reproducible, would not have the simple physical interpretation of the difference in mean log tolerance, which estimates log ρ.

Again, to take an extreme case, suppose that the experiment in Example 2.2 happened to fall into two roughly equal parts:

(*a*) with an average proportion diseased of 10 per cent, T_1 giving 8 per cent and T_2 12 per cent on the average;

(*b*) with an average proportion diseased of 50 per cent, T_1 giving 40 per cent and T_2 60 per cent on the average.

Then an averaging of the proportions diseased would give a difference between T_2 and T_1 of $36 - 24 = 12$ per cent. But this is clearly an artificial figure that depends on the particular incidence of disease encountered in the experiment; it is in this case much more revealing to say that T_1 gives a proportion $\frac{2}{3}$ of that corresponding to T_2.

Of course this is an extreme and oversimplified example, but it has been discussed to emphasize that the importance of the additive assumption is not essentially connected with details of statistical technique. However it would often happen that, if the experiment falls into sections with different treatment effects, the amount and distribution of the uncontrolled variation would be different in the different sections. A full statistical analysis will involve differential weighting of the sections; this will not be considered here.

Fortunately the complications that we have been considering are frequently unimportant because, as remarked above, if the variations involved are relatively small, the additive law (1), the multiplicative law (3), and other similar laws are nearly equivalent. In many applications it is probably enough to consider which of (3) and (1) is likely to be the more appropriate and to take or not take logarithms accordingly.

2.3 CONSTANCY OF TREATMENT EFFECTS

In the previous section we discussed the assumption that the observations are measured on a scale on which the effect of treatments is represented by the *addition* of appropriate quantities rather than by some other functional law, such as multiplication. In this section we in effect continue that discussion by considering other ways in which the treatment effects can fail to be constant.

First note that an additional completely random component added to the treatment term in formula (1) is indistinguishable from a random component added to the first, or unit, term and so can be disregarded, provided that the distribution of the random component is the same for all treatments. This possibility will not be discussed further. We shall deal in detail with what happens when the treatment effects depend on the value of some supplementary measurement that can be made on each unit.

Example 2.4. Suppose that it is required to compare two alternative processes A and B for extracting a product P from a raw material containing P in small quantities. The experimental units are different batches of raw material and the observation is the yield y of product. A supplementary observation x is also made by obtaining for each batch before processing, an estimate of the percentage of P in the batch. Then it might happen that the difference between the processes depends on the amount of P, e.g., A may work relatively much better when the raw material is rich in P. Information that this was so might not only be important in deciding what practical action to take, but also might throw some light on the fundamental reasons for process differences. Further the information might help to link the results with previous work in which, perhaps, the content of P in the raw material was systematically different.

A comparison of the mean value of y for those units receiving process A with the corresponding mean for process B would, with correct design, always estimate the mean process difference over the raw material used in the experiment. Although this would usually be of some interest, it is clear from the previous paragraph that such an overall difference may be only a partial description of the difference between the processes. Unless there is good prior reason for expecting the process difference to be constant, the data would therefore be analyzed by plotting y against x, distinguishing between the results for the two processes. This graphical analysis would be supplemented, if necessary, by appropriate statistical calculations, such as the fitting of regression lines. Attention would be paid to any change with x in the random variation of y.

Another way of dealing with the results of this experiment would be to work with y/x, which is proportional to the fraction of P in the raw material that is extracted in processing; if the difference between treatments in the ratio were expected to be constant, this would be the natural thing to do. However the general remarks on the constancy of treatment effects would still be relevant.

This example illustrates the use of a supplementary observation to examine whether a treatment difference is constant. A further use of supplementary observations is to increase precision, and this will be considered in detail in Chapter 4.

Example 2.5. Jellinek (1946) has described an experiment to compare three drugs A, B, C for the relief of headaches, with a pharmacologically inactive control D. Each subject used each drug for two weeks and one of the observations was the success rate, i.e., the number of headaches relieved divided by the number of headaches treated in the two-week period. Precautions, which need not be gone into here, were taken to remove any effect of the order in which the drugs were used. The first line of the table shows the mean success rates averaged over all subjects. They suggest that A, B, C are not appreciably different and all have appreciably higher success rates than D.

TABLE 2.1

MEAN SUCCESS RATES

	A	B	C	D
All subjects	0.84	0.80	0.80	0.52
Subjects not responding to D	0.88	0.67	0.77	0
Subjects responding to D	0.82	0.87	0.82	0.86

However the subjects fell quite sharply into two groups, those on whom D had no effect and those who did respond to D. The second and third lines of the table show the corresponding mean success rates. For subjects that do respond to D, the four drugs have practically the same success rates, whereas for those who do not respond to D, A has a higher success rate than C and a much higher rate than B. Comparisons based on averages for all subjects are thus quite misleading. The difference between the two groups in the response to the drugs is possibly due to a difference in type of headache.

In this example the response to D is used to divide the experimental units in a way similar to that in which the supplementary observations were used in Example 2.4.

The general conclusion to be drawn from these examples is the desirability of being able to detect variations in the treatment effects if these are likely to be important. This means making supplementary observations where appropriate and, in other cases, assigning the treatments to the units in such a way that the variations may be detected. Methods for doing this will be discussed later. In most of the book it will, however, be assumed, in accordance with (1), that the treatment effects are constant.

2.4 INTERFERENCE BETWEEN DIFFERENT UNITS

The last aspect of the assumption (1) to need discussion is the requirement that the observation on one unit should be unaffected by the particular assignment of treatments to the other units, i.e., that there is no "interference" between different units. In many experiments the different units are physically distinct and the assumption is automatically satisfied. If, however, the same object is used as a unit several times, or if different units are in physical contact, difficulties can arise and these will now be illustrated by some examples.

Example 2.6. In the textile process called carding, an entangled mass of fibers is passed over rotating cylinders carrying teeth, which straighten the fibers.

Consider an experiment to investigate the effect of various amounts of oil applied to the raw material. The treatments are, say, four percentages of oil and the experimental units are batches of raw material. Now when a batch with a high oil content is carded, some of the oil remains on the teeth, so that the following batch, or at any rate the part of it carded first, receives in effect a larger amount of oil than its nominal treatment implies. In other words the observation on any unit is likely to depend not only on the treatment applied to that unit but also on the treatment applied to the preceding unit and even, in certain cases, on the unit two before.

One way of avoiding this difficulty is to follow each experimental batch by a control batch sufficiently large to restore the amount of oil to a standard value or, alternatively, to use large experimental batches and to make observations only on the latter part of each batch, which is unlikely to be affected by the preceding treatment. However, both these procedures, and particularly the first, would very often not be economical ways of arranging the experiment. Instead it may be preferable to accept the overlap of the treatment effects and to deal with it in the design and analysis of the experiment. This is possible provided that it is reasonable to introduce a simple modification of (1), such as that the observation on any unit is

$$\begin{pmatrix}\text{a quantity depending}\\ \text{only on the}\\ \text{unit}\end{pmatrix} + \begin{pmatrix}\text{a quantity depending}\\ \text{on the treatment}\\ \text{used}\end{pmatrix} + \begin{pmatrix}\text{a quantity depending}\\ \text{on the treatment}\\ \text{applied to the}\\ \text{preceding unit}\end{pmatrix}. \quad (4)$$

This is plausible in the present example, provided that the oil contents investigated do not vary over too wide a range. If (4) is accepted it is natural to arrange that each treatment follows each other treatment (or each treatment) the same number of times. Then the systematic change caused by following the highest oil content affects all treatments equally. Such designs are discussed in Chapter 13.

Example 2.7. Similar problems arise in investigating the effect of different diets on the milk yield of cattle. If each animal is fed on a constant diet there is no difficulty, but it would often be preferable to change over the diets in the course of the experiment and, if possible, to use each diet once on each animal. This would eliminate the effect of systematic differences between animals.

Thus, with three diets, one animal might receive diet A for the first two weeks, diet B for the second two weeks, and diet C for the third. The main observation to be analyzed would be the milk yield determined as the average of two or three days' yield at the end of each two-week period. By thus taking observations at the end of each experimental period it would be hoped that a value would be obtained characteristic only of the treatment applied during the period; however, the overlap of the treatment effects might still occur and then difficulties like those of the preceding example would arise and in particular the assumption (4) might again be reasonable. It would also be necessary to ensure that for a group of animals each treatment occurred equally frequently in each period.

The interference between different units in the above examples can be coped with because it is of a simple form. Often, however, it is better to go to some trouble to arrange that the different units are isolated, rather than to allow interference and to attempt to deal with it by a more subtle design. For example, in agricultural field trials, guard rows are left between the different plots. Again, in an experiment in which some plots are inoculated with virus-carrying aphids, while other plots are untreated, it would be essential not only to leave substantial space between treated and untreated plots but also, as far as is possible, to check that there is no direct transmission of disease from one plot to another.

Competition may arise within an experimental unit, but this causes no difficulty provided that it is representative of the conditions under investigation. For example, in a poultry feeding trial, each unit might consist of a number of birds kept together and feeding in common. If the food is limited, large healthy birds may gain at the expense of others. However, this will not invalidate the assumption of no interference *between* different groups of birds, which is involved in (1).

In experimental psychology it is frequently required to use the same subject as an experimental unit several times. In this field, however, it often happens that the effect of one treatment on the subsequent observations is not represented by anything as simple as the addition of single constants as in equation (4). Babington Smith (1951) has described experiments on the "Muller–Lyer" illusion, which suggest that responses are dependent in a rather complicated way on the whole sequence of situations that have preceded them. Welford et al. (1950), in some experiments on fatigue in aircrew, noted that subjects who first met a task when tired continued to do it badly when fresh, whereas those who first met it fresh went on doing it well when tired. Other similar effects have been reported in the literature. In such cases either a special hypothesis to replace (4) must be set up appropriate to the problem, or the treatments must be taken as whole sequences of stimuli. These experiments are mentioned here to emphasize that the simple law (4) may not be adequate.

In the remainder of the book it will be assumed, unless explicitly stated otherwise, that interference between different units is absent. If it is suspected that such interference may arise, as when the same object is used as an experimental unit more than once, or when different units are in physical contact, either experimental precautions should be taken to prevent the interference or special allowances should be made in the design and analysis of the experiment.

SUMMARY

In most cases we estimate treatment differences by averaging observations over the whole experiment. There are three points to be watched if this is done, namely (*a*) that the observations should be analyzed on a scale on which the treatment *differences* are relevant;

(*b*) that either only average treatment effects are required, or that the treatment effects are constant. Special precautions should be taken if the treatment effects are expected to depend in an important way on the value of some supplementary observation, or to be different for different groups of units;

(*c*) that the observation obtained on one unit should not be affected by the treatment applied to other units.

In the ordinary way the second and third complications are assumed absent, but if it is suspected that they may arise, they should be allowed for both in the design and in the analysis of the experiment.

REFERENCES

Babington Smith, B. (1951). On some difficulties encountered in the use of factorial designs and analysis of variance with psychological experiments. *Brit. J. Psychol.*, **42,** 250.

Goulden, C. H. (1952). *Methods of statistical analysis.* 2nd ed. New York: Wiley.

Jellinek, E. M. (1946). Clinical tests on comparative effectiveness of analgesic drugs. *Biometrics*, **2,** 87.

Welford, A. T., R. A. Brown, and J. E. Gabb. (1950). Two experiments on fatigue as affecting skilled performance in civilian aircrew. *Brit. J. Psychol.*, **40,** 195.

CHAPTER 3

Designs for the Reduction of Error

3.1 INTRODUCTION

In this chapter we consider some ways of reducing the effect of uncontrolled variations on the error of the treatment comparisons. The general idea is the common sense one of grouping the units into sets, all the units in a set being as alike as possible, and then assigning the treatments so that each occurs once in each set. All comparisons are then made within sets of similar units. The success of the method in reducing error depends on using general knowledge of the experimental material to make an appropriate grouping of the units into sets. This method, and various generalizations of it, will be introduced mainly by examples.

3.2 PAIRED COMPARISONS

We begin by considering experiments for the comparison of just two treatments.

Example 3.1. Fertig and Heller (1950) have discussed an experiment for comparing the effect on sewage of two treatments, T_1 and T_2. Both treatments involved 100 per cent chlorination; with T_2 there was no special mixing and with T_1 there was an initial 15-sec period of rapid mixing. The observation made on each unit after processing was the logarithm of the coliform density per ml, and it was required to estimate any additional reduction in coliform density due to the rapid mixing in treatment T_1.

The main source of uncontrolled variation, other than random sampling errors in the determination of the coliform density, arose from variations in the sewage before processing. Therefore to obtain pairs of units as alike as possible, it was natural to take batches of sewage on the same day and as close together as possible in time. This was done on several days giving a series of pairs of similar units and it was then arranged that T_1 and T_2 both occur on each pair. This involves a series of choices between the orders $T_1 T_2$ and $T_2 T_1$. In the present case there was no reason for expecting a systematic difference between the first and second units in the pairs and the appropriate procedure is then to *randomize* the order of the treatments, i.e., to use an objective device such as a table of random numbers to choose, independently for each pair, between $T_1 T_2$

and $T_2 T_1$, giving each equal probability. The full discussion of this process of randomization is deferred to Chapter 5.

A typical arrangement of treatments resulting from such a randomization is shown in Table 3.1 together with some fictitious observations. For each pair of units the difference between the observation on T_2 and the observation on T_1 is calculated. The treatment effect is estimated by $\bar{d}$, the mean of these differences, and the estimated standard error of $\bar{d}$, and a test of the statistical significance of $\bar{d}$ can be obtained by simple standard statistical calculations (Goulden, 1952, p. 51), the amount of the uncontrolled variation being estimated from the observed dispersion of the differences in the last column of Table 3.1.

TABLE 3.1

PAIRED COMPARISON EXPERIMENT

Day	First Unit	Second Unit	Difference, d
1	T_1:2.8	T_2:3.2	0.4
2	T_2:3.1	T_1:3.1	0.0
3	T_2:3.4	T_1:2.9	0.5
4	T_1:3.0	T_2:3.5	0.5
5	T_2:2.7	T_1:2.4	0.3
6	T_2:2.9	T_1:3.0	−0.1
7	T_2:3.5	T_1:3.2	0.3
8	T_1:2.6	T_2:2.8	0.2
		Mean, $\bar{d}$ =	0.262
		Estimated standard error =	0.078

It is clear in this design that the variation from one day to another has no effect on the experiment, i.e., that if both the observations on one day are changed by the same amount, the estimate of the treatment difference, and its error, are unaffected. The elimination in this way of the effect of part of the uncontrolled variation is, of course, the object of the pairing of the units.

Notice that we take differences between observations which are themselves logarithms. This implies the assumption that T_1 achieves a constant *fractional* change in coliform density over what would be obtained on the same material with T_2.

A natural objection to the randomization used in obtaining the design in Table 3.1 is that T_2 has occurred five times in the first position and three times in the second, and that it would have been better to have arranged for each treatment to occur equally often in each column. This point will be discussed fully later, but in the meantime it should be noted that the objection is really only cogent if there is reason to expect a systematic difference between the first and second units and this was not so in this experiment.

This example could be paralleled from many applied fields. The general method is simply to obtain a number of pairs of experimental units, where the two units in each pair are expected to give as nearly as possible identical observations in the absence of treatment differences. The treatments T_1 and T_2 are then assigned in random order to each pair. The method will give a comparison of treatments free of systematic error whatever pairing of units is used, but the success of the method in reducing error depends on a skilful grouping of units.

The following are a few examples of methods that can be used to obtain a suitable pairing. Often, as in Example 3.1, the general fact that experimental units close together in some natural arrangement in space or time will tend to be alike suggests an appropriate pairing. Thus, plots close together in a field tend to give yields more alike than plots far apart, the products from one machine at two times nearer together tend to be more alike than products at times far apart or than products from different machines, and so on. If more explicit information is available about differences between units, this should, of course, be used. In experiments with rats, the pairs would probably be taken of the same sex, of approximately the same weight, and, so far as possible, from the same litter. In some experiments on animals it is possible to use twins, especially identical twins, for the pairs. In other animal work it is possible to use paired organs (kidneys, eyes, etc.) from the same animal. A somewhat similar idea in plant experimentation has been put forward by James (1948); he split clover plants through the middle of the tap root and used the two halves as paired units. Another device that is sometimes valuable is the use of the same physical object as a unit twice. This frequently happens in experimental psychology and in those clinical experiments in which the treatments are of a comparatively minor nature, so that each subject can be treated more than once. In such experiments, it may happen, even if there are no complications due to the overlap of treatment effects, that there is a systematic difference between the first and second units. In this case some restriction on the randomization is desirable to balance out the systematic difference; this will be discussed later.

The use of inbred lines of animals or plants has often been advocated in biological work as a method of ensuring uniform material. This use has been questioned, for example by Biggers and Claringbold (1954), who give examples where inbred lines are not more homogeneous than randomly bred material. They suggest that F_1 hybrids between inbred lines may be more suitable than the inbred lines themselves.

A final method depends on using a supplementary observation made on each unit before the experiment starts. For example, in an experiment on animals, the supplementary observation could be the initial weight. In this case the two animals with lowest weight are put in one pair, the two with the next lowest weight in the next pair, and so on. Provided that animals with extreme weights are omitted and that the final observation is highly correlated with initial weight, this provides a satisfactory grouping into pairs. The methods to be used if two or more supplementary measurements are available will be dealt with later.

One general warning is necessary in connection with the use of artificially uniform material. It may, by the use of such material, be possible to obtain a substantial increase in precision, but sometimes only at the cost of getting conclusions that are not representative of a wider class of units (see also § 9.2). What should be done in such cases depends on the purpose of the investigation; for example, if it is desired to obtain conclusions of immediate practical applicability in industry or agriculture, it will be desirable to use representative material.

3.3 RANDOMIZED BLOCKS

(i) Introduction and Example

If we have more than two treatments to be compared, the method just described can be extended in a straightforward way. If there are t alternative treatments, we group the units into sets of t, the units in each set being expected to give as nearly as possible the same observation if the treatments are equivalent in their effect. It is usual to call each set of t units a *block*. The order of treatments is then independently randomized within each block, arranging that each treatment occurs once in each block. Just as in § 3.2 the effect of variations between pairs is eliminated, so in the present case the effect of variations between blocks is eliminated, so far as treatment comparisons are concerned.

An experiment in which block differences are removed from the error in the way just described is said to be arranged in *randomized blocks*.

Example 3.2. In an experiment discussed by Cochran and Cox* (1957, § 4.23), the treatments were five levels of application of potash, 36, 54, 72, 108, and 144 lb K_2O per acre applied to a cotton crop. One observation analyzed was a measure of single-fiber strength in arbitrary units, obtained as an average of a number of tests on the cotton from each plot.

There were three blocks each containing five plots. The observations are given in the above reference but not full particulars of the arrangement of treatments within blocks, etc. In accordance with the general principle for grouping plots into blocks, the five plots in a block should be chosen to minimize the uncontrolled variation from plot to plot within blocks, and this is usually achieved by arranging the plots within a block in a compact approximately square area. This and the randomization of treatments within blocks allows statistical assessment of uncontrolled variation in the results arising from

* This book is referred to frequently. Section rather than page numbers are given because the section numbering is the same in both editions.

variation between plots. But this is certainly not the only way error can arise; three other possible sources of erratic variation are associated with

(*a*) the cultivation and harvesting of the crop;
(*b*) the selection of fibers for test;
(*c*) the strength testing;

and these will be discussed briefly.

Variations connected with the order in which the plots are cultivated or harvested would ordinarily be assumed negligible; however, if for instance the harvesting takes more than one day a useful precaution would be to harvest all the plots in one block on the same day. In this way constant differences between days become identified with block differences and do not contribute to the error of the experiment.

Only a minute proportion of the fibers on a plot are used in the strength testing; the use of a reliable method of sampling in selecting the fibers is a vital part of the method of testing but will not be discussed here.

There may be uncontrolled variations connected with the behavior of the testing machine, with the temperature or humidity of the testing room, and with the testing operative. The best procedure here is usually to test the cotton from one block in random order in as short a time period as possible. If several operatives or several testing machines are used in the whole experiment it is usually desirable that the results for each block should be obtained by one operative on one machine, i.e., possible differences within blocks that could arise from operative or machine differences should be eliminated.

To sum up, at each stage of the experiment, from the initial planting to the final testing, sources of uncontrolled variations are either identified with blocks and in effect eliminated from the treatment comparisons, or randomized, or possibly assumed negligible. The last course is avoided as far as possible, since, as discussed in Chapter 1, it is usually best to avoid assumptions about the nature of the uncontrolled variation.

The observations are given in Table 3.2(*a*); the five treatments have been denoted in order of increasing amount of K_2O, $T_1, \ldots, T_5$. (The detailed arrangement of treatments within blocks as shown has been obtained by randomizing the values given in Cochran and Cox and is presumably not the order actually used in the experiment.)

To analyze the observations* they are first rearranged as in Table 3.2(*b*) and the totals and means for each treatment (and block) calculated. Thus for the first treatment $7.62 + 8.00 + 7.93 = 23.55$, and this divided by 3 gives the treatment mean of 7.85. The differences between treatment means are the best estimates of the true treatment differences, provided that the basic assumption of Chapter 2 holds and that the amount of the uncontrolled variation does not vary appreciably from block to block.

To estimate the precision of these estimates we use formula (1) of § 1.2, i.e.,

$$\begin{pmatrix}\text{standard error of}\\ \text{difference of two}\\ \text{means of 3}\\ \text{observations each}\end{pmatrix} = \sqrt{\frac{2}{3}} \times \text{standard deviation.} \tag{1}$$

* The following account of the analysis may be omitted at a first reading.

TABLE 3.2

EXAMPLE OF RANDOMIZED BLOCK EXPERIMENT

(*a*) *Original Design and Observations*

Block 1	T_5:7.46	T_4:7.17	T_1:7.62	T_2:8.14	T_3:7.76
Block 2	T_2:8.15	T_1:8.00	T_5:7.68	T_4:7.57	T_3:7.73
Block 3	T_3:7.74	T_2:7.87	T_1:7.93	T_4:7.80	T_5:7.21

(*b*) *Rearranged Observations*

	T_1	T_2	T_3	T_4	T_5	Total	Mean
Block 1	7.62	8.14	7.76	7.17	7.46	38.15	7.63
Block 2	8.00	8.15	7.73	7.57	7.68	39.13	7.83
Block 3	7.93	7.87	7.74	7.80	7.21	38.55	7.71
Total	23.55	24.16	23.23	22.54	22.35	115.83	7.72
Mean	7.85	8.05	7.74	7.51	7.45	7.72	

(*c*) *Residuals*

	T_1	T_2	T_3	T_4	T_5
Block 1	−0.14	0.18	0.11	−0.25	0.10
Block 2	0.04	−0.01	−0.12	−0.05	0.12
Block 3	0.09	−0.17	0.01	0.30	−0.23

Estimate of standard deviation $= \sqrt{(0.3496/8)} = 0.2090$ (8 degrees of freedom)

Standard error of difference between two treatment means $= 0.2090\sqrt{(2/3)} = 0.171$

Estimated increase in strength per 18 lb K_2O increment is -0.090 with standard error 0.0251

We have first to estimate the standard deviation, that is the amount of uncontrolled variation from unit to unit. This is usually done by an elegant technique called *analysis of variance*; its application to the present problem has been described in full by Cochran and Cox and general accounts of the method will be found in any textbook on statistical methods.

It is, however, worth indicating briefly an equivalent method for estimating the standard deviation which, while rather inconvenient numerically, does indicate the physical basis for the estimate. We require to measure that part of the variation that is not due to real treatment effects and that cannot be regarded as systematic variation between blocks. Therefore it is natural first to express each observation as a difference from the overall mean and then to remove the variation accounted for by block differences. This is done by subtracting

$$\begin{pmatrix}\text{mean observation for}\\ \text{the particular block}\end{pmatrix} - \begin{pmatrix}\text{overall}\\ \text{mean}\end{pmatrix}.$$

Next the variation accounted for by treatments is removed by subtracting

$$\begin{pmatrix}\text{mean observation for}\\ \text{the particular treatment}\end{pmatrix} - \begin{pmatrix}\text{overall}\\ \text{mean}\end{pmatrix}.$$

At the end of this process we get, corresponding to each original observation, a *residual*, which may be defined directly as

$$\text{observation} - \begin{pmatrix}\text{mean observation}\\ \text{for the}\\ \text{particular block}\end{pmatrix} - \begin{pmatrix}\text{mean observation}\\ \text{for the}\\ \text{particular treatment}\end{pmatrix} + \begin{pmatrix}\text{overall}\\ \text{mean}\end{pmatrix}. \quad (2)$$

These are given in Table 3.2(*c*). Thus for the first observation we have that $7.62 - 7.63 - 7.85 + 7.72 = -0.14$. Except for rounding-off errors, the residuals add up to zero for each block and for each treatment.

The standard deviation measures the magnitude of the residuals and is calculated by finding the average of the squared residuals and then square-rooting the answer. However, in averaging the squared residuals it turns out to be appropriate to divide not by the number of residuals (15) but by what are called the *residual degrees of freedom*, (number of blocks − 1) × (number of treatments −1), which in this case is 8. The reason for this is essentially that if the 8 residuals in the upper left hand section of Table 3.2(*c*) were assigned arbitrarily, the condition that the row and column sums must be zero would determine the remaining residuals uniquely; i.e., effectively there are 8 *independent* residuals. Thus, the required estimate of standard deviation is

$$\sqrt{\{\tfrac{1}{8}[(-0.14)^2 + (0.18)^2 + \cdots + (-0.23)^2]\}} = 0.2090,$$

and is said to have 8 degrees of freedom. This is exactly the value that is given more quickly by the analysis of variance; the detailed table of residuals is, however, very useful if it is required to check the assumptions underlying the analysis. For example the occurrence of a single very large residual suggests that the corresponding observation may be suspect, whereas the distribution of the residuals gives information about the frequency distribution of error. It may sometimes happen that some blocks are much more variable than others and in extreme cases this too can be detected from the residuals, although considerable caution is needed in doing this. Important work on the examination of residuals has been done by F. J. Anscombe and J. W. Tukey; their work had not been published when this book went to press.

We now use formula (1) to estimate the standard error of the difference of two means as $0.2090 \times \sqrt{(2/3)} = 0.171$. In accordance with the account of the standard error given in § 1.2, the interpretation of this figure is that, for example, there is only a chance of about 1 in 20 that the estimate of a single preselected effect is in error by more than $\pm 2 \times 0.171 = \pm 0.342$. However, as explained in § 1.2, this interpretation needs some modification when the standard deviation is itself only estimated from a small number of observations, and in fact the residual degrees of freedom determine what this modification should be. The 1 in 20 limits for 8 degrees of freedom are increased to 2.31 × standard error, i.e., to plus or minus 0.395. This increase from 2 to 2.31 to allow for the uncertainty in the estimate of error is explained in textbooks on statistics and is an example of the use of what is known as "Student's" t distribution. Further modification of the multipliers is needed if they are applied solely to differences suggested by the data, such as to the difference between the treatments with highest and lowest mean responses.

The essential general points in this calculation are first the estimation of the treatment effects by a straightforward process of averaging, and second the estimation of the variation of the observations when treatment and block differences are removed. The important principle here, which applies also to more complicated cases, is that when the effect of a source of variation, for example block differences, is eliminated in the planning of the experiment, it must also be eliminated in the analysis, if an appropriate measure of error is to be obtained.

In this particular experiment the five treatments bear a special relation to one another in that they represent different levels of a continuous quantity, the amount of K_2O per acre. It is, therefore, natural to consider not just the differences between different treatments, but also the curve of mean strength against the amount of K_2O and, in particular, to see whether this curve is effectively a straight line. Standard statistical methods of regression analysis (Goulden, 1952, p. 102) can be used to show that the curve does not depart significantly from a straight line representing a decrease in strength of 0.090 per 18 lb K_2O per acre increment. The standard error of the slope is 0.0251.

Finally it is often worth examining the block means, even though they do not bear directly on the estimation of treatment effects. First, it is possible to assess from the magnitude of the differences between blocks whether the grouping of the units into blocks has appreciably reduced the error and this information may be useful in planning similar experiments in the future. Second, particularly in experiments with more blocks than the present one, a detailed examination of the block differences may be very helpful. For example, if two operatives had been used in the strength testing, a comparison of block means to see whether there is evidence of a systematic difference between operatives might be interesting. No very cogent conclusions would normally be drawn because of the difficulty of disentangling operative differences from other sources of block variation. (However, we shall later consider *split plot experiments*, which are essentially randomized block experiments in which a further set of treatments are applied to whole blocks, and in these reliable conclusions can be drawn from block differences.) It must be repeated that the examination of block means tells us nothing about the treatment effects and is of interest only in adding to the general knowledge of the experimental material.

(ii) Missing Values

The relatively simple analysis just described depends in an essential way on the balanced nature of the randomized block design. For example, it is only because each treatment occurs the same number of times within each block that the mean observations on the treatments can be used to compare treatments in a way unaffected by constant block differences. If, for instance, treatment T_1 did not occur in the first block and if the first block happened to give systematically high results, the mean for T_1 would be depressed relative to the means for the treatments that did occur in the first block, so that the treatment means would no longer give a fair basis for comparing treatments, free of block effects.

It can happen, particularly in experiments with many units, that the results on one or more units are lost, do not become available, or have to be discarded. For instance, if the units are animals, some may die from causes unconnected with the treatments. This loss will destroy the property of balance, i.e., the pattern of observations will no longer be that of a randomized block design.

A special case of a general principle, called the method of least squares, can be used for the efficient analysis of observations grouped into blocks and subject to treatments arranged in an unbalanced scheme, but the calculations tend to be complicated. Fortunately a very simple method is available when observations are missing for only one unit. This is to calculate a so-called estimated missing value by the formula

$$(kB + tT - G)/[(k - 1)(t - 1)],$$

where k is the number of blocks, t is the number of treatments, B is the total of all remaining observations in the block containing the missing observation, T is the total of observations on the missing treatment, and G is the grand total.

Then we analyze by the straightforward method, just as if the estimated missing value is a genuine observation. A small modification is that the degrees of freedom for residual are reduced by one. This procedure gives the same estimated treatment effects as the method of least squares and also the same estimated standard error for comparing two treatments for which no observations are missing. The correct standard error for comparisons involving the missing treatment is slightly greater and is closely approximated by using formula (2) of Chapter 1, allotting the treatment with the gap one fewer observations than the other treatments.

The importance for experimental design of the missing-value formula is that it would have been a serious drawback to the use of randomized blocks, and of course also of more complex designs, had the analysis and interpretation been greatly complicated whenever an observation is lost, or more generally whenever it proves impossible to get data in exactly the form intended. The existence of the formula means that the randomized blocks design can safely be adopted even when the occurrence of occasional missing values is expected.

Extensions of the method can be used if there are several missing observations; it is then necessary to solve a set of simultaneous linear equations, the number of equations being equal to the number of missing values. Similar methods are available if by accident the treatments are not applied exactly according to plan and the pattern of treatments departs somewhat from the randomized blocks form.

Analogous formulas are available for the other designs described in the present book (see Cochran and Cox, 1957; Goulden, 1952).

(iii) Further Examples

Example 3.3. Another application of randomized blocks is in laboratory work with animals such as mice or rats. Suppose for definiteness that there are five treatments under comparison, their nature depending on the particular field of application, but being, for example, different diets, different amounts and types of drug, different diets fed to rats during pregnancy, etc. The final observations might be of the amount of a certain substance in an organ of the animal at the end of the experimental period or, for the last case, some characteristic of the offspring.

To make successful use of the idea of randomized blocks, we begin by grouping the animals into sets of five in such a way that the final observation that would be obtained under uniform treatment is expected to be as nearly as possible constant within each set. Any special knowledge of the animals, such as their performance in previous experiments, can be used. In the absence of special knowledge it is common to rely either on the correlation that often exists between the final observations and a suitable, easily measured, initial property of the experimental animal, such as body weight, or on the general fact that animals from the same litter tend to respond similarly.

To use the last property, five suitable animals are taken from a number of litters of five or more animals, numbering the animals in each litter $1, \ldots, 5$ in any convenient way. The order of treatments is then randomized within each block to give some arrangement such as

Litter 1. Animal 1, T_3: 2, T_1: 3, T_5: 4, T_2: 5, T_4
Litter 2. Animal 1, T_2: 2, T_5: 3, T_3: 4, T_4: 5, T_1,
etc.

To use a quantitative property, such as body weight, to form blocks, the animals are numbered in order of increasing body weight, say, 1 through 20 if four blocks are required. Animals 1 through 5 form the first block, 6 through 10 the second block, and so on, the order of treatments again being randomized independently within each block. The effect of this is that the animals within any one block have approximately the same body weight. It sometimes happens that the first or last blocks contain one or more animals with very extreme body weights, so that there is appreciable variation of body weight within these blocks. If practicable, this is best avoided, for example, by starting with a few more animals than it is intended to use and discarding those with very extreme weights.

The results from such a design can be analyzed by the method of Example 3.2, the value of initial body weight being ignored once the grouping into blocks has been determined. In the next chapter we shall consider an alternative method of using the initial quantitative variable, in which the actual value is used in the analysis.

Example 3.4. In certain textile investigations it is required to test a number of modifications in a process for producing a thin web of parallel fibers. One important property of the web is the number of fiber entanglements, say per mg of web, and this is measured by passing a section of web slowly over an illuminated strip, when individual entanglements can be noted and the total found. However it is difficult to define precisely what constitutes an entanglement so that, whereas one observer can get reasonably reproducible counts over a short period of time, there are liable to be large systematic differences between observers and between the same observer's counts on different days. This example is typical of an important class of experiments in technology in which the properties under investigation are either rather difficult to define precisely and so are subject to personal errors of measurement, or, in extreme cases, are essentially matters of subjective judgement.

The first step in planning this sort of experiment is to take all reasonable precautions to eliminate the sources of systematic variation, for example by displaying in front of the observer photographs or slides showing typical fiber arrangements, some to be counted as entanglements and some not. The remaining systematic variations can then be reduced by the randomized block principle, as follows:

Suppose for definiteness that we have six different batches $W_1, \ldots, W_6$ of web to be compared. Let us assume to begin with that they have been produced by six different processes under highly controlled conditions, so that any difference between $W_1, \ldots, W_6$ can be confidently attributed to the effect of processes. A similar point arose in connection with the preceding example and this is of course just the sort of assumption that it is so often desirable to avoid; this can be done by having several batches from each process, produced and tested independently.

TABLE 3.4

PLAN FOR COMPARING SIX WEBS FOR ENTANGLEMENTS

		Order of Measurement					
		1	2	3	4	5	6
Block 1	Observer 1 First period	W_4	W_1	W_2	W_5	W_6	W_3
Block 2	Observer 2 First period	W_5	W_6	W_2	W_1	W_4	W_3
Block 3	Observer 1 Second period	W_3	W_6	W_2	W_4	W_1	W_5
Block 4	Observer 2 Second period	W_3	W_1	W_4	W_5	W_6	W_2
Block 5	Observer 1 Third period	W_6	W_5	W_4	W_3	W_1	W_2
Block 6	Observer 2 Third period	W_5	W_1	W_4	W_3	W_6	W_2
Block 7	Observer 1 Fourth period	W_2	W_3	W_6	W_4	W_1	W_5
Block 8	Observer 2 Fourth period	W_2	W_6	W_4	W_5	W_3	W_1

In each block there will be observations on all six webs and we want it to be possible to complete the observations in a block within a fairly short time, so that time differences are eliminated. Therefore, we take as units small sections of web that can each be examined for entanglements in say 10–15 min, the sections being selected by a random-sampling procedure. The number of sections of each web that it would be desirable to measure depends on the final precision required and on the regularity with which the entanglements are distributed, and they would have to be determined from previous work or from the results of a preliminary experiment. Suppose that eight sections are judged adequate and that two observers are available, each measuring four times. Then an arrangement in randomized blocks is shown in Table 3.4.

The order of webs is randomized independently within each block. Subjective biases of measurement are minimized by concealing from the observer the identity of the section under analysis. In analyzing results it would be desirable to check the consistency of the observers in their comparison of the 6 webs. It is not necessary to randomize the allotment of observers to blocks, because the object of the experiment is not the comparison of observers; the only purpose that would be served by randomizing observers over blocks would be that of ensuring the absence of systematic differences in the external conditions during the two observers' measurements. Note that if the primary object had been an examination of the difference between observers, it would have been advisable to have had each section of web measured by both observers. However when the object is the comparison of webs, the more distinct sections that are taken from each web the better, provided that the main cost of the experiment is in the counting of entanglements and not in the selection of sections or in the cost of the material that is in effect destroyed in sampling the web.*

In this example the randomized block principle has been used to eliminate the effect of systematic variations arising in the actual measurement, rather than variations arising from the experimental material itself. There are other ways of achieving this end. For example we may insert into each series of sections of experimental webs, a section of standard web, which has been counted many times and may be considered to have a known number of entanglements. The observation actually recorded on the standard section is then used to adjust the remaining observations. Another possibility, which may well be the best if large observer and time differences seem unavoidable, is to abandon the idea of directly counting entanglements and instead to measure the amount of entanglement either by assigning a score to each section after a subjective comparison of it with standard sections showing varying degrees of entanglement, or alternatively by direct ranking of a series of sections in order of increasing apparent degree of entanglement. The discussion of the relative advantages of these procedures raises difficult general questions; they are dealt with briefly later (§ 9.4).

We have now had several examples of the use of randomized blocks. The grouping of the material into blocks eliminates the effect of constant differences between blocks and the randomization allows us to treat the remaining variation between units as random variation, so far as assessing treatment comparisons is concerned. The success of the method depends on a good grouping of the units into blocks. The general idea of grouping into blocks is of fundamental importance and is not only frequently used in simple experiments but also forms the basis for most of the more complicated designs.

Sometimes a generalized form of the randomized block design is useful. It may be required to make some treatment comparisons more precisely

* For example, if only a very limited quantity of each web is produced and it is required to leave as much as possible for further processing, it might be advisable to measure each section more than once. If the magnitudes of the different components of variation can be estimated and if the relative costs of the various stages of the experiment can be measured, the optimum distribution of effort can be determined.

than others. For example, we may have a control treatment C and a number of other treatments T_1, T_2, . . . and the main interest may be in comparing T_1, T_2, . . . individually with C rather than in comparing T_1, T_2, . . . among themselves. In such a case it is proper to devote more units to C than to each T treatment. The block principle can still be used with a simple analysis, provided that the number of times a particular treatment occurs in a block is the same for all blocks. Thus, in the above example C might occur four times in each block and T_1, T_2, . . . once in each block. The difference between two treatment means is again unaffected by constant differences between blocks.

3.4 ELIMINATION OF ERROR BY SEVERAL GROUPINGS OF THE UNITS

(i) Latin Squares

In the preceding section we have seen how a single grouping of the units into blocks can be used to reduce the error of an experiment. Sometimes two or more systems of grouping suggest themselves and it may be desired to use them simultaneously. For instance in a paired comparison experiment, like Example 3.1, it might happen that there is reason to expect a systematic difference between the first unit and the second unit in the pair. Then we should have two systems of grouping, into pairs and into order within pairs, and we would wish to balance out both the associated types of systematic variation. A discussion of this example would involve one or two special points, and it is convenient instead to introduce the basic design, the Latin square, with a somewhat different problem.

Example 3.5. Consider an industrial experiment in which four processes are under comparison and in which it is suspected that there will be systematic changes in external conditions from day to day and also between different times of day, e.g., observations on material processed in the early morning may on the whole be lower than on material processed in the afternoon, etc. Suppose that the number of units that can be dealt with on one day is limited and that four is a convenient number, say two in the morning and two in the afternoon. Suppose also to begin with that four observations on each process are considered likely to give sufficient precision. The sixteen experimental units can then be set out in the square array shown in Table 3.5(*a*). If we preferred to use "days" as blocks in a randomized block design, we should arrange that each process is used once on each day, the arrangement otherwise being random. If we preferred to use "times of day" as blocks in a randomized block design, we should arrange that each process is used once at each time of day. Therefore if we wish to eliminate both sources of variation simultaneously we must arrange the four processes P_1, P_2, P_3, P_4 in the 4×4 square of Table 3.5(*a*) so that each letter occurs once in each row and once in each column.

One such arrangement is shown in Table 3.5(*b*) and is an example of a 4 × 4 *Latin square*. In general an $n \times n$ Latin square is an arrangement of n letters in an $n \times n$ square, such that each letter occurs once in each row and once in each column.

The particular Latin square given in Table 3.5(*b*) has been obtained by randomization in a way to be described in Chapter 10. The procedures to be used if it is required to have more experimental units, for example to have eight units for each treatment, will be discussed later.

TABLE 3.5

A LATIN SQUARE DESIGN

(*a*) *General Arrangement of Experimental Units*

	Time of Day			
	Time 1	Time 2	Time 3	Time 4
Day 1	—	—	—	—
Day 2	—	—	—	—
Day 3	—	—	—	—
Day 4	—	—	—	—

(*b*) *A Latin Square*

	Time 1	Time 2	Time 3	Time 4
Day 1	P_2	P_4	P_3	P_1
Day 2	P_3	P_1	P_2	P_4
Day 3	P_1	P_3	P_4	P_2
Day 4	P_4	P_2	P_1	P_3

The analysis of the observations from a Latin square is done by a procedure exactly analogous to that for the randomized blocks design. The treatment effects are estimated by comparing the average observations for the different treatments, and the estimate of the error standard deviation is obtained either from the appropriate analysis of variance or by calculating residuals. In accordance with the principle stated in § 3.3, every source of variation balanced out in the design of the experiment must be removed in the analysis before the standard deviation is estimated. The definition of the residual corresponding to a given observation is thus

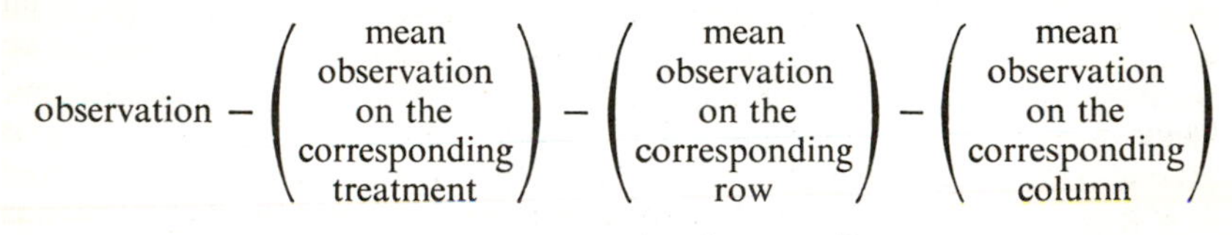

$$\text{observation} - \begin{pmatrix}\text{mean}\\ \text{observation}\\ \text{on the}\\ \text{corresponding}\\ \text{treatment}\end{pmatrix} - \begin{pmatrix}\text{mean}\\ \text{observation}\\ \text{on the}\\ \text{corresponding}\\ \text{row}\end{pmatrix} - \begin{pmatrix}\text{mean}\\ \text{observation}\\ \text{on the}\\ \text{corresponding}\\ \text{column}\end{pmatrix}$$

$$+ \text{ twice the overall mean,}$$

and the estimate of the standard deviation is

$$\sqrt{\left\{\frac{1}{(n-1)(n-2)} \times \text{sum of squares of the residuals}\right\}},$$

for an $n \times n$ square. The divisor $(n - 1)(n - 2)$, the residual degrees of freedom for the Latin square, is the number of independent residuals. The estimated standard error of the difference in mean observation is*

$$\begin{pmatrix}\text{estimate of} \\ \text{standard deviation}\end{pmatrix} \times \sqrt{\left(\frac{2}{\text{no. of observations per treatment}}\right)}.$$

Full details of the procedure of analysis are given in textbooks on statistical methods.

The example just discussed shows that the Latin square arrangement is a simple and natural extension of the randomized block design. The example is of wide applicability, since there are many types of work in which the rate at which experimental material can be dealt with is limited and in which it is worth balancing out certain time variations. Another common possibility arises when there are a number of observers or sets of apparatus or machinery, which can be used simultaneously. The Latin square arrangement can then be used with the rows standing for different times, and the columns for the different sets of apparatus, etc. In this way systematic time differences and systematic differences between sets of apparatus, etc., are eliminated.

A restriction, which clearly limits the use of the Latin square in its simple form, is that the number of rows, the number of columns, and the number of treatments must all be equal. Arrangements not restricted like this are discussed in Chapter 11. It is convenient now to consider some more examples of the use of the Latin square.

Example 3.6. In an agricultural field trial to compare a fairly small number of varieties or treatments, the best arrangement of plots depends in part on the shape of the plots, which is largely dictated by technical considerations. For example in a variety trial, particularly if it is required to have small plots, the plots would be long and narrow, only a few drills wide. In this case a natural grouping of plots for the use of a randomized block design is that shown for six varieties in Table 3.6(a), in which the blocks are approximately square and are, if possible, oriented to minimize the effect of the predominant fertility variations, if the best direction for doing this is known from previous experience of the field. If, on the other hand, the plots are more nearly square, a compact arrangement, such as that illustrated in Table 3.6(b), would usually be better. The exception is when it is confidently expected that fertility variations in one direction will be much larger than variations in the perpendicular direction; in this case the arrangement corresponding to Table 3.6(a) is likely to be successful. The objection to this design would ordinarily be that with wide plots, the whole block is of considerable extent and is therefore likely to contain excessive variation; all the variation of fertility between plots within the long block would contribute to the error. The design would be particularly bad if the predominant direction of fertility variation was misjudged and happened to lie parallel to the length of the block.

* In fact this formula, or its generalization (Chapter 1, Eq. (2)) applies for all designs in which the treatment effects are estimated by simple treatment means.

TABLE 3.6

AGRICULTURAL FIELD TRIALS

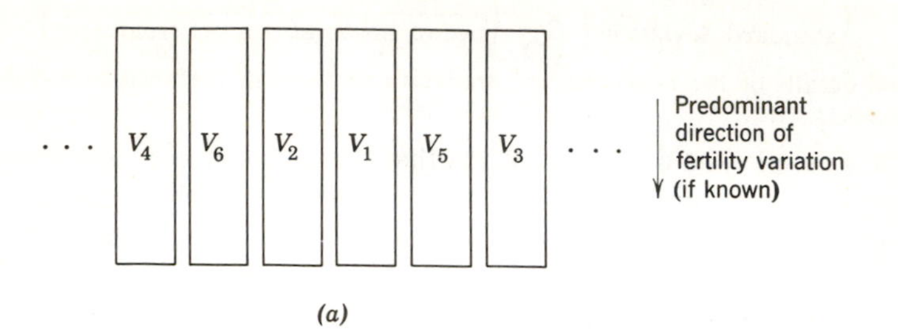

(a)

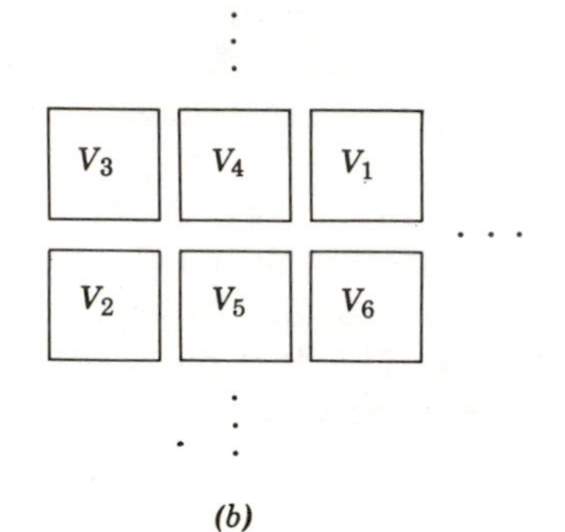

No predominant direction of fertility variation

(b)

V_2	V_1	V_6	V_4	V_3	V_5
V_1	V_5	V_2	V_3	V_4	V_6
V_4	V_3	V_1	V_6	V_5	V_2
V_3	V_2	V_4	V_5	V_6	V_1
V_6	V_4	V_5	V_2	V_1	V_3
V_5	V_6	V_3	V_1	V_2	V_4

(c)

Consider, however, the Latin square arrangement shown in Table 3.6(*c*). Here we are eliminating fertility variations in two directions and so there is a good chance that one or other, if not both, of the groupings will account for an appreciable portion of uncontrolled variation. Much depends on the particular circumstances and on special information that may be available in particular cases, but it seems that for many types of field experiment with approximately square plots and with not more than about ten to twelve treatments, the Latin square is a good design, and is to be preferred to randomized blocks, provided that it is reasonable to have the number of plots per treatment equal to, or a simple multiple of, the number of treatments. Exceptions to this have, however, been reported in the literature.

Example 3.7. The Latin square principle is employed frequently in experiments in which the same object or person is used several times. Thus for the cattle experiment discussed in Example 2.7, suppose that interference between different two-week periods can be ignored. Then we have two types of systematic error to try to eliminate, arising from variations between animals and from the common time trend between periods. Therefore the Latin square design is indicated. For each group of three similar animals a 3×3 Latin square is used, as shown in Table 3.7; the randomization of each 3×3 square is done independently.

TABLE 3.7

ANIMAL FEEDING TRIAL

	Two-week Period		
	1	2	3
Cow 1	*C*	*A*	*B*
Cow 2	*B*	*C*	*A*
Cow 3	*A*	*B*	*C*

The diets under comparison are denoted by *A*, *B*, *C*.

The three animals in a square should be chosen so that they are likely to have similar lactation curves and to be at corresponding points on the curve at the start of the experiment. The reason for this is that the balancing of columns in the Latin square removes the effect of a common trend in milk yield, but that if the trends are appreciably different for the three animals, the error of the treatment comparisons is inflated. The whole experiment will consist of several squares of the above type. It does not matter if the trends are different in the different squares; what is important is that as far as possible any trends that do exist should be the same for the three animals in any one square. If this state of affairs is not attained the randomized Latin square arrangement is, of course, still a perfectly valid experiment, giving treatment comparisons of measurable precision and free of systematic error. The point is that precision is lost.

The straightforward use of the Latin square applies when there is no carry-over of the treatment effect from one period to another. We shall see later that a special sort of Latin square is the design to use when there is a carry-over of treatment effect.

Example 3.7 could be paralleled from many applied fields. For example in some experiments on the fuel consumption of buses, described by Menzler (1954), four vehicles were used to compare four different tire pressures, four different tread thicknesses, or four different methods of operation, the experiment being repeated on four days. There are two types of systematic variation to be balanced out, that between days and that between vehicles, and a 4×4 Latin square is appropriate, with the rows representing vehicles and the columns, days.

Example 3.8. The following case of the misuse of the Latin square, quoted by Babington Smith (1951), illustrates the importance of considering the basic assumptions of Chapter 2. Four backward readers, Tom, Dick, Harry, and George undergo in succession four trainings in spelling, denoted by *A*, *B*, *C*, *D*, the treatments being arranged in a Latin square as in Table 3.8. An observation

TABLE 3.8

COMPARISON OF TRAINING METHODS

	Period 1	Period 2	Period 3	Period 4
Tom	*B*	*A*	*D*	*C*
Dick	*C*	*B*	*A*	*D*
Harry	*A*	*D*	*C*	*B*
George	*D*	*C*	*B*	*A*

is made on each subject at the end of each period, using a standard type of spelling test.

The justification for using the Latin square is that it aims at balancing out differences between subjects, and also any systematic effects accounted for by the order in which the methods of training are applied. However, we have an untenable assumption, namely that the observation obtained, say with method *D* for Tom in the third period, is unaffected by the particular choice of treatments for Tom in the preceding periods. The observation obtained on a subject in a particular period is likely to depend, probably in rather a complicated way, on all the training received up to that point. It is not difficult to conceive of situations in which the comparison of methods of training by a simple averaging of observations would give quite misleading results.

A list of standard Latin squares, from which designs can be constructed by randomization, is given in Chapter 10, where there is also a discussion of more complicated designs based on the Latin square principle. Two simple extensions of the Latin square design will be discussed briefly here; the first is that several Latin squares may be used simultaneously and the second is that instead of two sources of systematic variation to be dealt with, there may be three or even more.

(ii) Combined Latin Squares

In any experiment of the type we have been considering, for which an $n \times n$ Latin square is appropriate, it may well happen that more than n

units are required for each treatment to get estimates of adequate precision. This situation can be dealt with simply if one of the sides of the square, say the rows, represents extension in space or time, and if a multiple of n^2 units can be used.

Example 3.9. Consider again Example 3.5, where the rows of the square represent different days, the columns times of the day. If we want to extend the experiment over twelve days instead of four, two procedures are available and are illustrated in Table 3.9(*a*) and (*b*).

In the first design, Table 3.9(*a*), we have three independently randomized Latin squares placed underneath one another. In the second design, Table 3.9(*b*), we have completely randomized the rows of the previous design, so that, for example, the first four rows by themselves no longer necessarily form a Latin square.

The difference in practice between the designs is best seen by considering what types of systematic variation are eliminated from the error by the two designs. In both cases constant differences between days have no effect on the treatment comparisons. In the second design, the effect of constant differences between times of the day persisting throughout the whole experiment is likewise eliminated. In the first design, however, not only is this done, but also time of day effects are eliminated separately from each set of four days. This would be particularly useful if, as might be convenient, there is a considerable gap in time between the sets of four days, or if it were desired to introduce some external change in conditions, either of which things might mean that time of day effects would not be the same in all parts of the experiment.

Therefore in general we prefer the design (*a*) because it achieves all that (*b*) does and more. There are, however, two considerations which prevent (*a*) always being better than (*b*), although neither consideration is likely to be of any importance in the present case. First we would usually need to estimate the error standard deviation from the results of the experiment itself and, as we have seen above, the accuracy with which we do this, measured by the degrees of freedom for residual, affects somewhat the effective precision of the experiment. Now the residual degrees of freedom for the design in Table 3.9(*a*) may be shown to be 24 and the corresponding value for Table 3.9(*b*) is 30. In accordance with the full discussion in Chapter 8, this means that if the true standard deviations corresponding to the two designs were equal, the effective standard deviation for Table 3.9(*b*) would be about 1 per cent less than that for Table 3.9(*a*). This gain is negligible, but in smaller experiments the corresponding gain may be enough to justify the use of the design analogous to Table 3.9(*b*), in a case where there is good reason to expect any column effects to be constant throughout the experiment.

The second consideration, which is of merely academic interest in the present case, is that with very special patterns of uncontrolled variation Table 3.9(*b*) may give the more precise design, even apart from the consideration of the residual degrees of freedom. Suppose, to take the extreme case, the observations in the absence of treatment effects were of the form in Table 3.9(*c*), with only two possible values x, y distributed in the pattern shown. Notice that the systematic variation between days is zero, since for each day the observations add up to $2(x + y)$. Likewise if the columns are taken in sets 1–4, 5–8, 9–12 there is no systematic variation between columns. Of course, if it were known or suspected

TABLE 3.9

EXTENDED LATIN SQUARE DESIGNS

(*a*) *Separate Latin Squares*

	Time of Day				
	Time 1	Time 2	Time 3	Time 4	
Day 1	P_3	P_2	P_1	P_4	
2	P_2	P_3	P_4	P_1	Replicate
3	P_4	P_1	P_3	P_2	1
4	P_1	P_4	P_2	P_3	
5	P_4	P_1	P_2	P_3	
6	P_3	P_4	P_1	P_2	Replicate
7	P_2	P_3	P_4	P_1	2
8	P_1	P_2	P_3	P_4	
9	P_2	P_4	P_3	P_1	
10	P_3	P_1	P_2	P_4	Replicate
11	P_4	P_3	P_1	P_2	3
12	P_1	P_2	P_4	P_3	

(*b*) *Intermixed Latin Squares*

	Time 1	Time 2	Time 3	Time 4
Day 1	P_3	P_1	P_2	P_4
2	P_2	P_3	P_4	P_1
3	P_3	P_4	P_1	P_2
4	P_4	P_1	P_3	P_2
5	P_3	P_2	P_1	P_4
6	P_2	P_3	P_4	P_1
7	P_4	P_1	P_2	P_3
8	P_1	P_2	P_4	P_3
9	P_2	P_4	P_3	P_1
10	P_1	P_4	P_2	P_3
11	P_4	P_3	P_1	P_2
12	P_1	P_2	P_3	P_4

(*c*) *A Very Special Pattern of Uncontrolled Variation*

	Time 1	Time 2	Time 3	Time 4
Day 1	x	x	y	y
2	x	x	y	y
3	y	y	x	x
4	y	y	x	x
5	x	x	y	y
		and so on up to		
12	y	y	x	x

that the uncontrolled variation was of this form, neither of the designs we are considering would be at all appropriate. However, if the pattern of variation in Table 3.9(*c*) did occur, the standard deviation for design (*a*) can be shown to be $\sqrt{(33/27)} = 1.11$ times that for design (*b*). This, the largest factor in favor of (*b*) that can occur, does not represent a large change in precision and in any case only arises in the exceptional circumstances of Table 3.9(*c*).

We can formulate an important general rule for experiments of this type; that it is better to keep separate the sections, e.g., Latin squares, from which the whole design is built, except possibly when the experiment is a small one with few degrees of freedom available for the estimation of error, or when very special patterns of uncontrolled variation may arise.

All the arrangements that we have considered so far give equal precision for the comparison of every pair of treatments. If, however, it is required, say, to have two observations on A for each observation on B, C, etc. this can be done most simply by taking a 5×5 Latin square for treatments A, B, C, D, E and applying the treatment A every time the letters A or E occur. If the conditions of the experiment do not allow the use of five units in one day, the more complicated "unbalanced" arrangements of Chapter 11 must be used, if the additional observations on A are to be obtained.

Example 3.10. As a final example of the simple two-way elimination of error, consider the paired comparison experiment, Example 3.1, in which it is required to compare two treatments T_1, T_2, eliminating from the effective error not only the variation between pairs of units, but also any systematic variation associated with the order in which the units are arranged within pairs.

If we look back at Table 3.1, which shows a particular arrangement of treatments with one-way grouping of the units, we see that T_1 occurs three times in the first position and five times in the second. What we want, if it is expected that the observation on the first unit in a pair will tend to be larger (or smaller) than the observation on the second unit in the pair, is that each treatment should occur four times in each position. This raises essentially the same problems as Example 3.9. The key design in the 2×2 Latin square:

$$\begin{matrix} T_1 & T_2 \\ T_2 & T_1 \end{matrix} \qquad \text{or} \qquad \begin{matrix} T_2 & T_1 \\ T_1 & T_2 \end{matrix}$$

and we want four of these to build up the requisite number of observations. There are three different arrangements that merit consideration. First we might consider taking four Latin squares kept separate, analogous to the arrangement in Table 3.9(*a*). It can be shown that this design has only three degrees of freedom for residual and this would ordinarily be insufficient (see § 8.3), so that this arrangement would be used only in very special circumstances. The second arrangement, analogous to Table 3.9(*b*), is obtained by choosing completely at random four pairs to receive the order $T_1\ T_2$ and assigning the order $T_2\ T_1$ to the remaining four pairs. This may be shown to leave 6 degrees of freedom for residual, and is the arrangement that would normally be used. The third method is an intermediate arrangement, which is worth considering when the pairs fall naturally into two equal sets in which the order effects are quite possibly different. In this design the treatments are randomized separately within each set, so that $T_1\ T_2$ and $T_2\ T_1$ both occur twice in each set. This leaves 5 degrees of freedom for residual. The reader should write out examples of the three methods and consider carefully the types of systematic variation balanced out by each.

In experiments like this in which the degrees of freedom for residual are inevitably small, it will be worth considering whether useful information about the error standard deviation can be derived from the results of previous similar experiments.

(iii) Graeco-Latin Squares

The randomized block design is useful when the experimental units are grouped in one way. The Latin square design is useful when the units are simultaneously grouped in two ways. It is natural to consider what can be done if the units are grouped in three (or even more) ways.

Example 3.11. Consider again Example 3.5 used to illustrate the idea of a Latin square. Suppose that the observations are made by four observers and that each experimental unit is to be measured by one observer. Then, unless the absence of systematic observer differences can confidently be assumed, which would not be often, each observer should measure one unit of each process.

This could be done by a further application of the randomized block principle. One unit could be selected at random from each process for observer 1, a further set for observer 2, and so on. However, it would normally be an advantage to be able, in the analysis, to separate out differences between observers, between days and between times of day. Although this separation is not essential for the immediate purpose of comparing the processes, it may give information about the uncontrolled variation, of value both in attaining a general understanding of the experimental set-up and in designing future experiments.

Therefore we want to superimpose on the Latin square in Table 3.5(*b*) the symbols O_1, O_2, O_3, O_4 for four observers in such a way that

(*a*) each observer occurs once in combination with each process;

(*b*) each observer measures once on each day and once at each time of day.

The second condition is satisfied if the O's, considered by themselves, form a Latin square.

One such arrangement, after randomization, is shown in Table 3.10(*a*). The

TABLE 3.10

EXPERIMENT IN A GRAECO-LATIN SQUARE

(*a*) *Arrangement of Processes and Observers*

	Time 1	Time 2	Time 3	Time 4
Day 1	P_2O_3	P_4O_1	P_3O_2	P_1O_4
Day 2	P_3O_4	P_1O_2	P_2O_1	P_4O_3
Day 3	P_1O_1	P_3O_3	P_4O_4	P_2O_2
Day 4	P_4O_2	P_2O_4	P_1O_3	P_3O_1

(*b*) *The Same with Latin and Greek Letters*

$B\gamma$	$D\alpha$	$C\beta$	$A\delta$
$C\delta$	$A\beta$	$B\alpha$	$D\gamma$
$A\alpha$	$C\gamma$	$D\delta$	$B\beta$
$D\beta$	$B\delta$	$A\gamma$	$C\alpha$

processes are arranged in the same way as in Table 3.5(*b*). Note that the *O*'s form a Latin square and that condition (*a*) is satisfied because, for example, O_3 occurs in combination with P_2 just once. An arrangement like this is called a *Graeco-Latin square*. The reason for this name is that it is a common convention to rewrite the square replacing one set of symbols, say $P_1, \ldots, P_4$, by Latin letters *A*, *B*, *C*, *D* and the other set of symbols, $O_1, \ldots, O_4$, by Greek letters $\alpha, \beta, \gamma, \delta$. This has been done in Table 3.10(*b*); the general definition is that an $n \times n$ Graeco-Latin square is an arrangement of *n* Latin letters and *n* Greek letters in an $n \times n$ square in such a way that each Latin letter (and each Greek letter) occurs once in each row and once in each column, and that each combination of a Latin letter and a Greek letter occurs paired just once.

The statistical analysis of the results of such an experiment is a straightforward extension of that for a Latin square. Process, day, time of day, and observer effects are estimated by averaging and the error standard deviation is estimated either by analysis of variance or, equivalently, by forming and squaring the residuals, defined as

$$\text{observation} - \begin{pmatrix}\text{mean obs. on}\\ \text{corresponding}\\ \text{process}\end{pmatrix} - \begin{pmatrix}\text{mean obs. on}\\ \text{corresponding}\\ \text{day}\end{pmatrix} - \begin{pmatrix}\text{mean obs. on}\\ \text{corresponding}\\ \text{time of day}\end{pmatrix} - \begin{pmatrix}\text{mean obs. on}\\ \text{corresponding}\\ \text{observer}\end{pmatrix} + 3\,(\text{overall mean}).$$

The degrees of freedom for residual in an $n \times n$ square are $(n - 1)(n - 3)$, so that a 4×4 Graeco-Latin square gives only 3 degrees of freedom for residual. This would not, by itself, lead to an adequate estimate of error and so it would, with one replicate of this design, be necessary to have a supplementary estimate of error.

The Graeco-Latin square, though an important design both in principle and as a basis for constructing further designs, is not itself used very frequently in practice. The same applies to the more complicated squares, in which, for example, a third alphabet is placed in Table 3.10(*b*) in such a way that any alphabet by itself forms a Latin square and any pair of alphabets a Graeco-Latin square. This would be relevant if there were four simultaneous groupings of the experimental units.

Examples of Graeco-Latin and higher-order squares for small and moderate values of *n*, and instructions for their randomization, are given in Chapter 10. Some ingenious practical applications of these squares have been described by Tippett (1935).

3.5 THE NEED FOR MORE COMPLICATED ARRANGEMENTS

The essential point of randomized block and Latin square designs is that the experimental units are grouped into sets, the grouping being chosen so that the uncontrolled variation within sets is as small as possible.

It quite often happens that if this last condition is to be satisfied the number of units in a block must be small.

Thus, if the experimental units consist of pairs of identical twins, we are restricted to two units per block in order to make effective use of the similarity of the twins. If we wish to make one day's work a block in a randomized block design, this will set an upper limit to the number of units in a block, depending on how many units can be dealt with in a day. In an agricultural field trial there is no such clear upper limit to the number of plots per block, but the more plots in a block, the greater the area of the block and the more likely it is to contain substantial heterogeneity. Hence, there is again reason for limiting the number of plots per block, and twelve to sixteen is usually regarded as a maximum satisfactory number.

Now the randomized block design has at least as many units in a block as there are treatments, and similarly the simple form of Latin square has the number of rows and columns equal to the number of treatments. But what if the number of treatments exceeds the allowable number of units per block or exceeds the permissible number of columns in a Latin square design? For example, suppose that we wish to use the pairs of twins to compare five diets. We need an arrangement similar to randomized blocks, eliminating differences between blocks from the error, but having fewer units per block than the number of treatments. Much of the mathematically advanced work connected with experimental design aims at providing designs that enable this elimination to be achieved efficiently and simply. Some special types are balanced incomplete blocks, lattices, confounded arrangements, and so on. Similarly there are designs, Youden squares, lattice squares, and quasi-Latin squares, which fulfil the purpose of the Latin square but have the number of rows or columns, or both, less than the number of treatments. These will all be described later; the point of the present discussion has been to see just how the need for these more complicated arrangements arises.

SUMMARY

Knowledge available to the experimenter about the probable nature of the uncontrolled variation can be used to increase the precision of the treatment comparisons. The procedures considered in this chapter are:

(*a*) randomized blocks, in which the units are grouped into blocks and the treatments arranged randomly within blocks, each treatment occurring once, or more generally the same number of times, within each block;

(*b*) Latin squares, in which a similar method is used, although with two groupings of the experimental units.

These methods depend for their success on a skilful grouping of the units.

REFERENCES

Babington Smith, B. (1951). On some difficulties encountered in the use of factorial designs and analysis of variance with psychological experiments. *Brit. J. Psychol.*, **42**, 250.

Biggers, J. D., and P. J. Claringbold. (1954). Why use inbred lines? *Nature*, **174**, 596.

Cochran, W. G., and G. M. Cox. (1957). *Experimental designs*. 2nd ed. New York: Wiley.

Fertig, J. W., and A. N. Heller. (1950). The application of statistical techniques to sewage treatment processes. *Biometrics*, **6**, 127.

Goulden, C. H. (1952). *Methods of statistical analysis*. 2nd ed. New York: Wiley.

James, E. (1948). Incomplete block experiment with half plants. *Proc. Auburn Conference on Applied Statistics*, 52.

Menzler, F. A. A. (1954). The statistical design of experiments. *Brit. Transport Rev.*, **3**, 49.

Tippett, L. H. C. (1935). Some applications of statistical methods to the study of variation of quality in cotton yarn. *J. R. Statist. Soc., Suppl.*, **2**, 27.

CHAPTER 4

Use of Supplementary Observations to Reduce Error

4.1 INTRODUCTION

The methods described in the previous chapter use a qualitative grouping of the units. For instance, even in Example 3.3, where a quantitative measurement, body weight, is used to group experimental animals into blocks, no use of the measurements is made once the blocks have been formed, so that in effect only a ranking of the animals in order of increasing weight is used. It is natural to consider whether effective use can be made of the actual value of weight, either in addition to, or instead of, the grouping into blocks.

We shall deal with this topic in the present chapter. It is convenient to reserve the term *concomitant observation* for a supplementary observation that may be used to increase precision. An essential condition has to be satisfied in order that after use of the concomitant observation, estimated treatment effects for the desired main observation shall still be obtained. This condition is that the concomitant observations should be quite unaffected by the treatments.

4.2 NATURE OF CONCOMITANT OBSERVATIONS

We shall, therefore, consider situations in which in addition to the main observations, for which we want to find the treatment effects, we have for each experimental unit one or more concomitant observations. The essential point in our assumptions about these observations is that the value for any unit must be unaffected by the particular assignment of treatments to units actually used. In practice this means that either

(*a*) the concomitant observations are taken before the assignment of treatments to units is made; or

(*b*) the concomitant observations are made after the assignment of treatments, but before the effect of treatments has had time to develop.

This case occurs, for example, in some agricultural field trials, and in some laboratory experiments with animals; or

(*c*) we can assume from our knowledge of the nature of the concomitant observations concerned, that they are unaffected by treatment differences. For example in comparing a number of textile spinning processes, a main observation might be the end breakage rate and a concomitant observation the relative humidity in the spinning shed during processing. Both are taken during processing, after the treatments have been assigned to units, but it is clear that the relative humidity for any particular period of processing is unaffected by the process applied during that period, and so this observation is a concomitant observation in our sense.

Examples of concomitant observations have already been mentioned. Some more are the yield of product on a plot in years previous to the experimental year (which is particularly useful in experiments on perennial crops), the purity of the raw material in a chemical process, the weight of a particular organ (for example, the heart) of an experimental animal used in a biological assay, the score attained by a subject in a preliminary test in a psychological experiment, and so on.

4.3 THE USE OF A CONCOMITANT OBSERVATION AS AN ALTERNATIVE TO BLOCKING

Consider a situation, such as that of Example 3.3, where one concomitant observation on each unit is available, and suppose for simplicity that the same number of units are to be devoted to each treatment. We have seen in Example 3.3 how such an observation can be used to effect a grouping into blocks; suppose now either that the concomitant observations are of types (*b*) or (*c*) and so are not available at the time the treatments are allotted, or that it is desired to use the observations quantitatively in the analysis of the results instead of in the formation of blocks.

Suppose first that no alternative system of blocking suggests itself, so that the treatments are assigned to experimental units completely randomly. That is to say if five units are to receive treatment T_1, these are selected completely randomly from all the units available, using the methods described in the next chapter. Five more units are selected at random from the remainder for the second treatment, and so on. This is in a sense the simplest and most flexible design that can be used. Its disadvantage, if there is no concomitant observation, is that no attempt is made to reduce the effect of uncontrolled variation.

Suppose, however, that a concomitant observation, denoted by x, is made on each unit and that the main observation is denoted by y. Thus the full set of observations consists of a series of pairs (x, y), one pair for each experimental unit. Thus if the experiment concerned alternative chemical processes, x might be a measure of the purity of the raw material and y the yield of product.

Consider first what would happen if there were no treatment effects, i.e., if the observation obtained in any unit did not depend on the treatment applied to it. Imagine the value of y plotted against the corresponding value of x. Qualitatively two things may happen. There may

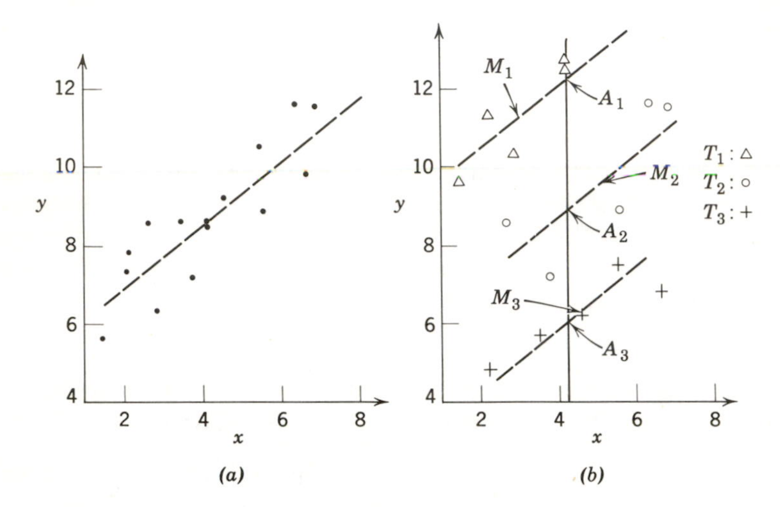

Fig. 4.1. The calculation of adjusted treatment means. (*a*) Plot of y against x in the absence of true treatment effects; (*b*) Corresponding plot after imposing treatment effects. The points M are unadjusted means, the points A are adjusted means, and the lines are fitted treatment lines.

be no appreciable relation between y and x, the points forming a random scatter. In this case no useful information about y can be obtained from the values of x. The interesting case is when the values of x and y cluster reasonably closely around a smooth curve. It is often convenient to assume that this curve is a straight line and we shall do this, but this assumption is not necessary; any simple smooth curve can be dealt with in a similar way.

Figure 4.1(*a*) shows a typical case for fifteen experimental units. The line that has been drawn through the points is known statistically as the regression line of y on x and its physical interpretation is that it gives, corresponding to any particular value of x, an estimate of what the mean value of y would be for a large number of experimental units similar to those used and all having the particular value of x. A detailed method for the calculation of the line is described in textbooks on statistics; for a rough estimate it is sometimes enough to divide the points into four or five sets in order of increasing x, to find, perhaps by eye, the centroid of each set and then to fit a line by eye to these five centroids, checking that there is no evidence of systematic departure from linearity.

Now consider the situation when there are treatment effects. Plot a graph corresponding to Fig. 4.1(a) but with the units receiving different treatments distinguished in some way. Figure 4.1(b) shows what will

TABLE 4.1

FICTITIOUS OBSERVATIONS TO ILLUSTRATE THE CALCULATION OF ADJUSTED TREATMENT MEANS

x	Imaginary Value of y before Applying Treatment	Treatment	Observed Value of y	x	Imaginary Value of y before Applying Treatment	Treatment	Observed Value of y
1.5	5.6	T_1	9.6	4.1	8.6	T_1	12.6
2.2	7.8	T_3	4.8	4.6	9.2	T_3	6.2
2.2	7.3	T_1	11.3	5.5	10.5	T_3	7.5
2.7	8.6	T_2	8.6	5.6	8.9	T_2	8.9
2.9	6.3	T_1	10.3	6.4	11.6	T_2	11.6
3.5	8.6	T_3	5.6	6.6	9.8	T_3	6.8
3.8	7.2	T_2	7.2	6.8	11.5	T_2	11.5
4.1	8.5	T_1	12.5				

	Unadjusted Treatment Means	Adjusted Treatment Means
T_1	11.26	12.20
T_2	9.56	8.87
T_3	6.18	5.94

result; the data on which this figure is based have been constructed from Fig. 4.1(a) by adding $+4$ to the y values of the five units selected at random for the treatment T_1, leaving five more unchanged for T_2 and adding -3 to the remaining units considered to have received T_3; the data themselves are given in full in Table 4.1.

What can we say in general about the form of the graph corresponding to Fig. 4.1(b)? First we expect to find that so far as the values of x are concerned, there are no systematic differences between treatments; for example we would have been surprised to find that the five units with lowest values of x all received T_1. This is because of our initial assumption that x is a concomitant variable and because the treatments were assigned to units randomly. If inspection of the results suggests that there are nevertheless systematic variations in x between treatments, there are three possible explanations;

(i) the effect is a chance one. Whether or not this is likely to be so can be assessed by a statistical significance test;

(ii) our belief that the allotment of treatments to units is random may be false. Now if the methods of objective randomization described in the next chapter have been used, this possibility can be disregarded, but it may happen that for some reason of convenience, or by an oversight, the assignment of treatments has been left systematic or determined subjectively by some procedure other than strict randomization. In such a case the assignment of treatments could be correlated with x.

(iii) for concomitant measurements of type (*b*) and (*c*), see pp. 48, 49, our assumption that x is unaffected by treatments may be false, i.e., the variations in x may represent a genuine treatment effect.

In certain applications (ii) or (iii), or both, can be ruled out. If the significance test indicates that it is very unlikely that the x differences are chance ones and (ii) is the only possible explanation, we may proceed with the methods to be described below, although the possibility should not be overlooked that correlation with x is not the only peculiarity in the allocation of treatments to units. However, if (iii) is a possibility, the methods must not be used in order to estimate the simple treatment effects of y; this case will be discussed in Example 4.6.

To sum up the discussion of this first point, we normally expect there to be no systematic differences between treatments in the values of x, but in certain cases it is all right to go ahead cautiously even if there are such differences.

The second general point follows from the basic assumption of Chapter 2 that the treatment effects are represented by the addition of constants. Therefore if, in the absence of treatment effects, we have a reasonably linear set of values such as Fig. 4.1(*a*), we shall find that the points for each treatment tend to cluster around a line and that the lines for the different treatments are parallel. This has happened in Fig. 4.1(*b*) and it will be clear why this is from the way the values for this diagram have been constructed in Table 4.1.

Therefore if the graph corresponding to Fig. 4.1(*b*) shows definite evidence of nonparallelism, the fundamental assumption about the treatment constants must be false. An apparent nonparallelism may arise from random fluctuations, i.e., from chance properties of the particular arrangement of treatments actually used, and the statistical significance of the nonparallelism can be tested by standard statistical methods. If real nonparallelism is established beyond reasonable doubt, it would usually be required to estimate the treatment effects separately for a number of values of x. Alternatively if the nonconstancy can be removed by a simple transformation of y, for example, to $\log y$ or to y/x, this would be done. These points will not be gone into in detail here, since they concern the analysis rather than the design of the experiment; the general effect on the interpretation of the experiment has already been discussed in §§ 2.2, 2.3 and will be mentioned again below. The main implication for experimental design is that if the treatment effects are suspected to vary systematically from unit to unit, it will be wise to record the values of suitable supplementary observations, in order that the variations in treatment effect may be detected and explained.

If the true treatment effects are constant but the initial values, in the absence of treatment effects, tend to cluster not around a line but around a curve, the final graph corresponding to Fig. 4.1(*b*) will consist of a series of parallel curves, one for each treatment. The nonlinearity makes no difference in principle to the argument, but complicates the procedure. It is therefore desirable to make the relation effectively linear if this can be done easily. The method is to use a transformed concomitant variable, such as $\log x$, $1/x$, $\sqrt{x}$, etc., selected after an initial graphical analysis in terms of the original variable, x. We shall not go into details.

To sum up, we usually expect to find that there are no systematic differences between treatments in the values of x, and that the points for different treatments lie along parallel lines (or curves). We can now see how to use the diagram to obtain improved estimates of the treatment effects. Fit parallel lines to the sets of points and call these *treatment lines*. Take any convenient value of x, say the overall mean value of x, and find the corresponding value of y on each treatment line. Call these values the *adjusted treatment means*. Then the differences among these quantities are estimates of the true treatment effects. This process can sometimes be done adequately by purely graphical methods, but if an objective answer is required, or if the standard error of the adjusted values is wanted, the whole procedure should be done arithmetically by the statistical technique called analysis of covariance (Goulden, 1952, p. 153). Again the details of this need not concern us; it is simply a question of fitting the parallel lines (or curves) arithmetically by statistically efficient methods,* rather than graphically.

The reason why the adjusted treatment means give a better estimate of the treatment effects than the uncorrected treatment means can be seen by thinking about the results for T_1 and T_2 in the numerical example.

* Some complications will arise if the scatter of y about the fitted line or curve varies appreciably with x.

By chance T_1 has been tested on units with on the whole lower values of x than those for T_2, and this means that the uncorrected mean M_1 of observations on T_1 will tend to be low. The adjusted value of A_1, obtained by sliding along the treatment line, is in effect an estimate of what the treatment mean for T_1 would have been had the units for T_1 had average values of x. The comparison of the adjusted treatment means for T_1 and T_2 therefore corrects the error arising from the differing values of x. Note that in fact the differences among the adjusted means are nearer the "true" values 4, 3 than are the corresponding differences for the unadjusted means.

The precision of comparisons based on adjusted treatment means depends on the standard deviation of y about the regression line, i.e., roughly speaking, on the standard deviation that y would have over a set of units all with the same value of x, treatment effects being absent. The formula for the standard error of the difference between two adjusted treatment means is complicated by the fact that the adjustments themselves have error, due to random error in estimating the slope of the regression line. This results in the standard error not being the same for all pairs of treatments. However for quick comparison the following approximate formula may be used, provided that there are no systematic differences between treatments in the value of x and that there are the same number of observations on each treatment;

$$\begin{pmatrix}\text{standard error of}\\ \text{difference between two}\\ \text{adjusted means}\end{pmatrix} = \begin{pmatrix}\text{standard deviation}\\ \text{about regression}\\ \text{line}\end{pmatrix}$$

$$\times \sqrt{\left\{\frac{2}{\begin{matrix}\text{no. of obs. per}\\ \text{treatment}\end{matrix}}\right\}} \times \sqrt{\left\{1 + \frac{1}{\begin{matrix}\text{no. of}\\ \text{treatments}\end{matrix} \times \begin{pmatrix}\text{no. of obs.}\\ \text{per treatment}\end{pmatrix} - 1)}\right\}}. \quad (1)$$

This should be compared with the formula (1) of § 1.2, namely that the standard error in an experiment not involving correction for a concomitant variable is equal to

$$\text{standard deviation} \times \sqrt{\left(\frac{2}{\text{no. of obs. per treatment}}\right)}. \quad (2)$$

The factor in the second square root in formula (1) is the contribution arising from the error in the slope of the fitted treatment lines.

There are several consequences of (1) and (2). First, if there is really no relation between y and x, the two standard deviations in (1) and (2) are equal, so that the precision is lower after adjustment. This is because for nearly every particular arrangement of treatments there will be some apparent dependence of y on x, so that a nonzero adjustment will be applied. This is an additional random term, inflating the error. However, if the number of units is large the quantity under the second square root in (1) will not be much greater than unity and the additional error is unlikely to be appreciable. The second, and more important, consequence is that, provided that this second square root can be ignored, the ratio of the standard error with and without adjustment for x is

$$R_s = \frac{\text{standard deviation of } y \text{ about regression line}}{\text{overall standard deviation of } y}, \tag{3}$$

both true standard deviations being calculated in the absence of treatment effects. The ratio R_s is a measure of the degree of relation between y and x. Readers familiar with the definition and meaning of the population correlation coefficient r may note that $R_s = \sqrt{(1 - r^2)}$. To give a rough picture of the strength of relation between y and x that will lead to a given value of the ratio R_s, Fig. 4.2 shows a number of scatter diagrams of y and x together with the corresponding values of the ratio. For example if, in the absence of treatment effects, y and x are related as in Fig. 4.2(c), the use of x as a concomitant variable would halve the standard error and thus be equivalent to a fourfold increase in the number of units.

4.4 ALTERNATIVE PROCEDURES

The procedure just given uses only the observations from the experiment and does not depend on assuming a completely specified relation between y and x; the relation is in fact estimated from the data and is not regarded as known a priori. All that we assume is that the relation is approximately linear. The disadvantage of the method is that it can be rather tedious to apply.

A simpler method, and one that in special cases is often used implicitly, is the construction of a suitable *index of response*, i.e., a combination of y and x which is treated as a new observation for analysis as a single quantity. This method is frequently used when the concomitant observation is of the same nature as the main observation, differing from it in being taken before the treatments are applied. Thus x may be the initial score in a spelling test, or the initial weight of an animal, y being the score or weight after treatment. An index of response widely used in such cases is $y - x$, the improvement or increase during test, a comparison of treatments being made in terms of this, the separate values of y and x being ignored.

It can be seen that if the slope of the treatment lines, fitted by the method of § 4.3, happens to be exactly unity, the index of response and the method of adjustments give identical estimates of the treatment effects. In other words the decision to use a simple analysis of the differences, $y - x$, amounts to assuming a special form for the residual relation between y and x, whereas the method of adjustments in effect finds in an objective way from the data, the most suitable linear combination of y and x for analysis.

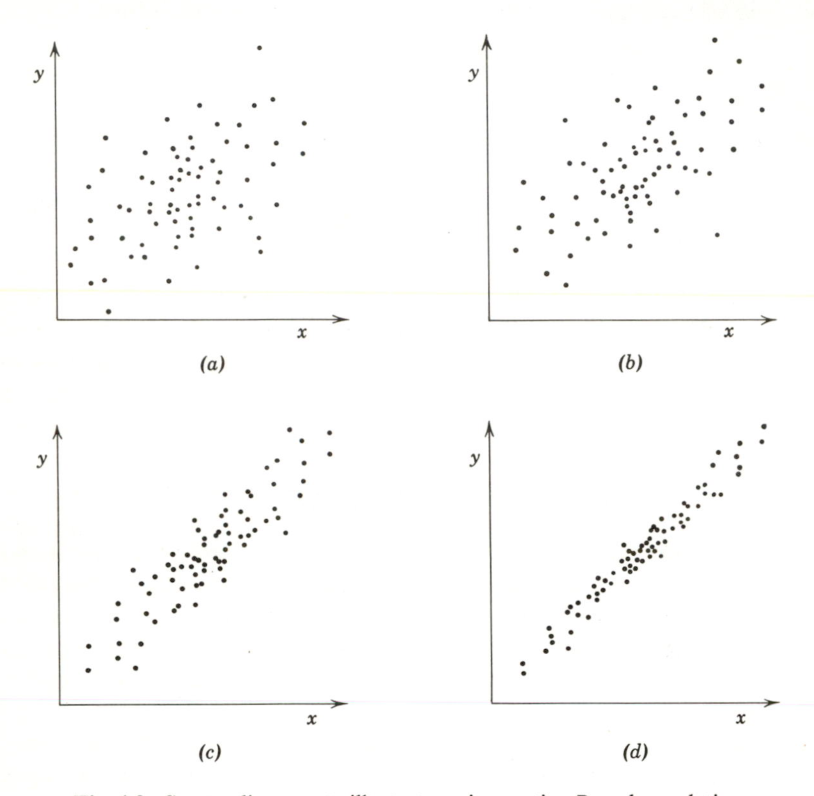

Fig. 4.2. Scatter diagrams to illustrate various ratios R_s and correlations r.
(*a*) $R_s = 0.9$, $r = 0.44$ (*b*) $R_s = 0.8$, $r = 0.60$
(*c*) $R_s = 0.5$, $r = 0.87$ (*d*) $R_s = 0.2$, $r = 0.98$

Quite generally, whenever theory or prior knowledge suggest a particular relation between y and x, an alternative to the method of adjustments is to assume that the relation is approximately true and to make a simple analysis of $y - kx$, where k is the expected slope of the relation between y and x (see Example 4.4 in the next section). In the previous paragraph k is taken to be unity. The objection is of course that if the value of k chosen is seriously different from the true slope, an appreciable loss of precision will result. The uncritical use of indices of response was criticized on these grounds by Fisher (1951, p. 161), who introduced the method of adjustment of § 4.3.

Gourlay (1953) and Cox (1957) have investigated how close k need be to the true slope in order to avoid a serious loss of information. Table 4.2, taken from the second of these papers, sums up the conclusions.

TABLE 4.2

LOSS OF PRECISION FROM USING WRONG INDEX OF RESPONSE

Correlation between y and x	Range within which Ratio of True to Assumed Slope must Lie to Avoid a Loss of Precision of		
	10%	20%	50%
0.4	(0.28, 1.72)	(−0.02, 2.02)	(−0.62, 2.62)
0.6	(0.68, 1.32)	(0.40, 1.60)	(0.06, 1.94)
0.8	(0.76, 1.24)	(0.67, 1.33)	(0.47, 1.53)
0.9	(0.82, 1.18)	(0.74, 1.26)	(0.59, 1.41)
0.95	(0.90, 1.10)	(0.85, 1.15)	(0.77, 1.23)

A loss of precision of say 50 per cent means here that the standard error with the assumed slope is $\sqrt{1.5}$ times what it would have been with the correct slope.

For example, suppose that the correlation between main and concomitant observations is 0.8; see equation (3) and Fig. 4.2 for an explanation of what this means. Then the loss of precision arising from using the wrong index of response is less than 20 per cent provided that the assumed slope k is not less than 0.7 times or more than 1.3 times the true slope; a precise definition of loss of precision is given in the footnote to the table. A reasonable general conclusion from the table is that the index of response does not have to be too close to the best one to give quite good results.

It is not possible to give a general rule about when to use an index of response rather than the method of adjustments. For the decision rests on, for example, the importance of saving time in analysis, the strength of belief in the assumed relation, and the importance of objectivity in the answer. However, if a number of similar experiments have to be analyzed, a sensible thing may be to analyze one or two experiments by the lengthier method before deciding what to do with the whole series. In any case a quick graphical analysis of all or part of the data will often show whether the assumed value of k is a reasonable one and whether the treatment effects appear to be independent of x, i.e., whether the treatment lines are parallel.

When the concomitant variable is available before the treatments are assigned to the experimental units the method used in Example 3.3 is a further alternative; that is, the units can be grouped in randomized blocks on the basis of the values of x, putting in one block those units with the lowest values of x, etc. If there is a nearly perfect relation between y and x, with negligible random scatter, the use of adjustments based on x will give nearly zero random errors, whereas the randomized block design will give appreciable error arising from the dispersion of x within blocks. However this is an extreme case and in many practical situations the randomized block method is adequate unless there are one or two very extreme values of x, or the number of units per treatment is small, or the correlation between the two variables is high, say 0.8 or more. The randomized block approach is of course simpler in analysis since it does not involve the calculation of adjustments. Some quantitative comparisons of blocking and adjustment have been given by Cox (1957).

The most profitable uses of the adjustment method are likely to be when there is no known index of response and when

(*a*) the value of x is not available until after the treatments have been assigned to the units; or

(*b*) the relation between y and x is of intrinsic interest; or

(*c*) it is important to examine whether the treatment effects vary with the value of x; or

(*d*) it is desired to block the experimental units on the basis of some other property, so that the information contained in x has to be used in another way. We consider this in the next section.

4.5 THE USE OF A CONCOMITANT OBSERVATION IN ADDITION TO BLOCKING

In § 4.3 we described a method for adjusting the treatment means to take into account a concomitant variable x, it being assumed that there is no grouping of the experimental units into blocks. Suppose now that the units are arranged in a randomized block, Latin square, or other similar design and that it is required to make an adjustment for a concomitant variable. Then the assumptions and general method discussed in § 4.3 above apply, with the single change that instead of plotting the observations of y directly against those of x, we plot a partial residual value of y against a corresponding partial residual value of x. These partial residuals differ slightly from the residuals defined in §§ 3.3, 3.4 in that, in order to get a more meaningful graph, the treatment means are not to be subtracted. For example in a randomized block experiment, the partial residual corresponding to a particular observation y is

$$\text{observation} - \begin{pmatrix} \text{mean observation} \\ \text{on the} \\ \text{corresponding block} \end{pmatrix}, \tag{4}$$

whereas for a Latin square the partial residual is

$$\text{observation} - \begin{pmatrix} \text{mean observation} \\ \text{on the} \\ \text{corresponding} \\ \text{row} \end{pmatrix} - \begin{pmatrix} \text{mean observation} \\ \text{on the} \\ \text{corresponding} \\ \text{column} \end{pmatrix} + \begin{pmatrix} \text{overall} \\ \text{mean} \end{pmatrix}. \tag{5}$$

The general idea is that before plotting, variation in y and x accounted for by the grouping of the experimental units should be removed.

The procedure is best explained in detail by an example.

Example 4.1. Pearce (1953, p. 113) has illustrated the statistical technique for calculating adjustments in a randomized block experiment. His data are reproduced in Table 4.3. The main observation, y, is the yield in pounds of apples over a four-year experimental period and the concomitant observation, x, is the yield in bushels in the preceding four-year period, during which no differing treatments were applied to the trees. Thus x is a proper concomitant observation of type (*a*). The six treatments under comparison are denoted by $T_1, \ldots, T_6$ and the observations in the table have been set out ordered by treatments, rather than randomly.

The first step in the analysis is to work out the mean values of y and of x by treatments and by blocks (the first part of Table 4.3(*b*)). The adjusted treatment means can be obtained directly by analysis of covariance without the calculation of partial residuals, but to do the adjustment semigraphically and to obtain a general understanding of what is being done, we proceed as follows.

The partial residuals denoted by Y and X are set out in Table 4.3(*c*), and have been calculated from equation (4). Thus for the observation y on T_2 in block III, the partial residual is $243 - 268.7 = -25.7$, since the mean of all observations, y, in block III is 268.7. Figure 4.3 shows the corresponding partial residuals Y and X plotted against one another for each treatment; this figure is analogous to Fig. 4.1(*b*). Inspection of the graph shows that within any one treatment there is a strong effectively linear relation between the values of Y and X and that the lines for each treatment are substantially parallel. Thus there is no evidence that the true treatment effects depend on x; a line for one treatment rising steeply from a relatively low value of Y for negative X to a relatively high value of Y for positive X would have suggested that the corresponding treatment was relatively good for "good" trees with a high initial yield and relatively bad for "poor" trees with a low initial yield. In the absence of such effects, we may fit parallel treatment lines. The slope as estimated graphically is about 30 units, that is an increase of 30 in Y for unit increase in X. The slope calculated from the analysis of covariance is 28.4 and this value has been used in what follows.

Treatment lines are drawn with this slope for the points corresponding to each

TABLE 4.3

THE CALCULATIONS OF ADJUSTMENTS IN A RANDOMIZED BLOCK EXPERIMENT

(*a*) *The Data*

	Block I		II		III		IV	
	y	x	y	x	y	x	y	x
T_1	287	8.2	290	9.4	254	7.7	307	8.5
T_2	271	8.2	209	6.0	243	9.1	348	10.1
T_3	234	6.8	210	7.0	286	9.7	371	9.9
T_4	189	5.7	205	5.5	312	10.2	375	10.3
T_5	210	6.1	276	7.0	279	8.7	344	8.1
T_6	222	7.6	301	10.1	238	9.0	357	10.5

(*b*) *Some Mean Values*

(*i*) *by blocks*

I 235.5 7.10 II 248.5 7.50 III 268.7 9.07 IV 350.3 9.57

(*ii*) *by treatments*

	Mean y	Mean x	Mean x − Overall Mean	Adjustment*	Adjusted Mean
T_1	284.5	8.45	0.14	−3.98	280.5
T_2	267.8	8.35	0.04	−1.14	266.7
T_3	275.2	8.35	0.04	−1.14	274.1
T_4	270.2	7.92	−0.39	11.08	281.3
T_5	277.2	7.48	−0.83	23.57	300.8
T_6	279.5	9.30	0.99	−28.12	251.4
Overall	275.75	8.31			

(*c*) *The Partial Residuals*

	Block I		II		III		IV	
	Y	X	Y	X	Y	X	Y	X
T_1	51.5	1.1	41.5	1.9	−14.7	−1.4	−43.3	−1.1
T_2	35.5	1.1	−39.5	−1.5	−25.7	0.0	−2.3	0.5
T_3	−1.5	−0.3	−38.5	−0.5	17.3	0.6	20.7	0.3
T_4	−46.5	−1.4	−43.5	−2.0	43.3	1.1	24.7	0.7
T_5	−25.5	−1.0	27.5	−0.5	10.3	−0.4	−6.3	−1.5
T_6	−13.5	0.5	52.5	2.6	−30.7	−0.1	6.7	0.9

(*d*) *Some Estimates of Precision*

The estimated standard error of the difference between two uncorrected treatment means is 28. The estimated standard error of the difference between two adjusted treatment means depends slightly on which pair of treatments is being compared; an average value is 12.

* The adjustment is equal to minus the slope (28.4) times the preceding column.

treatment. The position of each line is chosen so that it passes through the centroid of the appropriate points. Thus, for T_6 the mean values of x and y are 9.30 and 279.5. Therefore, in terms of residuals the mean point for T_6 has X equal to $9.30 - 8.31 = 0.99$ and Y equal to $279.5 - 275.75 = 3.75$, and this is the point M_6 in Fig. 4.3. To avoid complicating the figure unduly, the treatment lines are shown only for T_5 and T_6. The difference between the Y values for the centroids M_5 and M_6 is just the difference between the unadjusted treatment means, Table 4.3(*b*), column 1.

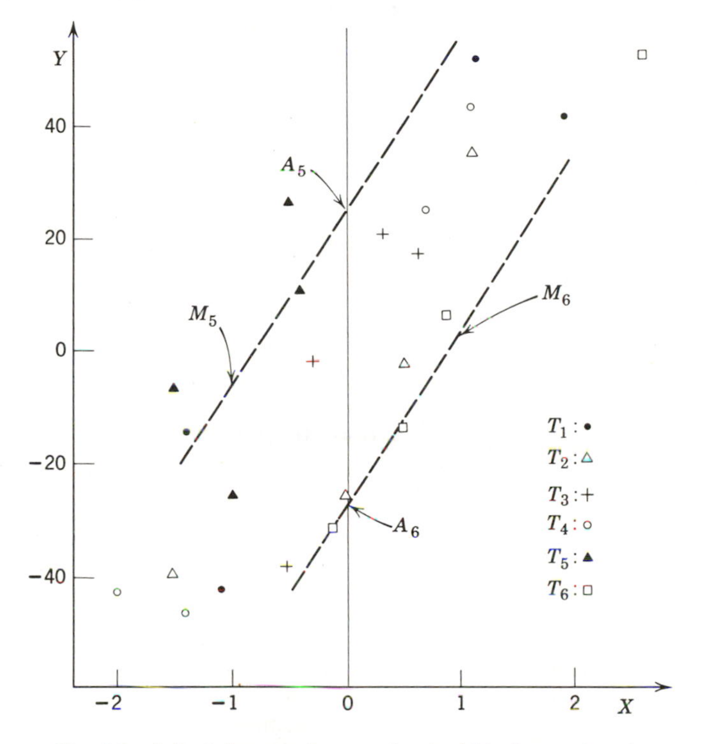

Fig. 4.3. Adjusted means in a randomized block experiment.

To find the adjusted estimates we take a standard value of X, say zero, in Fig. 4.3 and read off, or calculate, the corresponding values of Y on the treatment lines. These correspond to the points A_5 and A_6 in the figure; when the overall mean value of y, 275.75, is added to the Y values, we obtain the adjusted treatment means in the last column of Table 4.3(*b*). Thus the Y value for A_6 is about -28, agreeing with the value in the last column but one of the table. An arithmetical method for obtaining the adjustments is indicated in the table.

The common-sense justification of this procedure is exactly the same as that for the simple example in § 4.3. The only new point is the elimination of block differences prior to the plotting of the scatter diagram. The reasonableness of this step can be seen by considering that it should be possible to superimpose on the data arbitrary block effects for both x and y without affecting the conclusions.

The main effect of the adjustments has been on the treatments T_5 and T_6. The treatment T_6, for example, was applied to trees having on the whole high values of initial yield, so that the adjustment to y has lowered the yield. Some estimates of precision obtained from the full statistical analysis are given in Table 4.3(*d*) and show that the use of x has apparently appreciably increased the precision of the experiment.

The description of this example in terms of graphical methods is, of course, not to be taken as an implied criticism of the usefulness or appropriateness of the arithmetical technique of analysis of covariance.

Quite generally, in an experiment in which the concomitant observation x is available before the allocation of treatments to units, it would be possible to group the units into blocks on the basis of x and in addition to apply adjustments to remove the effect of variations in x not accounted for in the blocking. It would, however, very rarely be worth doing this solely for the purpose of increasing precision. The main value of the method of adjustments is when x represents some property of the experimental unit not directly connected with that used in grouping the units into blocks, or into the rows and columns of a Latin square.

4.6 SOME GENERAL POINTS

The following examples serve both to illustrate applications of the method and to bring out points of general interest connected with concomitant variables.

Example 4.2. In experiments on spinning textile yarn, it is sometimes difficult to ensure that each batch of yarn spun has the same mean weight per unit length. Yet one of the observations of most interest, the end-breakage rate in spinning, is quite critically dependent on the weight per unit length. Therefore it is natural to take the weight per unit length as a concomitant variable in an analysis of the type just described. Randomized block or Latin square designs are frequently useful to control systematic differences between machines, times, etc. This is an example of the use of a concomitant variable of type (*c*) (§ 4.2), in that the estimate of mean weight per unit length does not become available until after the completion of spinning. However, differences in mean weight per unit length are accidental in the sense that they could have been eliminated by minor adjustment of the machinery had sufficient initial information been available. Hence there is no sense in which treatments "cause" differences in yarn mean weight per unit length, so that use of this quantity as a basis for adjustment is in order.

This example illustrates the use of adjustments to correct for failure to control completely the experimental conditions.

Example 4.3. Pearce (1953, p. 34) has given an excellent account of the use of concomitant variables for adjustment in experiments on fruit trees and other perennial crops. He points out that when an experimental unit consists of one tree or bush, positional variations, which can conveniently be controlled by blocking, are likely to be relatively less important than in experiments on, say, wheat, in which each unit contains at least several thousand plants and in which individual differences are therefore likely to balance out. Consequently in the experiments Pearce is describing, it is worthwhile not only to control positional effects by blocking, but also to obtain concomitant variables for eliminating, as far as possible, the effect of variations arising from the peculiarities of individual trees. He lists suitable concomitant observations for various crops.

It would be possible, although inconvenient, to use these observations as a basis for blocking, following the method of Example 3.3. This would, however, mean that blocks would not be formed from adjacent units and this would make spatial variations more difficult to control, as well as probably making the experimental work more difficult to organize.

The general conclusion to be drawn from this example is that the use of concomitant observations is especially worth consideration whenever a major portion of the uncontrolled variation is associated with the particular objects, animals, plants, people, etc. forming the experimental units, rather than with the "external" conditions under which the experiment is carried out.* On the other hand variation associated with "external" conditions (observer, time, spatial, etc. differences) is often most conveniently controlled by the randomized block or Latin square devices. Experimental psychology and education are two fields where these remarks are particularly relevant, since in them a major source of uncontrolled variation lies in differences between subjects.

Example 4.4. Finney (1952, p. 45) has discussed an experiment of Chen et al. (1942) relating to the assay of digitalis-like principles in ouabain and other cardiac substances. The method was to infuse slowly a suitable dilution of drug into an anaesthetized cat and to record the dose at which death occurred. Twelve drugs were under comparison, ouabain being taken as a standard. There were three observers each testing four cats per day and the experiment was repeated on twelve days. A 12×12 Latin square was used with each column representing a day's work and each row a combination of observer and time of day. The details of this are not relevant to the discussion of adjustments, but the reader not too familiar with the use of Latin squares should think over this use of the design, consulting the above references if necessary.

The concomitant observation was the heart weight of the cat, determined, of course, at the end of each test, therefore being a concomitant variable of type (*c*). There were sound reasons for regarding the heart weight as unaffected by drug differences. Finney discusses the use of a second concomitant observation, the body weight, but we shall disregard this here. Notice how this use of the Latin square and a concomitant variable fits in with the general remarks at the end of Example 4.3.

There were physiological reasons for expecting the fatal dose of a particular drug to be proportional to the surface area of the cat's heart, which in turn is roughly proportional to the two-thirds power of heart weight. This power-law dependency is converted into a linear relation by working with logarithms, namely

$$\log(\text{fatal dose}) = \tfrac{2}{3}\log(\text{heart weight}) + \text{constant}. \qquad (6)$$

* Another method of controlling such variations is to use each object as a unit several times (Chapter 13). However, this may, as in the present example, be impossible from the nature of the experiment, and in other cases may lead to troublesome "interference" effects between different units.

Even if this relation does not hold exactly, it seems more reasonable to expect an approximately linear relation between logarithms than between original observations. Also, as we have seen in Example 2.1, it is natural to consider drug differences as affecting doses multiplicatively (e.g., one drug always requiring a 10% greater dose than a second drug would have on the same cat) and thus affecting log dose additively. Hence there are two reasons for converting the observations of dose and heart weight into logarithms. When this was done the fitted slope for the relation of log dose on log heart weight came to 0.676, in close agreement with the theoretical value of $\frac{2}{3}$; hence the use of an index of response (§ 4.4) log (dose) $- \frac{2}{3}$ log (heart weight) would have given excellent results.

This example illustrates the point that we should consider whether there is any prior reasoning to suggest the general form of the relation between the main observation and the concomitant observation, and exemplifies also that the relation between the two may be of some intrinsic interest.

Another general question, of analysis rather than of design, raised by the previous example concerns the desirability of transforming the observations mathematically, for example by taking logarithms, before analysis. This brings up some difficult issues. The concomitant variable x does not enter into the definition of what we are trying to estimate, namely the comparison of doses, and it is sometimes very helpful to transform x in order to get a linear relation between the variables. When we apply a mathematical transformation to the main observation, however, we are, if we work with means of transformed quantities, estimating differences between treatments on the transformed scale, not on the original scale.

In the example, the transformation to log dose is suggested both because ratios of doses provide the natural measure of relative potency and also because the theoretical relation between main and concomitant observations is simplified thereby. Things are not always so easy. If there is a definite reason for regarding treatment differences in terms of an observational scale z as particularly meaningful and for expecting treatment differences to be constant on this scale, then it seems wrong to estimate treatment effects on some other scale such as log z just for reasons of statistical or arithmetical convenience. In many applications, however, there may be no particular reason to expect that the scale on which the main observation is recorded is the one most helpful for analysis and interpretation of the results.

Common transformations usually affect the results materially only if the total fractional variation of results around the average is very appreciable, representing say a twofold variation or more.

Example 4.5. In all our examples so far the concomitant variable x has been a quantitative measurement. We can, however, sometimes usefully employ a "dummy" variable x to represent a qualitative division of the units into two classes. Thus, in an experiment with animals, a randomized block design might be used, taking all the animals in any one block from a single litter. In general, however, it would not be possible in doing this to ensure that all the animals in one block are of the same sex. The problem may therefore arise of adjusting the treatment means for the effect of any systematic difference between sexes, since in general each treatment will not occur equally frequently on males and females. To achieve this adjustment, introduce a concomitant variable taking the value 0 for males and 1 for females. Then the adjusted treatment means, calculated by the procedure of § 4.3, give the estimated treatment effects corrected for variation in the sex-ratio between treatment groups.

It might also be interesting in this application to examine whether the treatment effects for males are different from those for females, and this too can be done by an extension of the above procedure. Other groupings of the experimental units into two* sets not controlled in the blocking can be handled in the same way, provided of course, that the condition for a concomitant variable is satisfied, i.e., that the concomitant variable is unaffected by the treatments.

Throughout the discussion of adjustments it has been assumed that x is a concomitant variable, i.e., that the value of x for any unit is unaffected by the treatment applied to that unit. A further important aspect of the method is the examination of whether the treatment effects are constant. This is done, as explained in § 4.3, by looking for nonparallelism in the treatment lines or curves. We now consider a different type of application in which x is not a concomitant variable.

Example 4.6. The following example is suggested by Gourlay (1953) in his discussion of the analysis of covariance applied to psychological research. To compare a number of methods of teaching composition, the methods are assigned randomly to a number of experimental groups, with several groups for each treatment. After an appropriate time, scores are obtained for each group to measure (*a*) ability in composition and (*b*) knowledge of the mechanical aspects of English. Call these two scores y and x.

Then x is certainly not a concomitant variable, since it is quite likely to be influenced by methods of teaching composition. If, however, we go ahead with the method of adjustments, we are answering the question: what would the mean values of y have been had there been no differences in x? In other words we are asking whether any differences in ability in composition can be accounted for in terms of the effect of teaching on a knowledge of the mechanical aspects of English.

Figure 4.4 shows four cases that could arise with two treatments in a completely randomized experiment. In Fig. 4.4(*a*) differences between treatments in y are clearly not accounted for by differences in x. We may, in terms of the particular example, conclude that the treatment T_2 has improved ability in composition and that the improvement is not wholly accounted for by improvements in the knowledge of mechanical aspects of English. In Fig. 4.4(*b*) the difference between treatments in y is less than would be expected on the basis of the increase in x. In Fig. 4.4(*c*) differences in x account for the differences in y. In Fig. 4.4(*d*) the interpretation is in doubt because, although the fitting of treatment *lines* might suggest unaccounted differences in y the data are also reasonably consistent with

* Groupings into, say, three or four sets can be handled reasonably easily by the methods for several concomitant variables, to be described in § 4.6.

a single smooth curve. Similar difficulties are likely to arise whenever there are large differences in x, so that substantial extrapolation is involved. In practice, statistical analysis would usually be desirable to examine the precision of the conclusions suggested by the diagrams.

It should be noted that we talk about differences in y being "accounted for" by differences in x. That this is the right thing is best seen from the consideration that x and y are variables of the same nature and that, from a statistical point of view, any arguments that purported to show that differences in x caused

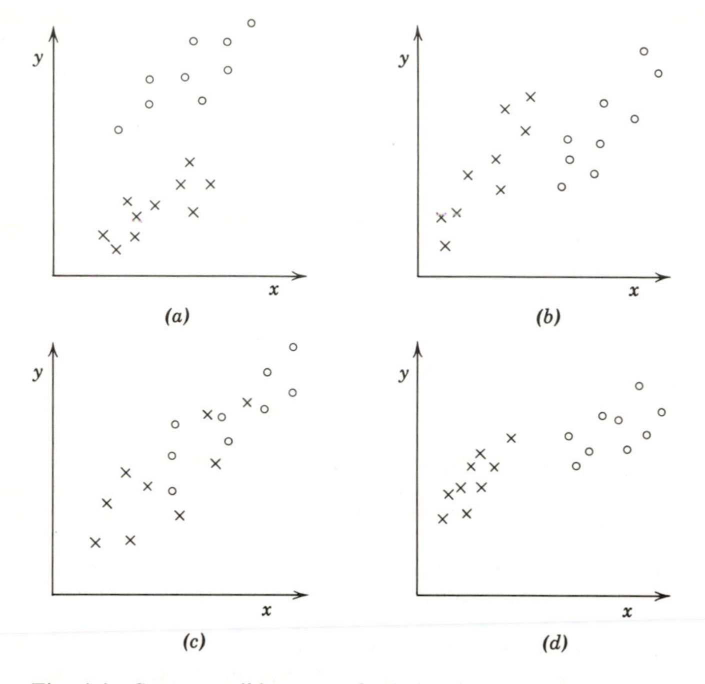

Fig. 4.4. Some possible types of relation between y and x when there are two treatments T_1, ×; T_2, ○.

differences in y would equally prove that differences in y caused differences in x. If we were to conclude that differences in x cause differences in y, this could only be because of hypotheses about the nature of the variables x and y extraneous to the experimental results themselves. Such hypotheses may be perfectly in order, but the scientist should always be aware when they are being appealed to.

This application, when x is not a concomitant variable, is clearly quite different from the use of a concomitant variable to increase the precision of treatment comparisons. The types of variable that are legitimate for increasing precision have been set out in § 4.2 and it is important to make sure in using the method that the conditions given there are satisfied. In particular it is in principle quite wrong to use as a concomitant variable a quantity x which there is no strong prior reason to regard as unaffected by treatments, but which has happened, in the particular experiment, to show nonsignificant treatment effects.

4.7 SEVERAL CONCOMITANT VARIABLES

It may happen that instead of one concomitant observation on each experimental unit there are several, that we think that they all give useful information about the main observation y and that we therefore wish to use them all to adjust the estimated treatment effects for y. For example, in an animal experiment we may have for each animal the initial weight and also other measurements that are considered important. In a psychological experiment we may have for each subject initial scores in several tests and other measurements on the subject relevant to the performance of the experimental task. In an industrial experiment we may have several different measurements on each batch of raw materials. We shall assume that all these observations are genuine concomitant observations.

A simple way of dealing with this problem is to combine the concomitant observations into one. Thus, if we had percentage scores in several tests, we could take the total score in all tests as a single concomitant variable. Similarly, if the concomitant variables $x_1, x_2, \ldots, x_k$ are not commensurable, we could construct a combination of them by some intuitively reasonable process. Thus, if $s_1, s_2, \ldots, s_k$ are the standard deviations of $x_1, x_2, \ldots, x_k$ measuring the variation between units and if $w_1, w_2, \ldots, w_k$ are rough measures* of the probable importance of $x_1, x_2, \ldots, x_k$ we could take a new preliminary variable $w_1x_1/s_1 + w_2x_2/s_2 + \ldots + w_kx_k/s_k$. If we are fortunate enough to have previous data linking y to $x_1, \ldots, x_k$, the best regression formula for predicting y from $x_1, \ldots, x_k$ gives a good single variable.

These methods are useful but are, except for the last one, open to obvious drawbacks. We may, by using an inappropriate single variable, sacrifice a lot of information; if we make a really bad choice we may do worse than using just one of the original variables. Also we lose objectivity in that there is no guarantee that another worker, faced with the same experimental data, even though asking the same questions, would not have reached somewhat different conclusions. This is not always important.

If we decide to use all the concomitant observations, without prior combination into a single quantity, the appropriate numerical technique,

* If $x_1, x_2, \ldots, x_k$ represent roughly independent properties of the experimental units, we should take w_1 proportional to the correlation coefficient we expect to find between y and x_1, etc.

multiple analysis of covariance, is a direct extension of the corresponding technique with one concomitant variable. The general interpretation of what is being done is the same too. Thus, with two concomitant variables x_1, x_2, we are effectively plotting a three-dimensional diagram of y against x_1 and x_2, separately for each treatment, and then fitting parallel treatment planes to each set of points. The values of y at some standard values of x_1 and x_2 give the adjusted treatment means. The same idea applies when there are three or more preliminary variables, although more than three dimensions are required to visualize the procedure geometrically.

We shall not go into details of the numerical method here, but it should be noted that the arithmetic, while straightforward, gets rapidly more laborious as the number of concomitant variables is increased. For this reason the method is not recommended as a routine tool, although there is no doubt that it is on occasion valuable and is worth knowing about. It would be possible to find the adjustments semigraphically, but the procedure is rather involved and not of much interest. The numerical calculations when we use a single concomitant variable, but fit treatment curves instead of lines, are very similar to those with several variables. This method of increasing precision by adjustment for several concomitant variables suffers from the general disadvantage that the gain is obtained by indirect calculation, i.e., by obtaining quantities that are comparatively remote from the original observations.

Sometimes, it may be required to use two or more concomitant observations in order to group the units into blocks, just as the single concomitant observation, body weight, was used in Example 3.3. If we have, from previous work, a regression formula for the main observation in terms of the concomitant observations there is no difficulty. If we have no such prior quantitative information, a sensible procedure with two variables x_1 and x_2 is the following. Plot a scatter diagram of x_2 against x_1 choosing a scale such that the dispersions in the two directions are approximately equal. Thus, with sixteen experimental units we should have a diagram with sixteen points, one for each unit. To group into four blocks of four units each, we take that division which makes each block of four points as compact as possible in the scatter diagram; if x_1 is thought to be a better variable than x_2, we pay more attention to dispersion parallel to the x_1 axis than to that parallel to the x_2 axis. Various ways of introducing a third variable, x_3, will occur to the reader.

SUMMARY

A concomitant observation is one whose value for any experimental unit is independent of the arrangement of the treatments under comparison.

Suppose that there is available on each unit a concomitant observation in addition to the main observation in terms of which it is required to compare the treatments.

The concomitant observations can be used to increase the precision of the treatment comparisons, provided that the concomitant and main observations on a unit would have been closely correlated in the absence of treatment effects. One method, which is essentially the statistical technique called analysis of covariance, is to obtain from the data adjustments that should be applied to the treatment means, in order to estimate what main observations would have been obtained had it been possible to make the concomitant variable the same for all experimental units. A second, simpler but usually less efficient, method is to analyze the data in terms of a suitable index of response formed by combining the main and the concomitant observations on a unit into a single quantity.

The skilful choice of concomitant observations can lead to an appreciable increase in precision, particularly when the main uncontrolled variation arises from the peculiarities of individual experimental units (animals, subjects, etc.).

A further important use of the concomitant observations is in the detection and explanation of variations in treatment effect from unit to unit.

REFERENCES

Chen, K. K., C. I. Bliss, and E. B. Robbins. (1942). The digitalis-like principles of *Calotropis* compared with other cardiac substances. *J. Pharmac. and Exptl. Therapeutics*, **74**, 223.

Cox, D. R. (1957). The use of a concomitant variable in selecting an experimental design. *Biometrika*, **44**, 150.

Finney, D. J. (1952). *Statistical method in biological assay*. London: Griffin.

Fisher, R. A. (1951). *The design of experiments*. 6th ed. Edinburgh: Oliver and Boyd.

Goulden, C. H. (1952). *Methods of statistical analysis*. 2nd ed. New York: Wiley.

Gourlay, N. (1953). Covariance analysis and its applications in psychological research. *Brit. J. Statist. Psychol.*, **6**, 25.

Pearce, S. C. (1953). *Field experimentation with fruit trees and other perennial crops*. East Malling, England: Commonwealth Bureau of Horticulture and Plantation Crops.

CHAPTER 5

Randomization

5.1 INTRODUCTION

In Chapter 3 designs called randomized blocks and Latin squares were introduced. Their object is to increase precision. In both cases some constraint was introduced into the allocation of treatments, for example by requiring each treatment to occur once in each block of the randomized block design. Subject to the constraint we said that the arrangement of treatments is to be randomized and we now must consider this process of randomization in detail.

The discussion falls into three main parts dealing with the practical details of carrying out the randomization, with the justification of the procedure and with various detailed points that occasionally arise in applications.

5.2 THE MECHANICS OF RANDOMIZATION

The basic operation is that of arranging in "random order" a series of numbered objects. In the more complicated designs this process has to be applied several times, but we shall begin with the simple cases. One of the essential features of randomization is that it should be an objective impersonal procedure; to arrange things in random order does *not* mean just to manipulate them into some order that looks haphazard.

One method of randomizing is to shuffle numbered cards or to draw numbered balls out of a well-shaken bag. Such methods are sometimes useful, but we shall not discuss them further. The main method, and the one we shall deal with, is the use of numerical random tables. Such tables for experimental design take two forms: tables of random permutations and tables of random digits. Short examples of both are given for illustration in the Appendix, Tables *A*.1 and *A*.2 of random permutations being taken from Cochran and Cox's book (1957, § 15.5) and Table *A*.3 of random digits from Kendall and Babington Smith's (1939) tables. These sources give much more extensive tables.

The use of tables will be illustrated by examples.

Example 5.1. Consider the randomization in Example 3.2; all that needs to be recalled is that this was a randomized block experiment with three blocks of five plots each and five treatments $T_1, T_2, \ldots, T_5$.

(*a*) Number the plots in each block 1, . . . , 5 in any convenient way.

(*b*) Use Table *A*.1, Random Permutations of 9, random permutations of 5, which are all that we need, not being available. Choose a starting point in a haphazard way without looking at the tables. For example write down a number (1 or 2) for the page, a number (1 to 5) for the row and a number (1 to 7) for the column block. Thus 2, 3, 6 gives the group beginning 9, 3, 4, 6, 2, 7, 5, 8, 1.

(*c*) Read off the first permutation, omitting the numbers 6 to 9 since there are only five treatments. This gives 3, 4, 2, 5, 1, and determines the allocation of treatments in the first block. Thus T_3 goes on plot 1, T_4 on plot 2, and so on.

(*d*) For the next block use the next permutation in the Table, which is 7, 4, 6, . . . and leads to the order $T_4\ T_2\ T_5\ T_3\ T_1$. Similarly for the third block, using, of course, different tabular permutations for each block.

A useful alternative device for selecting a starting point is to begin, on the first application, at the beginning of the table and to mark the last permutation used with a light pencil mark. At the next application carry on from where the last application finished, and so on. This assumes that recollection from a previous reading of the table is unimportant.

Example 5.2. Suppose that we have a randomized block experiment with 10 units per block and 7 treatments, T_1 occurring four times in each block and $T_2, \ldots, T_7$ each once. In this case let the digits 1, 8, 9, 10 represent T_1 and the digits 2, . . . , 7 represent $T_2, \ldots, T_7$ in order. The whole process of randomization is now analogous to that in Example 5.1, except that Table *A*.2, Random Permutations of 16, is used.

Thus if the first permutation is 7, 12, 1, 5, 16, 4, 11, 8, 2, 9, 10, 13, 15, 3, 6, 14, the numbers above 10 are rejected and the remainder replaced by the appropriate treatments to give

$$T_7\ T_1\ T_5\ T_4\ T_1\ T_2\ T_1\ T_1\ T_3\ T_6.$$

The randomization of Latin square designs is done by similar methods but since one or two special points are involved the discussion is postponed to Chapter 10.

A design that was only mentioned implicitly in Chapter 3 is the completely randomized arrangement, in which no grouping of the units is made and the treatments assigned at random subject only to the condition that each occurs the required number of times in all. The method, which is very simple and flexible, may be used in very small experiments, in order to get the maximum number of degrees of freedom for estimating error (Chapter 8), in experiments in which no reasonable grouping into blocks suggests itself, or when all attempts to increase precision are to be made by adjustment for concomitant variables.

Example 5.3. Consider the randomization of such an experiment over 21 units with three treatments each occurring seven times. There are several ways of proceeding; we cannot use the tables of random permutations in their simple form since more than sixteen objects are involved. (If sixteen or fewer were involved, we would write down a random permutation of the units and assign the first so many to T_1, etc.)

The following method is as quick as any.

(*a*) Number the units in any convenient way 00, 01, 02 through 20.

(*b*) Select haphazardly a starting point in the Table of Random Digits, Table *A*.3, and write out pairs of digits as they occur, subtracting 30 from two digit numbers from 30 to 59 and 60 from those that are between 60 and 89. Numbers 90 through 99 are rejected. Thus, 53 would be recorded as $53 - 30 = 23$. If the starting point chosen is the block in row 24 and column 12 on the first page of Table *A*.3, the first numbers are 5, 10, 6, 10 and 5 again, omitted, 21, and so on. When this device is used it is essential to check that each of the final set of numbers, in this case 00 through 29, has equal chance of selection.

(*c*) The first seven numbers determine the units to receive T_1, the next seven are to receive T_2 and the remainder T_3.

The process can be modified in various ways; for example, when 5 have been selected the remaining units could be ordered by a random permutation of 16. Another method (Cochran and Cox, 1957, § 15.3) is to produce a random permutation of $1, \ldots, 21$ by writing under each of the 21 figures a 3-figure random digit. The arrangement of the numbers is then changed until the random digits are in order of increasing magnitude.

Other tricks will occur to the reader when he has some experience of the tables. The randomization of some of the more complicated designs will be described when we come to them.

The tables in the Appendix are given primarily for illustrative purposes. They may be used for the randomization of small experiments but on no account should they be used for experiments so large that the same permutation or digit would have to be employed twice. A more extensive table is necessary in such cases.

5.3 NATURE OF RANDOM NUMBERS AND RANDOMNESS

A table of random digits is a series of digits $0, \ldots, 9$ in which each occurs approximately equally frequently and in which there is no recognizable pattern. A recognizable pattern means, for example, a tendency for some digits to follow let us say a 5 more frequently than others. Some readers may feel that no amplification of this statement is called for, but there are in fact a few general points worth making, although the reader satisfied with the statement may omit this section. It is simpler to discuss tables of random digits, although the same remarks would, with minor changes, apply to tables of random permutations.

A completely random sequence of digits is a mathematical idealization in which we think of a mechanism capable of producing an infinite sequence of digits. The sequence is to conform completely to the mathematical laws of probability, as applied to a set of mutually independent events, each equally likely to be 0, . . . , 9. That is to say, if we make any calculation of probability connected with the sequence, e.g., the probability of five adjacent pairs 01 occurring in a block of fifty, then the resulting probability is to equal the proportional frequency of times with which this event occurs in the infinite sequence. The main properties of such a completely random series would be that

(*a*) each digit would occur equally frequently in the whole sequence;
(*b*) adjacent digits, or adjacent sets of digits, would be completely independent of one another, so that, for example, if we knew one digit, we would have no basis for predicting the next one;
(*c*) moderately long sections of the whole would show substantial regularity, e.g., the number of 1's in a set of 1000 digits would not deviate much from 100, and so on.

A table of random digits is a finite collection of digits which

(*a*) is produced by a process which it is reasonable to expect will give results closely approximating to the above mathematical idealization;
(*b*) has been tested to check that in several important respects, e.g., in the relative frequencies of 0's, 1's, . . . , and in the simple independence properties, it does behave as a finite section from a completely random series should.

The first conclusion from this is that randomness is a property of the table as a whole; thus to be accurate we should talk about permutations produced by a random method, rather than about random permutations, as if the individual permutations were random. Thus, any permutation of 1, . . . , 12 is a possible random permutation and any two such, for example

$$1 \quad 2 \quad 3 \quad 4 \quad 5 \quad 6 \quad 7 \quad 8 \quad 9 \quad 10 \quad 11 \quad 12 \qquad (a)$$

$$8 \quad 3 \quad 9 \quad 7 \quad 6 \quad 11 \quad 10 \quad 1 \quad 12 \quad 4 \quad 2 \quad 5 \qquad (b)$$

are equally likely to occur. Whether or not they are legitimate random permutations is to be decided by the properties of the methods by which they were produced and not by inspecting them as individuals.

This, of course, conflicts with the every-day usage of the word random, and there are two related reasons for this. First, if we come across (*a*) in an application, we can usually think of good physical hypotheses that will explain the precise ordering; a hypothesis that will explain (*b*) is likely to be difficult to find. The second point is that the great majority of the permutations of 12 are disordered like (*b*) and not highly ordered like (*a*). These remarks have some bearing on the problem of the rejection of "unsatisfactory" randomizations (§ 5.7).

The second point to note from the general discussion is the independence of different numbers in the table. This is important where several randomizations have to be done in one experiment. Thus, we might have three Latin squares forming one experiment; we randomize these separately and then the random errors in the treatment estimates from the three squares are independent of one another. It would, of course, be wrong to randomize a Latin square and then to use it three times in the same form on each occasion. The reason is that if a very similar pattern of uncontrolled variation occurs in the three squares, the chance that it will produce a serious distortion in the treatment comparisons is much less if the squares are randomized independently than if a single common randomization is used.

5.4 JUSTIFICATION OF RANDOMIZATION

(i) Introduction

Having considered how to randomize, and briefly what a random series is, we must now consider why we randomize. Consider any of the designs we have discussed so far, for example, the randomized block, the Latin square, or one of its generalizations. When we have fixed the general type of design we are going to use, say a randomized block design with a certain number of blocks and treatments, we could determine the precise arrangement of treatments

(*a*) by adopting a particular systematic arrangement that seems unlikely to fit in with a pattern in the uncontrolled variation;

(*b*) by subjectively assigning the treatments in a way that seems haphazard;

(*c*) by randomization.

The dangers of (*a*) and (*b*) will be illustrated by examples.

(ii) Systematic Arrangements

Example 5.4. Greenberg (1951) has discussed an experiment in parasitology which illustrates the drawbacks of systematic arrangement. The experimental units were mice arranged in pairs of the same sex, one member of each pair receiving a series of stimulating injections, T, and the other member acting as a control (untreated, U). The observation consisted in challenging each mouse with 0.05 cc of solution, supposed to contain a standard number of larvae, and noting any response.

The point of the discussion depends on this. We have a series of pairs of mice *T*, *U*; *T*, *U*; *T*, *U*; ... In what order are the mice to be taken for inoculation? Greenberg reports that it was common to use the systematic order *TU*; *TU*; ... in the hope of cancelling variations in dosage. He produces data and experimental reasons, however, to show that during the course of the experiment, the number of larvae per injection increases steadily, and that therefore the ordering gives a systematically greater injection to the untreated mice. The consequences of this are:

(*a*) there is a systematic error in the estimated treatment effect, which would persist in a long experiment and even over several different experiments, if the above order were always used;
(*b*) the estimate of the error, based on the false hypothesis of the randomness of uncontrolled variations, is misleading.

When a systematic uncontrolled variation, such as this, is discovered in the experimental technique, it is, of course, important to take steps to eliminate the variation. However the aspect that concerns us now is the effect of such variations that we do not know about when the experiment is planned.

It is easy to be wise after the event and to say that a steady trend is a priori quite likely and that therefore some other pattern such as *TU*; *UT*; *TU*; *UT*; ... should have been used. However, whatever such pattern is chosen, there is the possibility that it coincides with some pattern in the uncontrolled variation, maybe one of obscure origin, producing a systematic error persisting even in a long experiment. To put the point another way, if a systematic arrangement of treatments is chosen, the presumption that it does not coincide with a pattern in the uncontrolled variation is a statement of the experimenter's opinion, which may well be justified, but which cannot be assessed quantitatively and which it is difficult for others to check on. If a surprising result is obtained, the experimenter may begin to doubt the validity of the systematic arrangement; if the results appear surprising to a later worker in the field, he will probably have no way of checking on the reasonableness of the pattern used.

Randomization, on the other hand, is an objective procedure, equally convincing to all and dealing equally with any pattern of uncontrolled variation that may present itself. The disadvantages of the systematic arrangements do not apply.

Example 5.5. When Latin squares were first introduced into experimental design, there was some discussion on whether a randomized square should be used or a square chosen deliberately for its balanced properties. An example of such a systematic square is the so-called knight's move 5×5 square,

$$\begin{array}{ccccc} A & B & C & D & E \\ D & E & A & B & C \\ B & C & D & E & A \\ E & A & B & C & D \\ C & D & E & A & B \end{array}$$

which has the treatments evenly spread out with respect to the diagonals of the square.

The disadvantages of this design are less striking than those of the systematic arrangement of Example 5.4. However there would certainly be objections to repeating the design unchanged in a series of trials and also, as Yates (1951) has pointed out, difficulties could arise from the fact that in four cases out of five *E*, for example, is immediately to the right of *D*. Thus if the experiment is an agricultural field trial and *D* is a "tall" variety, the effect would be, with certain orientations of the square, to depress the yield of *E*.

However the main objection to the systematic design in this case is not the appreciable possibility of serious systematic error in the treatment estimates but the difficulty of estimating the amount of random error (Fisher, 1951, p. 74). In the discussion of the analysis of the randomized block and Latin square designs in Chapter 3, the principle was mentioned that whenever a source of uncontrolled variation is eliminated from the error in the design of the experiment, it must also be eliminated in the analysis, if a correct estimate of random error is to be obtained. Now in the systematic square some variation parallel to the diagonals of the square is eliminated in virtue of the balanced property of the design, but the ordinary analysis of the Latin square takes no account of this. Therefore a biased estimate of error is to be expected and Tedin (1931) confirmed this by examining uniformity data from field experiments. The bias he found was small, although there is, of course, no guarantee that this would always be so.

The bias in the estimate of error could probably be removed by a modification of the method of analysis, but several different, although plausible, ways of doing this are available and it is not clear which should be used, so that there is some loss of objectivity.

To sum up, the objection to lack of randomization in this case is mainly connected with the estimation of error. If the design is not randomized, and particularly if it is chosen for its "balanced" nature, it is quite likely that the estimate of error from the conventional method of statistical analysis will not be appropriate, even if it is very unlikely that there is appreciable systematic error in the treatment estimates themselves. This consideration of the correctness of the estimate of error applies also to Example 5.4, but there it is rather overshadowed by the occurrence of substantial bias in the treatment estimates themselves.

The conclusion from these two examples is that systematic arrangements suffer from the disadvantages that

(*a*) the arrangement of treatments may combine with a pattern in the uncontrolled variation to produce a systematic error in the estimated treatment effects, persisting over a long experiment or even over a series of experiments. We may begin by thinking this possibility sufficiently unlikely to be disregarded, but this is a matter of personal judgement which cannot be put on an objective basis;

(*b*) there is likely, even in the most favorable cases, to be difficulty connected with the estimation of error from such designs.

Randomization removes these disadvantages and hence is, other things being equal, to be preferred to systematization. That is, we aim, by controlled grouping of the units (as in randomized blocks or the Latin square) to eliminate the effect of as much of the uncontrolled variation as possible and then to randomize the remainder.

It should not be thought, however, that these remarks mean that systematic designs are never to be tolerated. If we have good knowledge of the form of the uncontrolled variation and if a systematic arrangement is much easier to work with, as when the different treatments represent ordered changes of a machine, it may be right not to randomize. For example, suppose that there are at some stage of an experiment 24 test tubes of solution, to which are to be added x cc of reagent for treatment T_1, $2x$ cc for treatment T_2, and $3x$ cc for treatment T_3. Imagine further that this operation needs to be completed as quickly as possible. Clearly, the procedure that is quickest and least likely to lead to gross errors is to set the work out in systematic order, dealing first with all units receiving T_1, etc. If it is known that negligible error is involved in pipetting and if the whole set of 24 units can be completed in such a short time that the first and last tubes can be considered as dealt with simultaneously, it would be quite wrong to attempt randomization. If a number of repetitions are involved, a sensible precaution, however, would be to change the order of treatments for each repetition and, if practicable, to include some check on the assumptions.

Another reason for not randomizing is that in certain very special short experiments there is an appreciable gain in precision in using a special systematic arrangement (§ 14.2). What is important, however, is to randomize except when there is a very good reason not to, to understand that the conclusions from a nonrandomized experiment depend on the correctness of what is assumed about the uncontrolled variation, and to state this explicitly in reporting the experiment. The reader interested in a further discussion of systematic arrangements should read the papers by "Student" (1938), Yates (1939), and, for an account of some more recent work, Cox (1951). If a systematic design is adopted for an experiment to be repeated several times with the same design, the names of the treatments should be randomized independently in each repetition.

(iii) Subjective Assignment

We now consider the second alternative, the assignment of treatments not by strict randomization but in a subjective way that seems haphazard. The following is an example of an experiment spoiled by a procedure of this sort.

Example 5.6. In 1930 a very extensive experiment was carried out in the schools of Lanarkshire, in which 5000 school children received 3/4 pint of raw milk per day, 5000 received 3/4 pint pasteurized milk, and 10,000 children were selected as controls to receive no milk. The children were weighed and measured for height at the beginning and end of the experiment, which lasted four months. The discussion below is based on "Student's" critique of the experiment ("Student," 1931).

The following method was used in determining, for each school, which children should receive milk and which not, only one type of milk being used in any one school. A division into two groups was made either by ballot or using an alphabetical system. If this appeared to give a group with an undue proportion of well-fed or ill-nourished children, others were substituted in order to obtain a more level selection. In other words a random, or nearly random, assignment of treatments was made and then "improved" by subjective assessment.

This resulted in the final observations on the control group exceeding those on the treated group by an amount equivalent to three-months' growth in weight and four-months' growth in height. The explanation of this is presumably that the teachers were unconsciously influenced by the greater need of the poorer children, and that this led to the substitution of too many ill-nourished among the feeders and too few among the controls.

This had a particularly serious effect on the comparisons of weight, because the children were weighed in their indoor clothes in February at the beginning of the experiment and in June at the end. Thus the difference in weight between their winter and summer clothing is subtracted from their actual increase in weight. Had the control and treated groups been random this difference in weight due to the clothing would have decreased the precision of the results but would not have introduced bias; however there was the suggestion that the treated group contained more poor children, who probably lost less weight from this cause, so that the experiment was biased.

Although it was possible to draw certain conclusions, these were of a very approximate and tentative nature, even though the number of children taking part was large. The failure of the experiment to yield clear-cut conclusions was due to the failure to adopt an impersonal procedure in allocating treatments to experimental units.

"Student" pointed out that a much more economical and precise way of comparing two treatments, say the two types of milk, would have been to work with pairs of identical twins, in a randomized paired comparison experiment (§ 3.2). Very probably a comparatively small number of such pairs would give high precision and it would then have been practical to make detailed and carefully controlled measurements on each child.

The conclusion from this is that an experiment is in danger of being very seriously affected if the personal judgement of people taking part is allowed to determine the allocation of treatments to units. There is abundant evidence that observer biases occur even in apparently unlikely circumstances, and moreover, even if the arrangement chosen is in fact satisfactory, there is always the suspicion that it may not be, and this will detract considerably from the cogency of the experiment if surprising conclusions are found. The time taken to carry through a process of objective randomization by the methods of § 5.2 is trivial under all ordinary circumstances, so that there is no argument for subjective assignment on the grounds of simplicity.

(iv) Randomization as a Device for Concealment

In most of the previous examples, randomization has been used to deal with variations in space, in time, between different animals or subjects, and so on to ensure that any patterns of variation that may exist in the experimental material cause no systematic error in treatment comparisons. A very important further use of randomization, however, is in situations where a substantial amount of the uncontrolled variation arises from subjective effects due to personal biases of the people taking part in the experiment, including the experimenter himself. In such applications randomization achieves its aim by concealing from the persons involved which treatment is applied to each unit.

Consider first an application where bias may enter the selection of units to take part in the experiment.

Example 5.7. Suppose that a clinical trial is set up to compare two or more methods (drugs, surgical treatments, etc.) of treating a disease. Experimental units, i.e., patients, are included in the experiment as suitable individuals appear at the various centers taking part. There will often be doubt as to whether a particular person should, in fact, be included.

If the doctor responsible for the decision about inclusion knows that the patient, if included, will receive say treatment A, this may easily influence, consciously or unconsciously, the decision reached in doubtful cases, and if this happens, the groups of experimental units receiving different treatments will not be genuinely comparable.

If the allotment of treatments is determined by a systematic pattern, this may soon become apparent; equally, if the order of treatments is determined by an initial randomization and if the full key is available to the doctor concerned, the necessary concealment will not be achieved. The satisfactory method is either to do the randomization after the patient has been chosen for inclusion, or to arrange that the treatment a particular patient is to receive is named in a sealed envelope that is not opened until after the patient has definitely been selected. The order of treatments in successive envelopes is randomized by the controller of the experiment and not revealed.

This device is now widely used in the design of clinical trials. Similar remarks apply to any experiment in which a subjective element enters into the selection of experimental units for inclusion in the experiment.

Randomization to achieve concealment may also be necessary in applying the treatments, particularly where the units are people likely to be influenced in an irrelevant way if they knew the treatment which they have actually received.

Example 5.8. Consider an experiment on school children to assess the effect of a new tooth paste, say one with an added fluoride. We need a group of children with whom to compare the children who receive the experimental tooth paste, F. A quite unsatisfactory way of obtaining such a control group would be to give F to half the children chosen by a randomized method and to give the other half no special treatment. In order to obtain worth-while results, steps to encourage correct and frequent use of F would be necessary and any relative improvement in the experimental groups' teeth might well be due to the additional attention given to the cleaning of teeth rather than to the particular merits of F.

A better, but still unsatisfactory, procedure would be to issue the control group with a standard brand of tooth paste. The objection here is that the experimental group, knowing that they are receiving special treatment, may tend to be more diligent than the control group. The only satisfactory way of ensuring that no such effects occur, is to have identical tubes of control and experimental tooth pastes, so far as is possible differing only in the absence or presence of the special ingredient, and if possible indistinguishable in flavor, etc. The treatments are randomly assigned to the children, the key to the randomization being available only to the controller of the experiment. The children, their parents, and the staff responsible for instruction in the use of the tooth paste and for assessing the children's teeth at the end of the trial, should know neither which treatment any particular child has received, nor which groups of children have received the same treatment. The final observation on each child at the end of the experimental period would be some such index as the number of defective and missing teeth. This would also be determined for each child before the start of the experimental period and this initial value would enable adjusted treatment means to be calculated, eliminating from the error much of the variation connected with the initial state of the teeth (see § 4.3). It would also be possible to examine whether any difference between the treatments was more or less marked with children with good teeth.

These considerations are important in any experiment in which the application of one or more treatments may tend to be influenced by personal attitudes towards the treatments, these attitudes being considered irrelevant to the purpose of the experiment.* Thus in comparing a new and an old experimental technique or a modified and an unmodified industrial process, bias may arise due say to devoting greater attention to the running of the modified process. If, owing to the nature of the processes, this possible bias cannot be eliminated by concealment, whatever steps that may be practicable should be taken to remove such a bias. For example, the biases may tend to disappear if the experiment is spread over an appreciable time or is preceded by a practice period, or if the people taking part are deliberately misinformed as to the object of the investigation.

In some applications it may be required to conceal the nature of the treatment from the person who has to apply the treatments to a large number or all of the units. For example, one treatment might require an impure chemical, another an analytically pure source of the same substance. It would not in such cases be satisfactory to have two supplies one labelled A, and the other B, and to conceal which was which, since the necessary

* Note that for some purposes we might consider such attitudes as part of the treatments and would then not wish to eliminate their effect.

requirement of independence from unit to unit would not be satisfied. Thus a guess, possibly incorrect, might be made as to which treatment *A* was and a systematic error introduced. A better arrangement would be to have say at least six sources of material labelled *A* through *F*, three of which are impure, three not. For each experimental unit the source to use is given in the experimental instructions, having been determined by appropriate randomization.

The final stage in which concealment may be advisable is in the making of the observation itself. There are many fields where substantial personal biases may arise and in all these randomization of the order of presentation of the units for measurement is desirable. Thus in taste-testing experiments, where a judge is asked to state his preference among a number of products, it is most inadvisable that the judge should know the treatment to which each of the objects has been subjected. Again, in experiments in which the reproducibility of, say, an analytical technique is of interest, it is best for the analyst not to know which of the items submitted for analysis have received identical treatment. Quite generally, in any experiment in which personal judgement enters to a considerable extent into the determination of the final observation, concealment is desirable. Sometimes this is impracticable, but quite often randomization does achieve concealment in a simple and satisfactory way.

(v) Summing Up

To sum up, it seems fair to say that subjective allocation of treatments to units should never be used, because the method has serious disadvantages and no compensating advantages when compared with objective randomization. Of course, subjective allocation may work out perfectly well in some applications, but this is no argument for using it, since randomization is just as simple and has definite advantages.

Therefore, our general conclusion is that, with the minor exceptions noted in the discussion of systematic arrangements, randomization is to be preferred to alternative methods. This conclusion has been reached in a rather negative way by showing the disadvantages of other methods. In § 5.6 there is a brief discussion of the positive advantages of randomization from a more statistical point of view. That section may be omitted if desired. First, however, we deal with an important general matter concerned with the scheme of randomization to be used.

5.5 ERRORS ARISING IN SEVERAL STAGES

In many, if not most, experiments important uncontrolled variation may arise from several sources, notably in the experimental material, in the various stages of applying the treatments, and in the taking of the observations. It is important that the randomization should cover all important sources of variation connected with the experimental units and that, so far as is practicable, the different experimental units receiving the same treatment should be dealt with separately and independently at all stages at which important errors may arise, one such stage being in the application of the treatments.

This point has already been discussed to some extent in Example 3.2. It will be dealt with now in more detail using a somewhat fictitious example connected with the shrinkage of socks. The example should be studied carefully because it can be paralleled in many fields, particularly in that type of laboratory work in which a whole sequence of operations has to be carried out on each batch of experimental material.

Example 5.9. In an experiment to compare four treatments applied to knitted socks to reduce shrinkage, something like the following might be done. Forty-eight socks are divided into 4 sets of 12, each set to receive one of the 4 treatments, say a control and 3 different chlorination processes. The treatments are applied and the socks measured. Normal wear and washing is then simulated in a controlled way in a machine that can take from 1 to 12 socks at a time. At the end the socks are remeasured and the percentage shrinkage calculated.

The treatment comparisons can be affected by (*a*) the variation of intrinsic properties from sock to sock, (*b*) measurement errors, (*c*) variations arising during the application of the chlorination processes, (*d*) lack of complete uniformity in the simulation of wear. We may decide, after investigation, that measurement errors can be treated as completely random and are in any case small compared with other sources of variation. If this is done, the measurements may be obtained in any convenient order. We shall consider several randomization procedures in the light of the remaining three sources of variation.

Method I. The socks are divided randomly into 4 sets of 12. Each set is processed as one batch and after measurement, dealt with in one run of the simulation machine. There are, thus, 4 runs of the simulation machine, each run dealing with socks that have all received the same treatment.

Method II. The socks are divided randomly into 4 sets as before, but the chlorination processes are applied independently to single socks, for example by including each sock in a separate batch for processing. After measurement the simulation of wear and washing is carried through as in Method I.

Method III. This is the same as Method II up to the simulation stage. Here the socks are grouped into blocks of 4, one from each treatment, each block being used for one run of the machine.

Method IV. The socks are divided into 12 sets of 4, 3 sets per process. The chlorination processes are applied independently to each set, so that 3 separate batches need to be run for each process. The runs of the simulation machine are arranged by Method III.

Method I is adequate only if negligible variation arises from sources (*c*) and (*d*), the chlorination and simulation stages. If, for example, there is some variation in the performance of the simulation machine from run to run, this will appear as systematic error, since all the socks having one treatment are dealt with in a single run. If there happens to be appreciable variation between the conditions appertaining in different runs of the same chlorination process, the conclusions from Method I will apply solely to the particular runs of each process used and this will be a serious restriction. For example, it will be impossible from the experimental results alone to distinguish between real differences in the processes and variation from one application to another of one process.

Randomization is no help for treatment errors and the right procedure is to have independent applications of the treatment for each unit. This is Method II; any variation connected with the simulation process is still inadequately dealt with. Method III gives one way of dealing with this, by the randomized block principle, each run of the machine forming a block.

Method IV is a compromise version of Method III which meets the practical objection that would often be made that an independent run of a chlorination process for each sock is uneconomic. The method here is essentially to take experimental units consisting of 4 socks each and to set up a randomized block design of 3 blocks each with 4 treatments.

There are numerous further possibilities. Moreover, if the measurement process, instead of being fairly straightforward, involved a substantial subjective element, it would be necessary to measure the socks in randomized order, taking the sort of precaution discussed in the preceding section to conceal the treatment applied to the sock being measured. Of course, if this stage of randomization can be omitted, the work of measurement may be much simplified.

This discussion can be summarized as follows. Variation may arise from several sources, and randomization should cover all those at which the variation cannot be assumed negligible or completely random. It is frequently not good enough to randomize just at one stage of the experimental procedure and to leave the treatments systematically arrayed at other stages.

5.6 STATISTICAL DISCUSSION OF RANDOMIZATION

In this section, which may be omitted at a first reading, the statistical consequences of randomization are discussed. We start from the assumption,* § 2.2, equation (1), which says that the observation obtained when a particular treatment is applied to a particular experimental unit is

$$\begin{pmatrix} \text{a quantity} \\ \text{depending only} \\ \text{on the unit} \end{pmatrix} + \begin{pmatrix} \text{a quantity} \\ \text{depending on the} \\ \text{treatment} \end{pmatrix}. \tag{1}$$

Consider for definiteness a randomized block experiment, although the

* An analysis based on a more general assumption, allowing variations in treatment effect from unit to unit, has been made by Wilk and Kempthorne (1956).

following remarks apply with minor changes to nearly all the designs described in this book.

If we randomize the treatments within blocks, the unit quantities associated with a particular treatment T_1 consist of a random sample of one from the set of unit quantities for the first block, a random sample of one from the unit quantities for the second block, and so on. Similarly for the other treatments, the only complication being that the samples for T_2, T_3, . . . are drawn "without replacement," since no unit receives more than one treatment. Hence we may apply the mathematical theory of random sampling to the behavior of our observations, and moreover the theory is rigorously applicable, provided that the assumption (1) holds and that the table of random numbers, or random permutations, used in randomizing the treatments, is adequate. The latter point need cause us no trouble.

In this way we reach the following conclusions, without further assumptions about the nature of the uncontrolled variation.

(*a*) The estimated treatment effects are *unbiased*, in the sense that the average of the estimates over a large number of independent repetitions of the experiment would be equal to the true treatment effects defined from (1).

(*b*) In a single experiment with a fixed amount of uncontrolled variation, as measured by the standard deviation, the error in the estimated treatment effects would almost certainly be very small if the number of units were sufficiently large. That is, there is a negligible chance of appreciable error persisting in a very long experiment; this state of affairs should be contrasted with the situation for a nonrandomized experiment, such as Example 5.6, where there was appreciable systematic error even though the number of units was very large.

(*c*) The square of the standard error of the estimated treatment effects, calculated by the method described in § 3.3, is unbiased, in the sense that, averaged over a number of independent repetitions of the experiment, it would equal the average of the square of the actual error, i.e., estimated effect minus true effect, squared.

(*d*) In principle it is possible (Fisher, 1951, p. 43) to make exact significance tests concerning the treatment effects and to calculate limits within which the true effects lie at any assigned level of probability. Thus we can build up a distribution, inferred from the data, for the magnitude of the true treatment effect. In practice these calculations are almost always done, not by the "exact" method, but by introducing certain assumptions about the shape of the distribution of the unit quantities in (1). This enables the significance calculations to be made very simply by the t test and related methods. It is known that, except in small experiments, results obtained in this way agree satisfactorily with those based on the "exact" argument. In any case the assumptions about the uncontrolled variation concern the shape of the overall distribution of the unit quantities and not the nonexistence of patterns.

To return to a less statistical description, the positive advantages of randomization are assurances

(*a*) that in a large experiment it is very unlikely that the estimated treatment effects will be appreciably in error. In other words a randomized experiment may be more accurate than a corresponding nonrandomized one in which an unskilful assignment of treatments to units has led to systematic bias. Randomization achieves this mechanically;

(*b*) that the random error of the estimated treatment effects can be measured and their level of statistical significance examined, taking into account all possible forms of uncontrolled variation subject to (1).

Thus, to take a simple case to illustrate (*b*), we might conclude from a randomized experiment that there is a difference between two treatments that is statistically significant at a very high level. The corresponding conclusion for an experiment laid out in a systematic arrangement might be that the difference is very unlikely to be due to random uncontrolled variation (this is shown by the significance test) and that it is considered very improbable that the systematic arrangement is responsible for the apparent effect. This last statement has no measurable uncertainty, nor is there any guarantee that the standard error and significance test measure anything very relevant about the system. It is not that the systematic arrangement is necessarily less precise than the randomized one, but that the assessment of the results is on a less objective basis.

One or both points (*a*) and (*b*) may apply in any particular case.

This concludes the general discussion of the arguments for randomization in the allotment of treatments to experimental units. There is a second very important use for randomization in experimental work, namely in sampling, i.e., in selecting from a given bulk a portion for detailed study and measurement, the portion to be representative of the whole. The arguments for randomization in sampling are parallel to those developed above, but will not be discussed here.

5.7 SOME FURTHER POINTS

There are some difficulties that arise in the application of randomization, particularly to small experiments, and these will now be discussed.

The first point concerns the rejection of an arrangement produced by the randomization when it seems particularly unsuitable. As an example, consider the paired comparison experiment, Example 3.1, with eight pairs of units. Suppose that, as in our first account of this experiment, the units are arranged in a definite order within each pair, but that it is decided that this ordering is not of sufficient importance to warrant balancing it in the design of the experiment by the method of Example 3.10. Now it will happen, actually about once in 128 times in the long run, that the ordering of treatments is the same for every pair, either $T_1 T_2$ every time or $T_2 T_1$ every time. Further, once in about 14 times the arrangement is either of this type or has just one pair showing a different ordering from the remaining 7.

It is clearly undesirable to use these arrangements. Even though we think that there is probably not an important order effect, there are likely to be various things, connected say with the experimental technique, that could produce such an effect. In other words a pattern of uncontrolled variation with a substantial systematic difference between the first and second unit in the pair, is a priori considerably more probable than other *particular* patterns we can think of.

Similar considerations apply in other experiments where the randomization produces an arrangement that fits in with some physically meaningful pattern in the experimental material, even though this pattern is thought probably unimportant. Other examples are if a Latin square on randomization has a line of treatment T_1, say, down a diagonal, or if a randomized block experiment gives the same order of treatments within each block. The chances of these particular arrangements occurring are extremely small, except in experiments with a small total number of units.

There are three ways of dealing with the difficulty, all depending on curtailing the randomization. The first method is to incorporate a condition about order into the formal design of the experiment, as was done in Example 3.10, where T_1 and T_2 each occurred four times in the first position. This is probably the best solution in the present case, but it is certainly not a general answer to the problem, since there are various reasons why it may be impracticable or undesirable to introduce further constraints into the design. For example we lose degrees of freedom for residual in eliminating a source of variation that is probably not important, we make the experiment more complicated and there may already be several different systems of grouping in the design, making the introduction of further conditions difficult or impossible.

The second method is to reject extreme arrangements whenever they occur, i.e., to rerandomize. For example in the paired comparison experiment, we may decide to reject all arrangements with seven or more pairs in the same order. A highly desirable condition in using this method, if observer biases like those of Example 5.6 are to be avoided, is that if any arrangement is to be rejected, so must all other arrangements obtained by permuting the names of the treatments. Thus if the arrangement with eight $T_1 T_2$'s is rejected, so must the arrangement with eight $T_2 T_1$'s. There would be little likelihood of disagreement over such an extreme case, but since the decision as to what arrangements to regard as unsatisfactory is arbitrary, there could be disagreement with less extreme cases. The best plan is, if possible, to decide which arrangements are to be rejected before randomization. It is difficult to give general advice about which arrangements to reject, but the best rule is probably to have no hesitation in rejecting any arrangement that seems on general common-sense grounds to be unsatisfactory. Fortunately this matter is not nearly so important in practice as might be thought, since, as remarked above, extreme arrangements occur with appreciable chance only in very small experiments.

The third method is to use a special device, known technically as restricted randomization (Grundy and Healy, 1951; Youden, 1958). This is a very ingenious idea, in which a design is selected at random from a very special set of arrangements. The set is chosen to exclude both the extreme arrangements and the very balanced arrangements, in such a way that the full mathematical consequences of ordinary randomization follow. The method is probably of most value for a special design called the quasi-Latin square (Chapter 12), for which the method was first introduced, and otherwise in a series of small experiments, each of some interest in itself, but which also need to be considered collectively. The method is however too specialized to discuss here and its full implications have not yet been worked out; the nonstatistical reader requiring more information about it should consult a statistician.

The reader may object that the second method, the rejection of extreme arrangements, will falsify the mathematical consequences of randomization described in § 5.6. This is true of the estimation of error, although not of the absence of bias in the treatment estimates themselves. The estimate of error will only be unbiased if there is in fact no systematic order effect. However in single small experiments the estimate of error is very inaccurate anyway. More importantly we have here a mathematical interpretation of randomization: that it leads to desirable properties in the long run, or on the average, and on the other hand a practical problem—namely the designing and drawing of useful conclusions from a particular single experiment that we are now in the process of considering. Usually the concept that our procedures will work out well in the long run is a very helpful one, both qualitatively and in giving a vivid physical picture of the meaning of probabilities calculated in connection with a particular experiment. However to adopt arrangements that we suspect are bad, simply because things will be all right in the long run, is to force our behavior into the Procrustean bed of a mathematical theory. Our object is the design of individual experiments that will work well: good long-run properties are concepts that help us in doing this, but the exact fulfillment of long-run mathematical conditions is not the ultimate aim.

The second general matter is closely related to the first. Suppose that we design and carry out a randomized experiment, and that when we come to analyze and interpret the results we realize either that the arrangement we have used is probably an unfortunate one and should have been rejected, or, by inspection of the results, that there is some particular form of uncontrolled variation. For example, we might have the above paired comparison experiment with, say, six pairs receiving the order $T_1 T_2$ and two receiving the order $T_2 T_1$. Inspection of the results may suggest a substantial order effect comparable to the treatment effect. Another example would be if an agricultural field trial arranged in randomized blocks shows a systematic trend from one end to the other of the experimental area. What do we do in such situations?

In some cases, possibly in the first, we may decide that the data should be regarded with suspicion. Suppose, however, that we do wish to draw what conclusions we can. The previous discussion shows that it is not good enough to say that the long-run properties are valid whatever the form of the uncontrolled variation and on those grounds to analyze the experimental results by the usual methods. On the other hand, to introduce modifications into the analysis based on inspection of the results and on personal judgement about the design must lead to some loss of objectivity. The following procedure is suggested.

(*a*) Work through the conventional analysis of the observations ignoring the suspected complication.

(*b*) Make a special statistical analysis of the observations taking account of the complication in whatever seems the most reasonable way. The reader who is not familiar with fairly advanced statistical methods will probably need statistical advice in this. The method will usually involve the analysis of what is known technically as a nonorthogonal least-squares situation.

(*c*) If the conclusions of the two analyses are for practical purposes equivalent there is no difficulty. If the conclusions do differ, care is needed. The assumptions underlying the second analysis should be carefully thought over, and if they seem reasonable, the second analysis should be regarded as correct.

(*d*) In reporting on the experiment, conclusions from both analyses should be given, at any rate briefly. If the first analysis is rejected, reasons should be outlined. The general idea should be to make it clear to the reader what has been done and to give him the opportunity of forming his own conclusions as far as practicable.

Fortunately these difficulties tend to occur infrequently in practice.

Another difficulty that occasionally arises is that there is some practical reason why certain treatment arrangements are not allowable. One example arises in raspberry variety trials (Taylor, 1950). The point here is that additional canes spring up near many of the canes originally planted and it is necessary to remove these new canes from each plot. For this to be possible varieties that resemble each other closely must not occur close together, thus restricting the randomization. Another example occurs in carpet wearing trials, in which dyed and undyed carpets are under comparison. An experimental carpet is formed by sewing together squares of carpet of different types and the whole carpet placed say in a busy corridor. It would often be desirable that the carpet should look presentable and this would preclude full randomization of the dyed and undyed sections. The procedure in such cases is either to do as much randomization as possible or to use a systematic arrangement taking whatever steps are practicable to avoid bias.

SUMMARY

When a particular type of design, say a Latin square, has been chosen as likely to give precise treatment comparisons, the arrangement of treatments should be determined by impersonal randomization. This is done by shuffling cards, etc. or, much more usually, by tables of random permutations or of random digits.

Systematic arrangement is very occasionally to be preferred to randomization, for example on the grounds of simplicity; subjective assignment of treatments in a haphazard way should never be done. The justification for randomization is that it makes the chance negligible that systematic differences between units receiving different treatments will persist in a long experiment and that it enables the error to be estimated whatever the form of the uncontrolled variation. In effect the randomization rearranges the experimental units into random order and converts uncontrolled variation of whatever pattern into completely random variation. It is very important that randomization should cover all stages at which major errors may arise.

Care is needed, particularly in very small experiments, whenever unsatisfactory arrangements are produced in the randomization.

REFERENCES

Cochran, W. G., and G. M. Cox. (1957). *Experimental designs.* 2nd ed. New York: Wiley.

Cox, D. R. (1951). Some recent work on systematic experimental designs. *J. R. Statist. Soc.*, B, **14**, 211.

Fisher, R. A. (1951). *The design of experiments.* 6th ed. Edinburgh: Oliver & Boyd.

Greenberg, B. G. (1951). Why randomize? *Biometrics*, **7**, 309.

Grundy, P. M., and M. J. R. Healy. (1951). Restricted randomization and quasi-Latin squares. *J. R. Statist. Soc.*, B, **12**, 286.

Kendall, M. G., and B. Babington Smith. (1939). *Tables of random sampling numbers.* Tracts for computers, No. XXIV. Cambridge: Cambridge University Press.

"Student". (1931). The Lanarkshire milk experiment. *Biometrika*, **23**, 398. Reprinted in "*Student's*" *Collected Papers.* Cambridge, 1942.

——— (1938). Comparison between balanced and random arrangements of field plots. *Biometrika*, **29**, 363. Reprinted in "*Student's*" *Collected Papers.* Cambridge, 1942.

Taylor, J. (1950). A valid restriction of randomization for certain field experiments. *J. Agric. Sci.*, **39**, 303.

Tedin, O. (1931). The influence of systematic plot arrangement upon the estimate of error in field experiments. *J. Agric. Sci.*, **21**, 191.

Wilk, M. B., and O. Kempthorne. (1956). Some aspects of the analysis of factorial experiments in a completely randomized design. *Ann. Math. Statist.*, **26**, 950.

Yates, F. (1939). The comparative advantages of systematic and randomized arrangements in the design of agricultural and biological experiments. *Biometrika*, **30**, 440.

——— (1951). Bases logiques de la planification des experiences. *Ann. de l'Institut H. Poincaré*, **12**, 97.

Youden, W. J. (1958). Randomization and experimentation. *Ann. Math. Statist.* To appear.

CHAPTER 6

Basic Ideas about Factorial Experiments

6.1 INTRODUCTION

Up to now we have concentrated attention on experiments in which there is one set of treatments under comparison, the treatments in general not being arranged or ordered in any particular way. However it frequently happens that we wish to examine together the effect of several different types of modification to our system, for example, the effect of changes of pressure, temperature, and proportions of reactants on a chemical process. In this example each treatment consists of a combination of a temperature, a pressure, and a set of concentrations for the reactants, and, in general, in the type of experiment that we are going to consider now, each treatment consists of a combination of what will be called *factor levels*.

The methods discussed in previous chapters for increasing precision by grouping of experimental units and for the control of remaining variation by randomization apply without change, but there are numerous special problems that arise because of the particular relations holding among the treatments. In the present chapter we shall deal with some general concepts connected with these systems and then in the next chapter go on to discuss actual problems of design. Details of the present chapter may be omitted at a first reading; §§ 6.2, 6.3, 6.10 are the most important sections.

6.2 GENERAL DEFINITIONS AND DISCUSSION

It is convenient to start by introducing some definitions and then discussing in a general way the advantages and disadvantages of including all "factors" together in one experiment, rather than investigating them separately in separate experiments.

We shall call each basic treatment a *factor* and the number of possible forms of a factor the number of *levels* for that factor. A particular combination of one level from each factor determines a *treatment*. The experiment as a whole is called a *factorial experiment* if all, or nearly all, factor combinations are of interest. Some examples should make these definitions clear.

Example 6.1. A classical example is a fertilizer trial with three factors, nitrogeneous fertilizer N, phosphate P, and potash K. In the simplest case each type of fertilizer is either absent or present at a standard rate. There are then eight treatments:

no N,	no P,	no K	:	treatment 1
no N,	some P,	no K	:	treatment 2
no N,	no P,	some K	:	treatment 3
no N,	some P,	some K	:	treatment 4
some N,	no P,	no K	:	treatment 5
some N,	some P,	no K	:	treatment 6
some N,	no P,	some K	:	treatment 7
some N,	some P,	some K	:	treatment 8

This is an experiment with three factors, each at two levels. It is convenient to say that it is a $2 \times 2 \times 2$ (or 2^3) experiment. If all eight fertilizer combinations are tested on each of three varieties we should have a $3 \times 2 \times 2 \times 2$ experiment.

Example 6.2. Consider an experiment on the effect on conception rate of the addition of diluents to bull semen. Suppose that there are three substances which may be added (e.g., sulfanilamide, streptomycin, penicillin), each of which is either present or absent. If all eight combinations are used, the experiment is again a $2 \times 2 \times 2$ factorial experiment. If, however, we regard these substances as essentially alternatives to one another, there would be four treatments, a control and the above three. We should not then consider the experiment to be a factorial one, because an appreciable number of factor combinations are excluded from the experiment.

The design of a factorial experiment for this problem (Campbell and Edwards, 1954) illustrates some of the more advanced techniques and will be discussed in § 12.3.

Example 6.3. In experiments on the effect of fire-retardant treatments applied to wood (van Rest, 1937), factorial experiments of the following type are encountered. There are three different treatments and a control, and these are to be compared on material of two species. Each species is to be tested with a rough surface and with a smooth surface two days, three months, and one year after treatment. This is formally a 4 (treatments) $\times$ 2 (species) $\times$ 2 (rough and smooth) $\times$ 3 (times) factorial experiment. However, although for many purposes it is very useful to regard the experiment in this way, we have here an important distinction between factors that represent a treatment applied to the units (treatment factors) and factors that correspond in part to a classification of the units into two or more types (classification factors).

To elaborate on this point a little, consider the introduction in Chapter 1 of the idea of a treatment. An essential point, and one that has been appealed to implicitly throughout, is that any experimental unit is capable of receiving any treatment and that the allocation of treatments to units is under the experimenter's control. Clearly species is not a treatment in this sense; the species of a particular unit is an intrinsic property of the unit and not something assigned to it by the experimenter. Similarly the difference between rough and smooth surface does not represent a treatment factor if the experimenter has no control over which pieces are planed and which not, but is simply presented with a series of rough boards and a series of smooth boards. If however the experimenter has a supply of rough boards and can decide which are to be made smooth, we have a genuine treatment.

Therefore the present experiment is more accurately described either as a 4×3 factorial experiment with a 2×2 classification of the units or as a $4 \times 3 \times 2$ factorial experiment with a classification of the units into two sets.

Example 6.4. Another example may help to clarify the distinction between treatment factors and classification factors. Suppose that the artificial insemination experiment of Example 6.2 is repeated in say four centers, using cows arising in the day-to-day work of the centers. Then centers has to be regarded as a classification factor, since the fact that one cow occurs in one center's results rather than in another's is quite outside the experimenter's control.

Suppose now that, in addition to any treatment effects, there are appreciable differences between centers in the conception rate. Then we are unable, from the experimental results alone, to say whether these differences arise because

(*a*) the cows serviced in different centers are systematically different—this could very well happen; or

(*b*) there are some differences in technique between the different centers, or differences in the external conditions under which insemination is carried out.

In other words, so far as a comparison of centers is concerned, we are in the position of having an experiment with possible systematic differences between units receiving different treatments. This does not mean that useful information cannot be drawn from a comparison of centers, but that the conclusions will lack cogency unless supported by external evidence. On the other hand, when we compare two or more treatments, we know that a particular cow received one treatment rather than another in virtue of a purely random choice. Hence the effect of animal differences on the comparison of treatments is purely random, so that the resulting uncertainty can be measured objectively.

Purely to illustrate the logical point, imagine that it was possible to allocate each cow to a center in a way involving randomization. Centers would then represent a treatment factor, which could be examined in the usual way.

In much of the discussion that follows the distinction between the two types of factor does not need explicit mention, but in interpreting the results the distinction is very often important.

A factorial experiment in which each combination of factor levels is used the same number of times is called a *complete factorial experiment*. The experiments described in this and the next chapter are of this type. The reason for restricting attention to complete experiments is that the estimation of the separate effects of interest is simple only when the design is complete, or at any rate has a high degree of symmetry. Incomplete experiments are important in some fields, particularly when many factors are under study simultaneously (Chapter 12), but care is then needed in selecting suitable designs if very lengthy calculations are to be avoided.

We now discuss the advantages of factorial experiments, using for the most part the language of Example 6.1. Suppose for definiteness that we have this experiment arranged in 6 randomized blocks of 8 plots each, so that each of the 8 treatments occurs 6 times, there being 48 plots in all. There are two cases to be distinguished in the discussion:

(*a*) the additional yield obtained by changing from N absent to N present is, to a good approximation, the same for all levels of P and K, and that the same is true when P, K take the place of N. In this case we say that on this scale of measurement N, P, and K do not interact.

(*b*) there may be interaction between two or more of the factors. For example, there may be no gain from using N unless P is present too, and so on.

We shall discuss the definition and meaning of interaction carefully in the next sections; all that is needed at present is the general idea that interaction between two factors means that the effect due to one of them, say N, depends on the particular level of the other, say P.

Now in the first case, we can estimate the increase in yield caused by changing from N absent to N present by taking the difference of the mean of the 24 observations with N present, minus the mean of the 24 observations with N absent. We can do this because in the complete factorial system each set of 24 contains all combinations of P and K the same number of times, e.g., P absent 12 times and P present 12 times.

Therefore we get the precision that we would have got if the full 48 plots had been devoted to a test of N. Similar remarks apply to the corresponding comparisons for P and K.

On the other hand, if we had divided the plots into 3 equal groups of 16 and had devoted the first group to a test of N by itself, and the second and third groups to tests of P and K respectively, we should have been much worse off. Each comparison would have to be estimated from the difference of 2 means of 8 observations, not from means of 24. The more factors, the greater is the gain in using a factorial experiment to investigate the factors simultaneously.

The advantages of the factorial approach are even more marked when interaction is present. The averages considered in the previous paragraph estimate, for example, the mean increase in going from N absent to N present, averaged over all levels of the other factors. It is possible, in addition, to examine any particular factor combinations of interest, to estimate separately the effect of N, first with P absent and then with P

present, and so on. On the other hand, if we are to do separate experiments for N, P, and K we must in experimenting, say on N, fix levels for P and K. The estimate that we obtain of the effect of N may be quite misleading if the levels chosen for P and K happen to be appreciably different from those of final practical interest. Also, information obtained from the factorial experiment on how the treatments interact with one another is of value, not only in reaching a decision about which combination is best but also in reaching some insight as to how the treatments "work." In fact a factorial experiment, particularly one with factors at more than two levels, gives a much fuller picture of the relation between the observation and the factor levels than can be obtained from an experiment in which only a very restricted set of factor combinations is examined.

A further advantage of factorial experiments is that they enable the range of validity of the conclusions to be extended in a convenient way. For instance in Example 6.1 it would be common to include, as a further factor, a number of varieties, chosen to be typical of important different types of variety. The object of this is not a direct comparison of varieties, which would almost certainly be known to be very different, but is to examine in what way, if any, the fertilizer effects are different for different varieties. In particular, if it is found that the fertilizer effects are essentially the same for the three varieties, the conclusions can be applied to a new variety with much more confidence than if the experiment had been confined to one variety.

To sum up, factorial experiments have, compared with the one factor at a time approach, the advantages of giving greater precision for estimating overall factor effects, of enabling the interactions between different factors to be explored, and of allowing the range of validity of the conclusions to be extended by the insertion of additional factors.

The remarks in the previous paragraphs, if taken uncritically, may appear to suggest that if we have a particular situation under investigation it will be best to examine all conceivable factors together, using in one big experiment all resources that can be devoted to the study of the problem. There are several reasons why this is not so. First, there is the obvious consideration of simplicity; experiments with many factors and a large number of experimental units are, even if the resources for them are available, difficult to organize.

Second, and more importantly, it will often be a bad thing to commit oneself to a large experiment at the beginning of an investigation. Small preliminary experiments may indicate a much more promising line of enquiry and are, in any case, frequently useful for picking out the factors of most importance and for showing the ranges of these factors that should be investigated. Some of the methods useful in designing "preliminary" experiments are discussed in Chapter 12.

The third point raises an important general issue. Factorial experiments are particularly appropriate for describing the behavior of a system empirically by observing it under a wide range of conditions. In many problems, however, it is more profitable in the long run to aim, not just at a description of the effect of changing the system, but at a real understanding of the system in terms of more fundamental concepts. To do this a chain of simple experiments under special conditions is usually called for, each experiment suggested by and illuminating the results of the previous experiments. Simple factorial experiments with up to, say, three or four factors may very well be appropriate and, if appreciable uncontrolled variation is present, the other methods described in this book will be useful too. But the idea of a factorial experiment with very many factors is usually inappropriate, except for a rapid initial survey, because understanding is most often best obtained by starting with thorough investigation of simple cases and proceeding in stages from there. The situation is modified somewhat if, as in agricultural field trials, particularly with perennial crops, experiments take a long time to complete. In such cases the argument for using quite complex experiments is strengthened.

6.3 TYPES OF FACTORS

In addition to making the distinction between treatment and classification factors, it is convenient for some purposes to divide the factors encountered in applications into the following types:

(i) specific qualitative factors,
(ii) quantitative factors,
(iii) ranked qualitative factors, and
(iv) sampled qualitative factors.

These four types will now be illustrated by examples.

(i) Specific Qualitative Factors

These are factors for which there is no natural order established among the different levels and for which each level is of intrinsic interest. Very often we shall drop the adjective specific, which is included to distinguish this case from (iii) and (iv). Examples of specific qualitative factors are: varieties of wheat in a variety trial, qualitatively different treatments for a disease, different techniques for pruning in a fruit trial, different techniques of measurement in an interlaboratory comparative trial, and so on.

Factors which represent a classification of the experimental units are frequently qualitative; for example, different types (qualities) of wool, different groups of subjects in a psychological experiment, different laboratories or centers in an interlaboratory trial in which each laboratory provides its own experimental material, and so on.

(ii) Quantitative Factors

These are factors for which the different levels correspond to well-defined values of some numerical quantity called a carrier variable. Such factors are particularly common in experiments in the physical and chemical sciences and technologies. Some examples of quantitative factors are the following: different quantities of fertilizer in an agricultural field trial, different temperatures (or pressures or concentrations of a reactant) in a chemical experiment, different lengths of time for which a treatment is applied, different doses of a drug, different strengths of stimuli, and so on.

For instance, in the first example the carrier variable is the quantity of fertilizer in say cwt per acre and the levels of the factor might be zero and once, twice, and three times a standard dressing. We would then have a situation in which the factor levels used in the experiment correspond to equally spaced values of the carrier variable. It is often convenient to arrange for this to happen, but it is by no means essential and, as we shall see later, in dealing with certain types of curved relation it is desirable not to have equally spaced values.

In a specific qualitative factor, the different levels represent treatments of intrinsic interest to us. This is not usually the case with a quantitative factor; the precise levels used are selected more or less arbitrarily and we are interested in the curve giving the relation between the observation and the carrier variable, e.g., the curve of mean yield against amount of fertilizer. We shall call such a curve a response curve.

To put this more formally consider an experiment with just one factor and denote the carrier variable by v. Choose as a convenient reference point an arbitrary value v_0, somewhere in the middle of the range of values of the carrier variable of practical interest. Then if the basic assumption of Chapter 2 holds, we can consider the true treatment effect corresponding to any value of v within the range. This is defined as the difference between the observation obtained on a particular unit when a treatment is applied corresponding to v and what the observation would have been with a treatment corresponding to the reference point v_0. We call the curve of the true treatment effect against v the true response curve; a few common types of response curve arising in applications will be shown in Fig. 6.2 of § 6.8.

In estimating the curve from the results of an experiment there are two sources of uncertainty, one arising from the uncontrolled variation and the other from the fact that we can only make observations for a limited number of values of v. This means that if the range of values of v is continuous we can only estimate the whole form of the curve by assuming it to be sufficiently smooth. These two sources of uncertainty will be dealt with when we discuss the choice of levels, i.e., the choice of how many values of v to use and where to take them.

The distinction between qualitative and quantitative factors affects the analysis as well as the design of the experiment. However for a factor occurring at only two levels, no estimate can be formed of the detailed shape of the response curve and the only thing that can be estimated is the difference between the observations at the two factor levels. Hence in this case qualitative and quantitative factors are, from the point of view of analysis, equivalent.

(iii) Ranked Qualitative Factors

These arise less frequently than the previous two types of factors. It may happen that the factor levels are arranged in some order but that there is no natural quantitative variable that describes the factor levels, or alternatively that the different levels correspond to very coarsely grouped values of a quantitative variable.

Most ranked qualitative factors are classification factors. The following are examples: patients might be divided into those having a disease slightly, moderately, severely, and very severely. Severity of disease is then a ranked qualitative factor, since there is no natural numerical quantity defining the different levels. In a psychological experiment, it might be convenient to group the subjects into three sets of different ages, e.g., under 25, between 25 and 35, and over 35. Here the factor does correspond to a well-defined quantity, age, but the levels do not correspond to well-defined values; for example, the last group might contain subjects with a wide range of ages.

If in this last example it turned out that the observations depend in an important way on age, we should probably abandon the grouping into three sets and use the individual ages of the subjects in the analysis. However so long as we regard age as a three-level factor it is a ranked qualitative one.

In general, although the distinction between specific qualitative and ranked qualitative factors is often important in the analysis of observations, it is not very relevant in the design of an experiment, so that the distinction will not often arise in what follows.

(iv) Sampled Qualitative Factors

It sometimes happens, particularly with classification factors, that the levels used in the experiment are not of much interest in themselves but are to be considered as a haphazard sample from a larger collection, or population, of levels.

For example we might take as different levels of a factor different consignments of raw material for an industrial process. It might then be interesting to make individual comparisons among the different consignments, regarding each consignment as a thing of intrinsic interest and hence the factor consignments as a specific qualitative factor. Often, however, we would not regard the consignments as of particular interest individually and would be concerned solely with a population of consignments, of which the consignments used are a sample. If the consignments used in the experiment were selected from a larger set by a statistical sampling procedure, the population is well defined, namely as the larger set from which the selection was made. In other cases the population may be a rather nebulous concept, but the general idea is that we usually want our conclusions about other factors in the experiment to apply not just to the consignments actually used in the experiment, and we want to assess the additional uncertainty involved in this extension.

Similar remarks apply to agricultural experiments repeated in several years or in several centers. We often find it appropriate to consider the years or centers as a sample from a population of years or centers, but there is almost always some additional uncertainty in supposing that this population is equivalent to the one to which we want the conclusions of the experiment to apply (see Example 6.7).

The distinction between the different types of factor will often be referred to below, although for a good deal of the discussion the distinctions do not matter much.

6.4 MAIN EFFECTS AND INTERACTIONS IN A TWO-FACTOR EXPERIMENT

A very important idea that underlies much of the discussion of factorial experiments is that of main effects and interactions. This will now be explained using simple idealized examples.

Consider first an experiment with two factors A and B each say at four levels. For definiteness the levels of A may be thought of as four varieties of say wheat, and the four levels of B as four different kinds of nitrogeneous fertilizer. Suppose, to make the discussion simpler, that there is no uncontrolled variation and that the observations in Table 6.1a are obtained. These observations can be characterized in various equivalent ways:

(*a*) the difference between the observations corresponding to any two levels of A is the same for all levels of B. Thus, the change in going from A level 1 to A level 3 is plus 1 in each column;

(*b*) the difference between the observations for two levels of B is the same for all levels of A;

(*c*) the effects of the two factors are *additive*. The meaning of this statement is best understood by considering the row and column averages shown in the margins of the Table. Now the row averages show that the change in going from level 1 of A to level 3 of A is plus 1; also the change in going from level 2 of B to level 4 of B is plus 4. Consider now the change in going from the position determined by the first level of A and second level of B to the position determined by the third level of A and fourth level of B. It is $16 - 11 = 5$, which is equal to the sum of the average effects of A and B separately. If this rule

$$\begin{pmatrix}\text{change in}\\ \text{observation}\end{pmatrix} = \begin{pmatrix}\text{average effect of}\\ \text{change in } A\end{pmatrix} + \begin{pmatrix}\text{average effect of}\\ \text{change in } B\end{pmatrix}$$

applies to all pairs of cells in the table, we say that the factors A and B are additive. This will be found to be so in the present example and the reader should check a few cases;

(*d*) the residuals, obtained as in Chapter 3 by subtracting row and column effects, are all zero.

The conditions (*a*), (*b*), (*c*), and (*d*) are mutually equivalent; the reader may convince himself that this is so, either by thinking about the above special case or by constructing a formal proof using elementary algebra. In general if these conditions are satisfied we shall say that the factors A and B do not *interact*. If the conditions are not satisfied we say that there is *interaction* between A and B. These are very important definitions.

There are many ways in which interaction can occur, and a particular one is shown in Table 6.1(*b*). For comparison the row and column averages have been made the same as in Table 6.1(*a*), but now the difference between, for example, level 1 and level 2 of Factor B is not the same for all levels of A. In terms of the particular application, the effect of changing from one fertilizer to another is not the same for all varieties.

We define the main effect of the factor A (averaged over the four levels of B) to be the series of comparisons that can be made with the row averages at the right of the table; for example the main effect of the comparison of the second level of A with the first is $15.25 - 12.25 = 3$, and so on. We usually talk just about the main effect of A without mentioning B but, unless there is no interaction, the magnitude of the main effect of A depends on the particular levels of B used in the experiment.

TABLE 6.1

FICTITIOUS OBSERVATIONS IN SIMPLE FACTORIAL EXPERIMENTS

(*a*) *Absence of Interaction*

		Factor *B*				
	Level	1	2	3	4	Row Average
	Level 1	9	11	14	15	12.25
Factor *A*	2	12	14	17	18	15.25
	3	10	12	15	16	13.25
	4	13	15	18	19	16.25
Column Average		11	13	16	17	

(*b*) *Presence of Interaction*

		Factor *B*				
	Level	1	2	3	4	Row Average
	Level 1	9	11	14	15	12.25
Factor *A*	2	12	14	17	18	15.25
	3	11	11	14	17	13.25
	4	12	16	19	18	16.25
Column Average		11	13	16	17	

(*c*) *A Very Special Situation with Zero Main Effects*

		Factor *B*				
	Level	1	2	3	4	Row Average
	Level 1	14	16	14	16	15
Factor *A*	2	15	13	18	14	15
	3	12	15	16	17	15
	4	19	16	12	13	15
Column Average		15	15	15	15	

A very artificial case is shown for illustration in Table 6.1(*c*). Here the row and column averages are all equal so that the main effect comparisons are all zero. There is no average change in going from one *A* level to another, but, for any particular level of *B*, the observations corresponding to different levels of *A* are all different. We shall see that the cases most commonly arising in applications are not like this but have interactions smaller than the main effects.

The above account is in terms of a situation with no uncontrolled variation. In practice we have to distinguish between the true main effects and the true interaction both defined in terms of the true treatment effects, and our estimates of these obtained from the observations. That is, in virtue of the basic assumption of Chapter 2, there is for each combination of levels of A and B a true treatment constant. If these are entered in a table such as those in Table 6.1, they define the true main effects of A and of B and the true interaction. If the treatment means calculated from our observations are set out in a similar table we calculate estimates which are, of course, affected by the uncontrolled variation. In particular, if the true interaction is zero, there will almost always be some apparent interaction in the observations; a statistical test is usually required to find whether the data are consistent with the absence of true interaction.

We now consider the advantages of working with main effects and interactions. If there is no interaction there is considerable economy in the description of the results of the experiment, since only the main effects need to be thought about instead of the effects corresponding to all treatment combinations. The gain here is particularly marked in experiments with more than two factors. Secondly, there is sometimes a gain in understanding the experimental situation, both in that the absence of interaction may mean that the two factors act independently of one another and this may throw light on what is happening, and in that particular patterns of interaction may have special physical significance.

To illustrate this last point consider a nutritional experiment in which the two factors A and B refer to different supplements to a basic diet, each being either present or absent, i.e., at two levels. We can distinguish three extreme cases:

(*a*) there is no interaction between the factors;

(*b*) each supplement produces, say, a unit increase in growth rate by itself, but there is no additional effect when both are used together;

(*c*) there is no change when the supplements are used individually, but there is an effect when the two are used simultaneously.*

Fictitious observations illustrating these situations are shown in Tables 6.2(*a*), (*b*), (*c*); the observations might be, for example, weight increases over a standard period of time. It seems clear that the three sets of observations correspond to three different types of biological action and that, although we could not expect any of the types of treatment response to apply exactly, a comparison of the observations with these types to see which fits most nearly may be helpful. However, in an experiment of that type with quantitative factors, it is usually advisable

* It may be noted that these situations are analogous to three classical epistatic models in genetics.

to have more than two levels for each factor; we shall discuss this more fully later.

TABLE 6.2

FICTITIOUS OBSERVATIONS IN A SIMPLE NUTRITIONAL EXPERIMENT

(*a*) *No Interaction*

		Supplement *B*	
		Absent	Present
Supplement *A*	Absent	10	12
	Present	13	15

(*b*) *No Additional Response with Both Supplements*

		Supplement *B*	
		Absent	Present
Supplement *A*	Absent	10	15
	Present	15	15

(*c*) *No Response without Both Supplements*

		Supplement *B*	
		Absent	Present
Supplement *A*	Absent	10	10
	Present	10	15

We now go on to consider how we can interpret the presence of interaction.

6.5 THE INTERPRETATION OF INTERACTIONS

Nearly all sets of observations from experiments of the type we are discussing will show some apparent interaction. Before claiming that our experiment shows that such an interaction is a real property of the system, we usually need to apply the appropriate significance test (Cochran and Cox, 1957, Chapter 5; Goulden, 1952, p. 94). The result of such a test measures the uncertainty involved in assuming the reality of the apparent interaction. The test may, however, indicate that the data are consistent with the absence of interaction. This is not the same as saying that there is no real interaction.

In the present account we shall not deal with details of these tests of significance but will give a brief account of the interpretation of interactions, assuming that their statistical significance has been established. These details of interpretation are not directly relevant in the design of the experiment, and the section may be omitted on a first reading.

With a two-factor experiment the observations may be summarized in a two-way table similar to Table 6.1, each entry, however, being the mean of all observations on the corresponding treatment. Thus if the experiment has been done in five simple randomized blocks, each entry in the two-way table is the mean of five observations one from each block. Marginal means are calculated in a straightforward way.

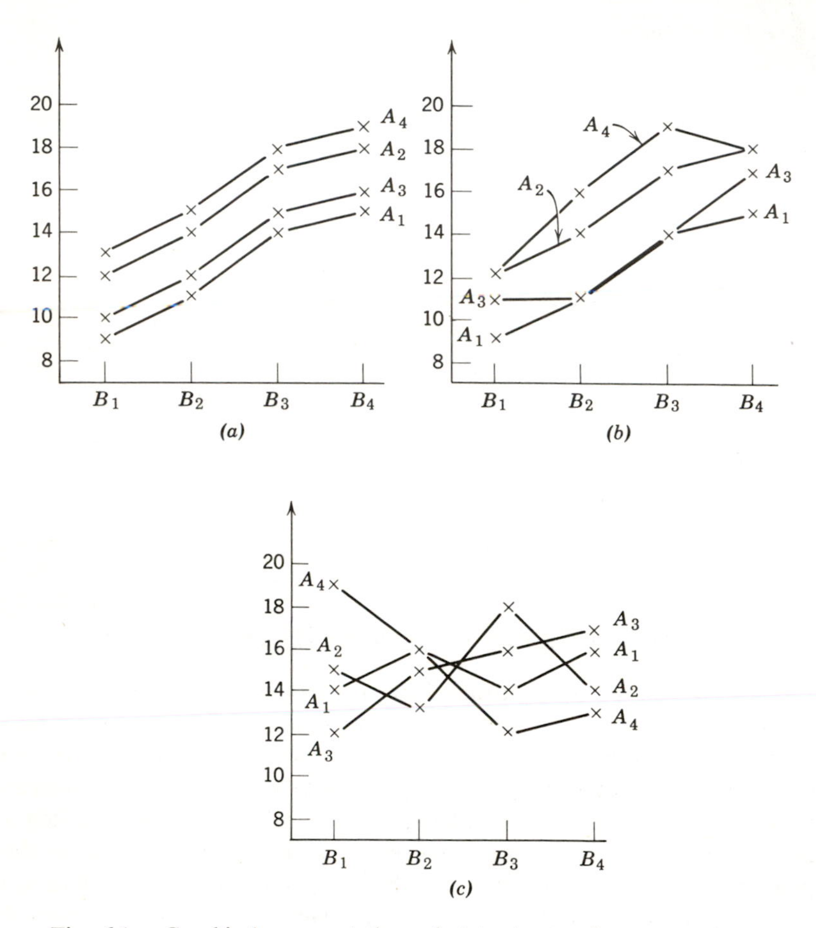

Fig. 6.1. Graphical representation of data in two-factor experiment, Table 6.1. (*a*) without interaction; (*b*) and (*c*) with interaction.

It may be possible to grasp the meaning of such a table as it stands; however it is often useful to present it graphically in the following way. We choose one of the factors to be represented along the x-axis of our diagram and plot a series of graphs, one for each level of the other factor. For example, take the numbers in Table 6.1(*a*). In Fig. 6.1(*a*) we have represented the levels of B by four points along the x-axis and each set of points joined by a light line corresponds to a different level of Factor A. The four curves are parallel; this is the condition for zero interaction. This graph should be compared with Figs. 6.1(b) and (c) for the observations in Tables 6.1(b) and (c), where interaction is present.

In this experiment, in which A represents varieties and B different types of fertilizer, both factors are specific qualitative factors. Either could be chosen to be plotted along the x-axis. It is often, however, clear which is the more informative way to plot the graph; for example, if one factor is a sampled qualitative factor, or a classification factor, this would probably be taken as A in the diagram. Similarly if one factor is quantitative, or ranked qualitative, it would probably be taken as B.

We shall distinguish briefly three types of interaction. The first is one that can be removed by a transformation of each observation to, for example, its square, its logarithm, its square root, or some other appropriate function. Our definition of the absence of interaction is that the *difference* between the observations at two levels of say A is the same for all levels of B; if, for example, the proportional change were constant, not the absolute change, we should say that interaction existed. Table 6.3(a) illustrates this. The first set of data show interaction of the type just described; when the logarithms of the observations are taken there is no interaction (see the second half of the table).

Tables 6.3(b) and (c) show corresponding cases in which the interaction is removed: (b) by squaring and (c) by square-rooting the observations. The reader should plot for these examples graphs corresponding to Fig. 6.1 and should note the difference between them. These examples correspond to cases with no uncontrolled variation, i.e., the transformed observations are exactly without interaction.

Statistical methods by which removable interactions can be recognized and dealt with systematically have been studied (Tukey, 1949, 1955; Moore and Tukey, 1954), but at the time of writing no simple routine procedure is available, and it is sometimes best to proceed by trying out various transformations if the presence of a removable interaction is suspected. There seems to be an essential difference between removable and nonremovable interactions in that the latter indicate some complexity of behavior, whereas the former can be regarded as arising from having recorded the observations on an inappropriate scale. The advantages of having no interactions have been discussed above, but there is of course no compelling reason to transform the data if there are compensating practical or theoretical advantages to working with treatment differences on the untransformed scale. Transformations are in any case usually only effective when the fractional variation of the data is large.

We have just discussed interactions removable by transformation

TABLE 6.3

SOME EXAMPLES OF INTERACTIONS REMOVABLE BY A TRANSFORMATION

(*a*) *Logarithmic Transformation*

Original Observations

	Factor *B* Level	1	2	3
Level Factor *A*	1	2.00	3.16	6.31
	2	2.51	3.98	7.94
	3	3.98	6.31	12.59

Transformed Observations

Factor *B* Level	1	2	3
	0.3	0.5	0.8
	0.4	0.6	0.9
	0.6	0.8	1.1

(*b*) *Square Transformation*

Original Observations

	Factor *B* Level	1	2	3
Level Factor *A*	1	1.73	2.24	2.83
	2	2.00	2.45	3.00
	3	2.45	2.83	3.32

Transformed Observations

Factor *B* Level	1	2	3
	3	5	8
	4	6	9
	6	8	11

(*c*) *Square-Root Transformation*

Original Observations

	Factor *B* Level	1	2	3
Level Factor *A*	1	9	25	64
	2	16	36	81
	3	36	64	121

Transformed Observations

Factor *B* Level	1	2	3
	3	5	8
	4	6	9
	6	8	11

The first type of nonremovable interaction consists of those describable in simple verbal terms. These arise particularly with specific qualitative factors. For example, it may happen that interaction would be absent were it not for a single row's or column's results. Thus, we can reach a good understanding of Table 6.4(*a*) by saying that for levels 1, 2, and 3 of A there is no interaction with B and giving a table of means to show the corresponding main effects. The response for the fourth level of B is given separately and commented on. Table 6.4(*b*) shows the resulting sets of means. This is again a case with no uncontrolled variation; in practice such variation must be allowed for and this is usually done by quoting the standard error of the difference between two means.

Some care is needed in this sort of analysis because the data may be consistent with several simple descriptions; we usually use the description that is a priori most likely, but this introduces a subjective element. The full two-way table of treatment means should always be given in reporting on the experiment.

The scientific importance of an interaction like this is that the action of B is modified at the fourth level of A, and consideration of what distinguishes the fourth level of A from the others may throw light on the way the treatment B works. The practical importance of the interaction is that if we wish to make a technological recommendation about the

TABLE 6.4

FICTITIOUS OBSERVATIONS ILLUSTRATING THE INTERPRETATION OF INTERACTIONS

(*a*) *Full Two-Way Table*

		Factor B		
	Level	1	2	3
Factor A Level	1	9	10	12
	2	11	12	14
	3	13	14	16
	4	13	13	9

(*b*) *Supplementary Tables*

		Factor B		
	Level	1	2	3
Factor A Mean of Levels	1–3	11	12	14
Factor A Level	4	13	13	9
Mean of all Levels		$11\frac{1}{2}$	$12\frac{1}{4}$	$12\frac{3}{4}$

	Factor A			
	Level 1	2	3	4
Mean of all Levels of B	$10\frac{3}{4}$	$12\frac{1}{3}$	$14\frac{1}{3}$	$11\frac{2}{3}$

Alongside each row standard errors would usually be quoted for differences between pairs of means in the row.

level of B it will be necessary to consider at what level A is to appear: furthermore, extension of the conclusions about B to apply to a level of A not appearing in the experiment will be difficult until the nature of the interaction is well understood.

It may happen that no simple description of the interaction appears reasonable and in this case we may have to regard the different treatment combinations as distinct and in effect ignore the factorial structure. This is the third type of interaction.

In all cases, whether or not there is interaction, the main effects of A and of B are defined, as in § 6.4, in terms of the marginal means of the two-way table. The main effect of A, therefore, tells us about the effect of A averaged over the particular levels of B used in the experiment. If there is interaction, the main effect may, or may not, be a useful thing to work with. The following remarks are relevant:

(*a*) If the general trend with A is the same for all levels of B, as for example when the interaction is removable by a transformation, the main effect may be useful for indicating qualitatively the general trend with A for all levels of B and vice versa;

(*b*) If the averaging over the levels of B, giving each level equal weight, has a direct physical meaning, the main effect is useful. For example, if factor B is sex, an average over the two levels is often appropriate. If the factor B is a sampled qualitative factor, the main effect of A will estimate the A effect averaged over the population of levels of B and this is usually required.

(*c*) In other cases, when the averaging over the levels of one factor is artificial and the main effect of A gives little idea of the variation with A for individual levels of B, the use of main effects seems undesirable.

6.6 EXPERIMENTS WITH MORE THAN TWO FACTORS

The previous discussion has been in terms of an experiment with only two factors. Many of the remarks apply with minor changes to complete factorial experiments with more than two factors. If we select any two factors, say A and B, we can form a two-way table of mean values like Table 6.1; in the entry for each mean, all combinations of levels of the other factors are represented equally often. From such a two-way table we define the main effect of, say, A averaged over all levels of all other factors. Similarly, the presence or absence of interaction in such a table determines what we shall call the two-factor interaction $A \times B$ (averaged over the levels of the other factors). There is a two-factor interaction for each pair of factors.

In addition to the main effects and two-factor interactions, there are more complicated things to consider. The two-factor interaction $A \times B$ determines whether the effect of A is the same for all levels of B. Similarly, in an experiment with three factors, we may examine the interaction $A \times B$ separately for each level of C. If the interactions are of the same form for each level of A, in that the set of residuals eliminating A and B is the same for each level of C, we say that there is no three-factor interaction $A \times B \times C$.

This is illustrated, again with fictitious data, in Table 6.5(*a*). The left-hand side of this shows the results for a $3 \times 3 \times 3$ experiment set out in three two-way tables for A against B, one for each level of C. There is an interaction $A \times B$, as can be seen by inspecting Table 6.5(*b*), which gives the results averaged over the three levels of C. The pattern of the interaction $A \times B$ in each two-way table is found by calculating residuals by the formulas of § 3.3. The residuals would all be zero if there were no interaction. Thus, to calculate the residual for the observation at the second level of A, B, and C we take

$$\text{observation} - \begin{pmatrix}\text{row mean}\\ \text{in two-way}\\ \text{table}\end{pmatrix} - \begin{pmatrix}\text{column mean}\\ \text{in two-way}\\ \text{table}\end{pmatrix} + \begin{pmatrix}\text{overall mean}\\ \text{of two-way}\\ \text{table}\end{pmatrix}$$
$$= 6 - 6 - 6 + 8 = 2$$

and this value is entered in the table in the right-hand column.

The set of residuals is the same for each level of C and for the

TABLE 6.5

FICTITIOUS OBSERVATIONS ILLUSTRATING THE ABSENCE OF A THREE-FACTOR INTERACTION

(*a*) *Two-Way Table for A and B, Separately for Each Level of C*

Observations

First Level of C

	B_1	B_2	B_3	Mean	Residuals		
A_1	5	6	10	7	−2	−1	3
A_2	7	7	1	5	2	2	−4
A_3	6	5	7	6	0	−1	1
Mean	6	6	6	6			

Second Level of C

	B_1	B_2	B_3	Mean			
A_1	9	7	14	10	−2	−1	3
A_2	9	6	3	6	2	2	−4
A_3	9	5	10	8	0	−1	1
Mean	9	6	9	8			

Third Level of C

	B_1	B_2	B_3	Mean			
A_1	10	11	15	12	−2	−1	3
A_2	10	10	4	8	2	2	−4
A_3	7	6	8	7	0	−1	1
Mean	9	9	9	9			

(*b*) *Two-Way Table for A and B Averaged over C*

Observations

	B_1	B_2	B_3	Mean	Residuals		
A_1	8	8	13	$9\frac{2}{3}$	−2	−1	3
A_2	$8\frac{2}{3}$	$7\frac{2}{3}$	$2\frac{2}{3}$	$6\frac{1}{3}$	2	2	−4
A_3	$7\frac{1}{3}$	$5\frac{1}{3}$	$8\frac{1}{3}$	7	0	−1	1
Mean	8	7	8	$7\frac{2}{3}$			

observations averaged over all levels of C. When this happens, we say that $A \times B$ is the same at all levels of C and that there is no three-factor interaction $A \times B \times C$.

The three-factor interaction has been introduced in an unsymmetrical way by considering the interaction $A \times B$ separately for each level of C. We could equally well have considered $A \times C$ separately for each level of B or $B \times C$ separately for each level of A. The reader new to these ideas will find it a good exercise to reform the figures in Table 6.5 into three two-way tables for say A and C separately for each level of B and to check that there is no three-factor interaction looked at from this point of view. Quite generally the three-factor interaction $A \times B \times C$ is a symmetrical property of the three factors and could equally well be written $A \times C \times B$, $B \times C \times A$, etc.

The statistical test to see whether an apparent three-factor interaction can be accounted for by random variation is a simple generalization of the corresponding test for two-factor interactions and will not be discussed here. The interpretation of three-factor interactions also follows a similar line to that for two-factor interactions. For example, we may be able to describe verbally the two-factor interactions of, say, A and B separately for each level of C in such a way as to make the meaning of the three-factor interaction understandable.

In an experiment with more than three factors we can define four (and possibly more) factor interactions. For example, $A \times B \times C \times D$ will determine whether or not the three-factor interactions $A \times B \times C$ are the same for all levels of D or, equivalently, whether or not $A \times B \times D$ is the same for all levels of C, and so on. High-order interactions like this are difficult to interpret and as a general rule are important only in very complicated systems.

6.7 MAIN EFFECTS AND INTERACTIONS IN EXPERIMENTS WITH ALL FACTORS AT TWO LEVELS

There is one case when the definitions of main effects and interactions can helpfully be put another way. This is when all the factors are at just two levels. Consider first an experiment with two factors A and B and denote the corresponding treatment means by (a_1b_1), (a_1b_2), (a_2b_1), and (a_2b_2), where, for example, (a_2b_1) means the average of all observations or units for which the factor A occurs at its second level and factor B at its first level. We can reasonably introduce the following definitions which are special cases of those given above for factors with more than two levels.

The effect of changing A when B is at its lower level is $(a_2b_1) - (a_1b_1)$

and the effect of A with B at its second level is similarly $(a_2b_2) - (a_1b_2)$. The main effect of A (averaged over the levels of B) is defined to be the mean of these two, namely

$$\tfrac{1}{2}[(a_2b_1) + (a_2b_2) - (a_1b_1) - (a_1b_2)]$$
$$= \begin{pmatrix}\text{average observation with} \\ A \text{ at its second level}\end{pmatrix} - \begin{pmatrix}\text{average observation with} \\ A \text{ at its first level}\end{pmatrix}.$$

The interaction $A \times B$ is defined to be one-half the difference between the effects of A at the two levels of B, i.e.,

$$A \times B = \tfrac{1}{2}[\{(a_2b_2) - (a_1b_2)\} - \{(a_2b_1) - (a_1b_1)\}]$$
$$= \tfrac{1}{2}[(a_2b_2) + (a_1b_1)] - \tfrac{1}{2}[(a_1b_2) + (a_2b_1)].$$

If the numerical values are, say,

	B_1	B_2
A_1	6	8
A_2	7	9

the interaction is $\tfrac{1}{2}(6 + 9) - \tfrac{1}{2}(7 + 8)$ and is zero. The reader should check that the earlier definitions give zero interaction too.

Consider now a 2^3 experiment with three factors A, B, C. To define the three-factor interaction $A \times B \times C$ we first consider the two-factor interaction $A \times B$ with C, (*a*) at its second level and (*b*) at its first level. These are obtained from the formulas for $A \times B$ in a 2×2 experiment, and are

(*a*) $\tfrac{1}{2}\{(a_2b_2c_2) - (a_1b_2c_2) - (a_2b_1c_2) + (a_1b_1c_2)\}$,

(*b*) $\tfrac{1}{2}\{(a_2b_2c_1) - (a_1b_2c_1) - (a_2b_1c_1) + (a_1b_1c_1)\}$.

The two-factor interaction $A \times B$ for the whole experiment is the average of (*a*) and (*b*), i.e., the mean of the separate $A \times B$'s over the two levels of C. The three-factor interaction is defined to be equal to one-half the difference between (*a*) and (*b*), and therefore is zero if $A \times B$ is the same for both levels of C. Thus

$$A \times B \times C = \tfrac{1}{4}\{(a_1b_2c_1) + (a_2b_1c_1) + (a_1b_1c_2) + (a_2b_2c_2)$$
$$- (a_1b_1c_1) - (a_2b_2c_1) - (a_2b_1c_2) - (a_1b_2c_2)\}.$$

The reader should check that the same definition is obtained by considering $A \times C$ separately for the two levels of B, or $B \times C$ separately for the two levels of A. (For most purposes the numerical factor is unimportant; what matters is the grouping of the treatments into two equal sets, one set receiving one sign and one set receiving the other.)

The general definition of any contrast (main effect or interaction) in a two-level experiment with any number of factors can be obtained by continuing the same process. Thus consider the definition of $A \times B \times C$ in a $2 \times 2 \times 2 \times 2$ experiment. Calculate first the interaction $A \times B \times C$, (*a*) with D at its upper level and (*b*) with D at its lower level. Then $A \times B \times C$ is the average of (*a*) and (*b*), and $A \times B \times C \times D$ is one-half their difference.

The formal mathematical expression of these general definitions is recorded here for completeness. To define $A \times B$ in a $2 \times 2 \times 2$ experiment with factors A, B, C, we consider the expression

$$(a_2 - a_1)(b_2 - b_1)(c_2 + c_1),$$

writing $(-)$ whenever the factor does occur in the contrast $A \times B$ that we are defining and $(+)$ whenever it does not occur. If this expression is multiplied out by the ordinary rules of algebra, putting brackets round each term, we shall obtain $2^3 = 8$ terms, each representing the average observation on one treatment. Each term has a coefficient ± 1, the full expression being

$$(a_2b_2c_2) + (a_2b_2c_1) + (a_1b_1c_2) + (a_1b_1c_1)$$
$$- (a_1b_2c_2) - (a_1b_2c_1) - (a_2b_1c_2) - (a_2b_1c_1).$$

Finally we multiply this by $1/4 = 1/2^{3-1}$, so that it is the difference of two averages.

In general, in an experiment with n factors, we consider an expression

$$\frac{1}{2^{n-1}}(a_2 \pm a_1)(b_2 \pm b_1)(c_2 \pm c_1)\ldots$$

with $(+)$ for absence and $(-)$ for presence of the corresponding letter in the interaction under definition.*

To sum up, it is sufficient for many purposes to remember that

(*a*) a main effect tells us the change produced by varying one factor, averaged over all levels of the remaining factors;

(*b*) a two-factor interaction, $A \times B$, tells us whether, averaged over all levels of the remaining factors, the effect of varying A is the same for all levels of B;

(*c*) a three-factor interaction, $A \times B \times C$, tells us whether, averaged over all levels of the remaining factors, the interaction between any two of A, B, C, say between A and B, is the same for all levels of the third factor C.

* Mathematicians may note that for certain more advanced work with interactions it is useful to bring in some ideas from group theory and Galois field theory (Mann, 1949, Chapter 8).

In most applications main effects are rather more important than two-factor interactions which in turn are a good deal more important than three-factor interactions. Four and more factor interactions do not often need consideration.

It is, of course, important to distinguish between the apparent interaction, as estimated from the data, and the true interaction that would have been obtained in the absence of uncontrolled variation. Each main effect or interaction is estimated by the difference between the mean of one-half of the observations and the mean of the other half. The formula for the standard error of the difference between two means, § 1.2(ii), therefore shows that the standard error of the estimated main effect or interaction is the residual standard deviation times $2/\sqrt{N}$, where N is the total number of units in the experiment. The quantity N is equal to the number of factor combinations, 2^n, multiplied by the number of times each treatment occurs. The estimation of the residual standard deviation is considered in § 6.11.

6.8 A SINGLE QUANTITATIVE FACTOR*

(i) Types of Response Curve

The definition and account of interactions just given is valid for all types of factor, but is of most practical value for qualitative factors. This is because the definitions are tied to the levels of the factors actually used in the experiment and do not bear directly on the continuous range of levels of potential interest when we work with a quantitative factor.

To develop methods more particularly appropriate for quantitative factors, consider first a single-factor experiment with the different treatments corresponding to different values of a carrier variable v.

As explained in § 6.3(ii) the true effect of the factor is described by the response curve plotted against v. Sometimes it is adequate to describe the curve qualitatively, for example, by plotting the mean observation corresponding to each factor level against the corresponding value of the carrier variable, giving an indication of the effect of uncontrolled variation on these treatment means. In such cases there is little difference from the analysis of a qualitative factor. Often, however, it is profitable to assume that this curve can be represented by a mathematical formula of a suitable type holding to a reasonable approximation over the range of values of v of interest; this is done partly to facilitate interpolation between the different factor levels and partly to obtain a concise and meaningful description of the experimental results.

* Some of the following section requires more mathematical knowledge than is assumed in the major part of the book.

Some of the more important types of response curve are shown in Fig. 6.2 and will now be discussed.

The simplest and most important case is the linear relation shown in Fig. 6.2(*a*) and defined mathematically by

$$\text{response curve} = a + bv, \tag{1}$$

where a and b have to be estimated from the data and are constant for any one application. This is characterized most importantly by the

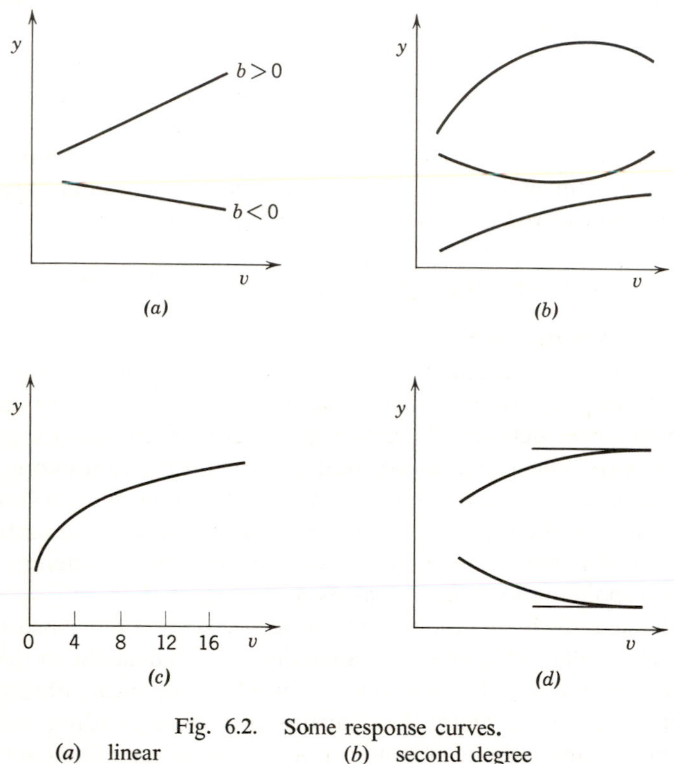

Fig. 6.2. Some response curves.
(*a*) linear (*b*) second degree
(*c*) linear against log v (*d*) rise (or fall) to a limit

slope b, which is the treatment effect for unit change in the carrier variable. If there are just two levels, b can be estimated from observations as

$$\begin{pmatrix}\text{difference between}\\ \text{mean observations}\end{pmatrix} \Big/ \begin{pmatrix}\text{difference between values}\\ \text{of } v \text{ at the two levels}\end{pmatrix}.$$

The next type of relation is shown in Fig. 6.2(*b*) and is given by the second-degree equation

$$\text{response curve} = a + bv + cv^2. \tag{2}$$

The new term, cv^2, represents a curvature; in Fig. 6.2(b) the top and bottom curves correspond to negative c, the middle curve to positive c. The maximum (or minimum) point of the curve is at a value of v equal to $-b/(2c)$. This point may, however, lie outside the range of values of v that are used, as in the bottom curve.

For any given response curve of this type, the numerical values of b and c depend on the units in which v is measured. If, however, we change the scale of v so that the three points $v = -1, 0, 1$ fall in the practical range of v we can give a simple picture of the meaning of b and c as follows:*

$$b = \tfrac{1}{2}[\text{value at } v = 1 - \text{value at } v = -1], \quad (3)$$

$$c = \tfrac{1}{2}[\text{value at } v = 1 + \text{value at } v = -1] - \text{value at } v = 0. \quad (4)$$

The reader not familiar with the properties of (2) should sketch a few cases.

We can extend the second-degree form (2) by including additional terms dv^3, ev^4, etc., and we can regard all the resulting expressions as particular cases of the general polynomial expression

$$\text{response curve} = a + bv + cv^2 + \ldots + kv^p, \quad (5)$$

where p is called the degree of the polynomial. Expressions (1) and (2) are the special cases $p = 1, 2$. The larger is the value of p, the greater is the range of curves that can be dealt with by suitable choice of the constants $a, b, \ldots, k$. On the other hand the meaning of relations with a large value of p is rather difficult to grasp physically and if it seems that a value of p of more than two is needed for a satisfactory fit, it is almost always worth considering whether some alternative type of expression would not be better; see the further discussion on p. 117.

For example we may consider polynomial expressions not in v but in some function of v such as $\log v$ or $\sqrt{v}$. Thus instead of (1) we may take

$$\text{response curve} = a + b \log v. \quad (6)$$

This has the effect of contracting the upper end of the scale of values of v and describes a curved response relation when plotted against v and not against $\log v$ [Fig. 6.2(c)]. The function of v to use may be suggested by prior considerations, by boundary requirements, or by inspection of the data. The use of a transformed carrier variable may or may not be combined with a transformation of the observation itself. In particular if on general grounds it is likely that there is a relation in which on the

* These formulas are given to give a geometrical interpretation to b and c; formulas for the analysis of data are considered in Table 6.6.

average the observation is proportional to an unknown power of the carrier variable, or equivalently

$$\text{log observation} = b \log v + \text{constant}, \tag{7}$$

it is natural to work with the log observation and log carrier variable, provided that this does not conflict with the basic assumption of Chapter 2. That is, it must be sensible to consider treatment differences on the transformed scale.

The common property of all the response relations so far is that they consist of the sum of a series of terms each of the form of an unknown constant times a known function of the carrier variable. We can write down a more general expression of this type, namely

$$b_1 f_1(v) + b_2 f_2(v) + \ldots, \tag{8}$$

where $f_1(v), f_2(v), \ldots$ are any *known* functions of v and $b_1, b_2, \ldots$ are unknown constants. Examples of this more general form are not common but are occasionally useful, for example in the analysis of periodic phenomena.

In practice, however, the simple first- and second-degree polynomials (1) and (2) are the most commonly used expressions of this type. The second broad class of relations consists of expressions that cannot be put in the forms (5) or (8). The two main types are the steady rise or fall to a limiting value illustrated in Fig. 6.2(*d*) and the rise from one limiting value to another. If the limiting values are known beforehand, modified logarithmic transformations of the observation are likely to make the relations approximately linear. Examples of mathematical curves to represent the two types of relation are

$$y = a + be^{-kv} \tag{9}$$

and

$$y = (1 + ae^{-kv})/(b + ce^{-kv}), \tag{10}$$

both with k a positive number and e the base of natural logarithms. The reader interested in these expressions should examine the geometrical meaning of the constants a, b, c, and k and should consider what transformations will make the relations linear. For example in the relation (9), a is the limiting value as v increases and the relation between $\log |y - a|$ and v is linear. If the limiting values are not known, but have to be estimated from the data, more complicated methods of analysis are needed. Also in a detailed analysis of this type of situation, allowance has often to be made for variation with v in the precision of the transformed observations.

A quantitative theoretical analysis of the system, if it can be made, will suggest a form for the response curve with parameters having physical significance, i.e., capable of being related to other phenomena. Such forms should, of course, be used wherever possible, unless interest is solely in the empirical response.

Limited sections of the curves in Figs. 6.2(*d*) and (*e*), and more generally of any reasonably smooth curve, can be represented by polynomials (5) of sufficiently high degree, but as a general rule it is better to take an expression of the right general form, if this is known, rather than to take a polynomial, even though polynomials are statistically simple to work with. The disadvantages of polynomials of high degree, particularly cubic and above, are

(*a*) the coefficients rarely have a simple interpretation;

(*b*) the coefficients are likely to change if the experiment is repeated with a somewhat different range of values of v;

(*c*) nonsensical results may be obtained if the polynomial is extrapolated, even only a short distance. Thus, suppose a second-degree equation is fitted to a section of the upper curve of Fig. 6.2(*d*) near the point where the limiting response is nearly attained. The fitted curve may agree well with the true curve, but will predict a fall in response if v is increased a small amount above the range considered.

The main use of the higher order polynomials is likely to be in constructing simple formulas for interpolation.

In many cases graphical inspection of the response curve will be enough and the reader should not think from the preceding discussion that the fitting of mathematical formulas is an essential element in the analysis of response curves.

(ii) Statistical Analysis of Response Curves

So far we have dealt with the mathematical forms that can be used to describe the response curve for a single quantitative factor. In practice our observations are affected by uncontrolled variation and we must make some allowance for this both in assessing whether a given set of observations are consistent with some particular form of response curve and also in measuring the precision of estimates of unknown quantities, such as slopes and curvatures. For this we shall consider only the simple polynomial curves such as (1) and (2).

The full statistical calculations are special cases of the method of multiple linear regression (Goulden, 1952, p. 134) and take a very simple form when the levels used in the experiment correspond to a series of equally spaced levels of the carrier variable, with equal numbers of observations at each level. In this case the appropriate formulas for estimating the slope and curvature of the response curve are set out in

Table 6.6. They are all based on expression (2), $a + bv + cv^2$, for the three response curves and also on the convention that v is measured in units such that the extreme levels of treatment are two units of v apart. For any particular number of levels, the levels are numbered in order so

TABLE 6.6

ESTIMATION OF SLOPE AND CURVATURE IN AN EXPERIMENT WITH ONE QUANTITATIVE FACTOR, WITH EQUALLY SPACED LEVELS AND EQUAL NUMBERS OF OBSERVATIONS PER LEVEL*

	Slope		Curvature	
Number of Levels	Estimate	Multiplying Factor for Standard Error	Estimate	Multiplying Factor for Standard Error
2	$\frac{1}{2}(\bar{y}_2 - \bar{y}_1)$	0.707	none possible	
3	$\frac{1}{2}(\bar{y}_3 - \bar{y}_1)$	0.707	$\frac{1}{2}(\bar{y}_3 - 2\bar{y}_2 + \bar{y}_1)$	1.225
4	$\frac{3}{20}(3\bar{y}_4 + \bar{y}_3 - \bar{y}_2 - 3\bar{y}_1)$	0.671	$\frac{9}{16}(\bar{y}_4 - \bar{y}_3 - \bar{y}_2 + \bar{y}_1)$	1.125
5	$\frac{1}{5}(2\bar{y}_5 + \bar{y}_4 - \bar{y}_2 - 2\bar{y}_1)$	0.632	$\frac{2}{7}(2\bar{y}_5 - \bar{y}_4 - 2\bar{y}_3 - \bar{y}_2 + 2\bar{y}_1)$	1.069

$$\begin{pmatrix}\text{standard error of}\\ \text{estimate of slope}\\ \text{or curvature}\end{pmatrix} = \begin{pmatrix}\text{multiplying}\\ \text{factor}\end{pmatrix} \times \frac{\text{residual standard deviation}}{\sqrt{(\text{no. of observations per level})}}$$

* The formulas in this table have been obtained from tables of orthogonal polynomials. A statistician will be able to give similar formulas for other cases.

that the lowest is number 1, the next lowest number 2, and so on. The mean of all observations on say the third level is denoted by $\bar{y}_3$. The use and interpretation of the formulas is illustrated by the following example.

Example 6.5. Suppose that we have the following observations on the yield, in arbitrary units, of a chemical process at four concentrations of a catalyst, the experiment being completely randomized with sixteen experimental units, four receiving each level of catalyst.

Concentration of Catalyst	1%	$1\frac{1}{2}$%	2%	$2\frac{1}{2}$%
	1.53	1.66	1.75	1.68
	1.63	1.58	1.57	1.74
	1.49	1.51	1.63	1.62
	1.56	1.56	1.68	1.66
Mean	1.5525	1.5775	1.6575	1.6750

There are four equally spaced levels, with an equal number of observations at each level, so that the formulas in the third row of Table 6.6 are applicable.

We first calculate the mean observation on each treatment; these are in order the mean values $\bar{y}_1$, $\bar{y}_2$, $\bar{y}_3$, and $\bar{y}_4$ used in the table. We get from the formulas of Table 6.6,

$$\text{estimate of slope} = \frac{3}{20}(3 \times 1.6750 + 1.6575 - 1.5775 - 3 \times 1.5525)$$

$$= 0.0671, \quad \text{and}$$

$$\text{estimate of curvature} = \frac{9}{16}(1.6750 - 1.6575 - 1.5775 + 1.5525) = -0.0042.$$

The next step is to estimate the residual standard deviation from the variation within the sets of units treated alike. We obtain, by a standard procedure (Goulden, 1952, p. 64) a value of 0.06265 (with 12 degrees of freedom). The formula at the foot of Table 6.6, together with the multiplying factors in the body of the table, now give that the estimated standard error of the slope $= 0.06265 \times 0.671/\sqrt{4} = 0.0210$ and that the estimated standard error of the curvature $= 0.06265 \times 1.125/\sqrt{4} = 0.0352$.

To sum up, the estimated slope is 0.0671 with a standard error of 0.021 and the estimated curvature is −0.0042 with a standard error of 0.035. The statistical interpretation of these results is as follows:

(*a*) The estimate of slope is just over three times its estimated standard error and a difference as great or greater would occur by chance only about once in one hundred times.* Hence it is very unlikely that the apparent increase in yield with increasing concentration of catalyst is spurious.

(*b*) We can work out limits within which the true slope will lie with any required probability. For example statistical tables show that there is only one chance in ten that the true and estimated slopes differ by more than 1.78 times the estimated standard error, i.e., that there is a chance of 0.9 that the true slope lies between $0.0671 - 1.78 \times 0.0210 = 0.0297$ and $0.0671 + 1.78 \times 0.0210 = 0.1045$.

(*c*) The slope measures the average increase in yield when the catalyst concentration is increased by one-half its total range of variation, which is $1\frac{1}{2}\%$. The estimated increase is therefore $0.0671/0.75 = 0.0895$ units of yield, per one per cent increase in catalyst concentration.

(*d*) The curvature is just over one-tenth of its estimated standard error. The estimated value deviates from the true value by more than this very frequently, and so the data are consistent with a true curvature of zero, i.e., with a linear relation between yield and concentration of catalyst. This statement must on no account be mistaken for an assertion that the true relation is linear. We are simply saying that *from the data under analysis* a statement that the relation is curved upwards, the direction of apparent curvature, would not be justified.

(*e*) It would often be worth having some idea of the maximum amount of true curvature consistent with the data. For example, just as in (*b*), the chance is 0.9 that the true curvature lies between $-0.0042 - 1.78 \times 0.0352 = -0.0669$ and $-0.0042 + 1.78 \times 0.0352 = 0.0585$. Limits at other levels of probability can be calculated by a similar process, using different multipliers to replace 1.78. The geometrical interpretations of a curvature of −0.07 would be that the average of the true yields at catalyst concentrations of 1% and $2\frac{1}{2}\%$ is 0.07 units

* This figure is found from tables of the statistical distribution known as the t distribution with 12 degrees of freedom.

lower than the true yield at the midpoint, $1\frac{3}{4}\%$ concentration. The yield-concentration curve would thus be concave to the concentration axis. The reader should sketch a curve with the average slope of 0.0671 and with this limiting curvature −0.07.

Formulas corresponding to Table 6.6 are available for more levels, for unequally spaced levels, for differing numbers of observations per level, and for the fitting of higher degree trends. A statistician should be consulted for these results.

The above discussion has been for one factor. Now suppose that we have several factors. If there is only one quantitative factor and this is a treatment factor, not a classification factor, the best procedure is usually to consider separately (*a*) the slope and its interaction with the qualitative factors and (*b*) the curvature and its interaction with the qualitative factors. That is we consider separately for slope and curvature whether there is evidence that they vary between the different levels of the qualitative factors. Much the most interesting case, however, arises when there are several quantitative factors and this situation we consider in the next section.

6.9 SEVERAL QUANTITATIVE FACTORS

Suppose first that there are two factors, both quantitative, with carrier variables denoted by v_1 and v_2. Then instead of a response curve of treatment effect plotted against the single carrier variable v, we have a response *surface* in which the treatment effect is plotted against the values of v_1 and v_2 in a three-dimensional diagram. To do this, two perpendicular axes are taken to represent the carrier variables v_1, v_2. If these axes are imagined to be in the plane of the paper, any point on the paper represents a combination of factor levels and hence a treatment. Now take a third axis perpendicular to the plane of the paper and along it measure the treatment response, i.e., the difference between the observation obtained on an experimental unit with a particular treatment and the observation that would have been obtained with some standard reference treatment. For each point in the plane of the paper, move parallel to the new axis a distance equal to the appropriate treatment response. If this is done for a large number of different treatment points, we build up the response surface showing how the treatment response depends on the factor levels.

Such a surface can be represented diagrammatically by a sketch of a surface in three dimensions. Usually, however, it is more convenient to use a contour map of the surface, constructed according to the ordinary principles of map-making. Figure 6.3 (*a*) shows the contours for a response surface that has a maximum at the treatment corresponding to levels

v_1' and v_2'. The response falls off steadily from the maximum, more steeply in the v_2 direction than in the v_1 direction. The actual values marked on the contours depend on the arbitrary reference treatment chosen in defining the response; in truly comparative experiments it is

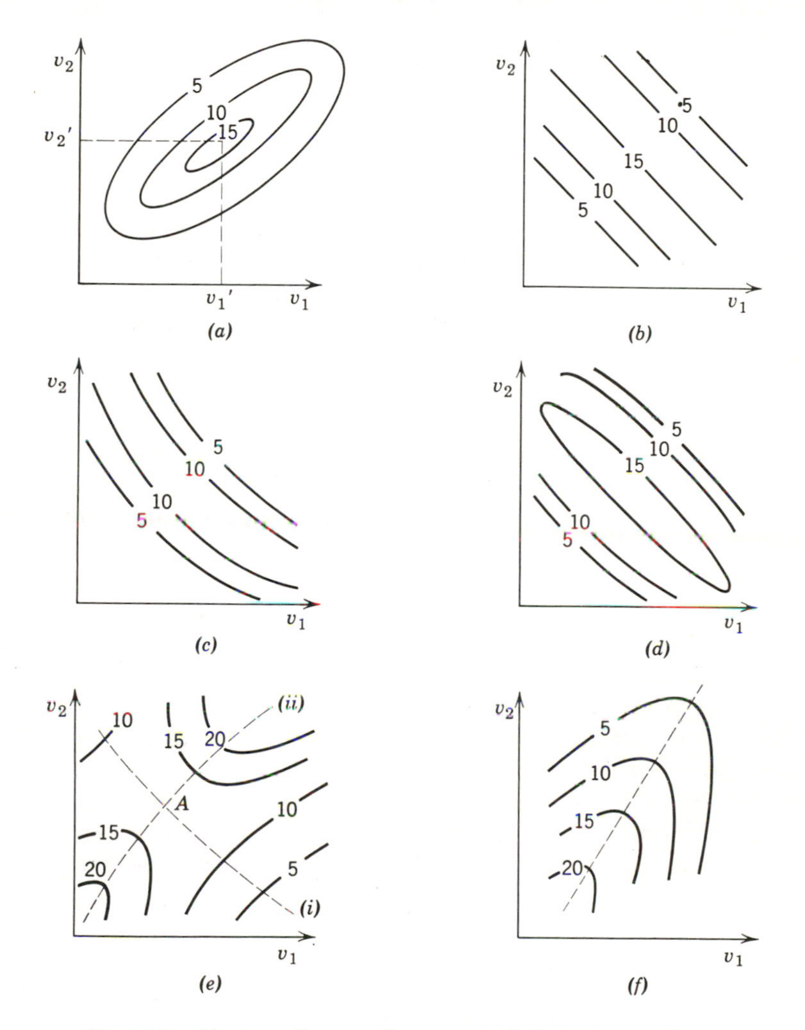

Fig. 6.3. Contour diagrams for some typical response surfaces.

only the differences between the heights of the surface at different points that are of interest.

Just as with a response curve for one factor, it may sometimes be sufficient to describe a response surface qualitatively, by giving the treatment means at the various factor combinations together with a measure of random error. For a complete factorial experiment, in which all combinations of a set of levels of each factor are investigated, the treatments form a rectangle of points in the plane of the axes of v_1 and v_2, and provided that this rectangle of points covers the region of treatments of interest, it may be possible to get a sufficiently good idea of the shape of the response surface without further analysis. There will be no interaction if the height of the response surface is the sum of a function of v_1 and a function of v_2. Absence of interaction does not necessarily mean that the surface is of a simple shape.

Some examples of interesting, if somewhat idealized, response surfaces are shown in Fig. 6.3. In (*a*), which we have already commented on, there is a maximum point on the surface with a decrease in response on all sides of the maximum. In (*b*) the maximum response is obtained at any point along a line or ridge of maxima. In (*c*) the ridge of maxima is a curve and not a straight line. Example (*d*) is like a limiting form of the general type (*a*) in which the maximum point is situated in a long narrow region of almost constant height; types (*d*) and (*b*) are likely to be indistinguishable in applications. In example (*e*) there is a col or saddle point at A; the point A is a maximum point for the path (i) and a minimum point for the path (ii). Finally, (*f*) is an example of a rising ridge. In the region shown the surface rises steadily, the position of the ridge being marked by the dashed line. Surfaces which fall to a minimum instead of rising to a maximum can be represented by numbering the contour lines in Fig. 6.3(*a*)–(*f*) in reverse order.

An interesting property of surfaces (*b*) and (*c*) is that we can take a single simple combination of the carrier variables v_1 and v_2 such that the height of the response surface is, within the range considered, determined by the new function. Thus in (*b*), where the contour lines are at 45° to the coordinate axes, the new function is $v_1 + v_2$, since every point with the same value of $v_1 + v_2$ lies on the same contour line and so has the same height. A similar fairly simple function could be found for (*c*).

If we know the appropriate composite carrier variable to take, e.g., $v_1 + v_2$, we could describe the variation of response by a curve plotted against this new carrier variable and the separate values of v_1 and v_2 would be irrelevant. This would not only simplify the empirical description of the surface but would also suggest that in some sense the new composite carrier variable may be physically more important for the system than the separate carrier variables. For example, the factors pressure and temperature may affect the yield of a chemical reaction only through that combination of pressure and temperature that determines the frequency of molecular collisions. In fact, the physical sciences contain many examples of systems in which some observation on the system depends only on a simple combination of factor values. Any theoretical information about the form such a combination of carrier variables might take, should be used in analyzing, and also in designing, the experiment. A remarkable example has been given by Box and Youle (1955).

It must be stressed, though, that all surfaces can be expressed in terms of a single carrier variable, provided that we are willing to take a sufficiently complicated combination of the separate v's. There is, thus, no fundamental distinction between (*b*) and (*c*) and the other surfaces; it is just that a more complicated combination of v_1 and v_2 would be required to describe say (*e*).

The general conclusion is thus that a study of the response surface, either qualitatively or by fitting equations, may suggest the form of the underlying factor combinations that are of most significance for the system, and hence may greatly add to one's understanding. One snag is, of course, that if there is appreciable uncontrolled variation, a very extensive experiment may be necessary to map out a complicated surface with sufficient precision for cogent conclusions to be drawn.

We shall not discuss the formulas for fitting mathematical equations to response surfaces although they are in fact very similar to the formulas for one factor explained in Example 6.5. It is, however, important to mention the two mathematical expressions most commonly used to represent response surfaces. These are

$$\text{height of response surface} = a + b_1v_1 + b_2v_2, \tag{11}$$

and

$$\text{height of response surface} = a + b_1v_1 + b_2v_2 + c_{11}v_1^2 + c_{12}v_1v_2 + c_{22}v_2^2. \tag{12}$$

The first equation (11) corresponds to (1) in the single factor discussion. There is a linear main effect for each carrier variable separately and there is no interaction. The response surface is a plane. This simple form is most useful in preliminary work where it is required to estimate the general direction in which the surface rises than to make a detailed investigation of shape. The unknown constants in (11) can be estimated from a two-level factorial experiment.

The surface (12) is analogous to the response curve (2) in which a square term is included to represent curvature. The practical interpretation of the quantities b_1 and c_{11} can be seen by supposing as before that v_1 and v_2 are measured in such a way that their extreme values are plus and minus one. The response curve against v_1 when v_2 takes the central value zero is $a + b_1v_1 + c_{11}v_1^2$, so that b_1 and c_1 are the slope and curvature of this curve, as defined previously. The only quantity in (12) for which we do not have a geometric interpretation is thus c_{12}. The reader may verify that when v_2 is 1 the slope of the response curve against v_1 is $b_1 + c_{12}$, and that when v_2 is -1 the slope is $b_1 - c_{12}$. Thus, c_{12} measures rate of change of the slope against v_1 as v_2 varies. In fact its interpretation is symmetrical in that it measures equally how the slope against v_2 varies with v_1. We call c_{12} the linear × linear interaction.

The curvatures and the linear × linear interaction can be estimated from experiments at three or more levels but not from two-level experiments.

The following example illustrates the occurrence of a linear × linear interaction.

Example 6.6. In a wool textile carding machine there is a device known as a burr remover. This has three pairs of rollers adjustable as to speed and spacing. To investigate the functioning of the burr remover a factorial experiment with six quantitative factors, three speeds, and three spacings would be called for.

The following observations refer to a fictitious simplified version of such an experiment having only two factors, the speed and setting at one point. Speed has been examined at three equally spaced levels 300, 400, and 500 rpm and spacing at four equally spaced levels of 1, 1.2, 1.4, and 1.6 units. There are thus twelve treatments forming a complete factorial experiment. Twelve short runs could be made on each day so that the experiment was arranged in randomized blocks, each day's results forming one block and the order of treatments being randomized within blocks, each treatment occurring once in each block. It was considered that four blocks would give sufficient precision.

The observation analyzed here is a measure of the efficiency of carding, taking into account fiber breakage and production of burr-free output. There are 48 observations in all and if they are analyzed by the method of Chapter 3 for a randomized block experiment with 12 treatments in 4 blocks, an estimate of the residual standard deviation with $3 \times 11 = 33$ degrees of freedom can be obtained. Its value in this case was 1.15.

The twelve treatment means are given in a two-way table in Table 6.8. Each entry in the table is a mean of four observations and, therefore, has a standard

TABLE 6.8

AN EXPERIMENT WITH TWO QUANTITATIVE FACTORS

		Speed			
		300 rpm	400 rpm	500 rpm	Mean
Spacing	1 unit	21.6	22.3	22.9	22.27
	1.2 units	18.7	19.1	21.6	19.80
	1.4 units	15.8	17.9	19.4	17.70
	1.6 units	13.2	16.7	19.5	16.47
	Mean	17.32	19.00	20.85	

error of $1.15/\sqrt{4} = 0.58$. The standard error of the difference between two means in the body of the table is $1.15\sqrt{2}/\sqrt{4} = 0.81$.

The general conclusion is that there is an approximately linear decrease in efficiency as spacing increases, but that the rate of decrease is much lower at the higher speeds. This steady change in the rate of change with one factor as the other factor is increased is a typical linear × linear interaction.

Many of the comments made in the discussion of a single factor apply when there is more than one factor. For example, we may use transformed carrier variables such as $\log v_1$ and $\log v_2$. A more detailed account of the fitting and analysis of response surfaces is given in a paper by Box (1954) and in the book edited by Davies (1954).

6.10 FURTHER DISCUSSION OF MAIN EFFECTS AND INTERACTIONS

In the previous sections we have discussed in some detail the description of the relation between observations and treatments in a factorial experiment. We now briefly summarize the main points of the discussion. To understand this section, the reader should be clear about the difference between treatment and classification factors (§ 6.2) and about the distinction between specific qualitative, quantitative, and sampled qualitative factors (§ 6.3).

(i) Two Specific Qualitative Treatment Factors

We usually begin by examining the two main effects, the differences between the mean observations at different levels of factor A averaged over all levels of factor B, and the similar thing for B. Then we look at the interaction; if this is too large to be accounted for by random errors, we either describe it qualitatively in what seems the most helpful way or occasionally try to remove it by a simple transformation of the observations. If one factor A is less important than B, it may be best to describe interaction by discussing the effect of B separately for each level of A. If appreciable interaction is present the main effects are not necessarily useful.

(ii) Two Specific Qualitative Factors, One a Treatment Factor and One a Classification Factor

We look first at the main effect of the treatment factor, and then at the interaction. Any interaction is dealt with either by transformation or, more commonly, by describing how the effect of the treatment factor varies from one level to another of the classification factor. Useful descriptive information about the nature of the experimental material may be obtained from the main effect of the classification factor.

(iii) One Specific Qualitative Treatment Factor and a Sampled Qualitative Classification Factor

We are interested primarily in the main effect of the treatment factor averaged over a population of levels of the other factor, of which the levels actually used are regarded as a random sample. The presence of interaction not removable by a simple transformation is therefore taken to mean variation in the treatment effect from level to level of the sampled factor. This variation should be explained physically if at all possible, for example, by correlating it with some supplementary observation taken at each level of the sampled factor. Failing this, the variation in the treatment effect must be taken as random, and then the treatment effect is estimated by the main effect of the treatment factor. The determination of the error of these estimates involves a statistical analysis of the interaction. It may be possible also to obtain useful information about the experimental material by individual comparisons within the main effect of the sampled factor.

(iv) Two Quantitative Treatment Factors

We consider a response surface in which the treatment effect is plotted against the corresponding values of the carrier variables. The form of the surface can be assessed either qualitatively, or by fitting a suitable mathematical equation. The two most common types of equations are (*a*) of the first degree and (*b*) of the second degree, in the carrier variables or some function of them, such as their logarithms, reciprocals, etc. The linear surface corresponds to the absence of interaction and to equally spaced parallel and linear contours. It is useful mainly in preliminary work to assess the general direction of the response surface in a particular region. The second-degree equation gives an approximate representation of a wide range of surfaces and allows for the possibility both of interaction and of curvature in the response curves for the separate factors. The linear surface can be estimated from two-level factorial experiments, but the estimation of the second-degree surface calls for observations on at least three levels of each factor. Investigation of the form of the response surface may shed some light on the nature of the phenomena under examination.

(v) Two Treatment Factors, One Quantitative and One Specific Qualitative

If the response to the quantitative factor is of the type that can be represented by a slope and a curvature, i.e., by a second-degree equation, it is often best to examine the slope and curvature separately. That is, we examine whether there are significant variations in slope between the different levels of the qualitative factor, and so on, and also carry through a parallel analysis on curvature. If the separate response curves are not of a second-degree type, a similar method would be adopted analyzing separately the different geometrical aspects of the curve.

Occasionally it may happen that the specific qualitative factor is much more important than the quantitative factor. Then the procedure is to examine first the main effect of the qualitative factor and then the effect of the qualitative factor separately for each level of the other factor. That is, we treat the quantitative factor just as if it was qualitative.

(vi) Quantitative Treatment Factor and a Specific Qualitative Classification Factor

This is treated in the same way as the first part of (v), namely by considering the geometrical aspects of the response curve separately for each level of the qualitative factor and averaged over all levels of the qualitative factor.

(vii) Quantitative Treatment Factor and a Sampled Qualitative Classification Factor

Much the same method usually works. We are interested in, for example, the slope of the response curve averaged over the population of levels of the sampled factor. The interaction of the slope with the qualitative factor determines the existence and amount of the variation in slope from level to level and hence affects the estimate of the precision of the final estimate of mean slope.

(viii) General Discussion

Although the detailed methods of interpretation vary in these seven cases, the general idea remains the same. Main effects give the change in the mean observation from level to level of one factor averaged over the levels of the other factor. Interaction is concerned with whether the effect of one factor is the same for all levels of the other factor. In many cases we need to make a statistical analysis to see whether an apparent main effect or interaction could reasonably be accounted for by random error, but the details of such statistical analysis, which have not been gone into here, should not be allowed to obscure the more general questions of interpretation that are involved.

If there are more than two factors in the experiment, there are many more possible combinations of types of factor, but the same general principles hold. With, say, three factors A, B, and C we have main effects showing the differences between the levels of A averaged over the levels of B and C and two-factor interactions: B with C, C with A, A

with B, averaged respectively over the levels of A, B, and C. Also there is a three-factor interaction which is concerned with whether the pattern of any of the two-factor interactions, say B with C, is the same for all levels of A.

Similarly in experiments with more than three factors, we have four-factor interactions, five-factor interactions, and so on. But these are infrequently of practical importance and it is enough for nearly all purposes to understand and to be able to work with main effects, two-factor interactions, and, less frequently, three-factor interactions.

6.11 THE ESTIMATION OF ERROR IN FACTORIAL EXPERIMENTS

We saw in Chapter 3 that the estimation of the precision of the comparisons in a randomized block or Latin square experiment depends on estimating a quantity called the residual standard deviation. This is done in practice by the statistical procedure known as analysis of variance, which in effect calculates the variation between observations on units receiving the same treatment, making due allowance for any systematic variation between block, rows, or columns. The estimate of standard deviation can equivalently be obtained by calculating residuals in the way explained in Chapter 3.

Often these methods apply directly to factorial experiments. For example we might have say a $2 \times 2 \times 2$ experiment arranged in six randomized blocks of 8 units each. The residual standard deviation can then be estimated just as for any other experiment with 8 treatments and 6 blocks; the degrees of freedom for residual would be 35.

Sometimes, however, it is necessary to use special methods for factorial experiments, and these methods will now be described briefly.

First it may happen, particularly with experiments with many factors, that one observation on each factor combination gives sufficient precision. Thus in a $4 \times 4 \times 3 \times 3$ experiment there are 144 treatment combinations, and it could very well happen that 144 experimental units are all that are available for the experiment, or that sufficient precision for main effects, etc., can be obtained from this number of units. But no estimate of error can be formed by the previous method, since with only one observation on each treatment combination, there are no sets of units receiving the same treatment on which to base the estimate.

To surmount this difficulty we must find some comparisons among the observations that are effectively equivalent to comparisons of units receiving the same treatment. Such comparisons are those forming the four-factor interaction. We have seen above that this interaction is likely to be of negligible importance, except possibly in systems behaving in a very complicated way. A standard deviation can be calculated based on the comparisons forming this four-factor interaction. It is the square root of the mean square for this interaction in the analysis of variance and would in the ordinary way be used for testing the statistical significance of the four-factor interaction. In using the four-factor interaction as a substitute for a residual standard deviation, there are two dangers. The four-factor interaction may actually contain a real effect of importance which will get overlooked. Also the presence of such an effect will increase the corresponding standard deviation, leading to an overestimate of the error of the main effects and low-order interactions. Neither of these points is often likely to be of much importance.

In the same way, in experiments with more than four factors, the five- and higher-factor interactions could be used to estimate the error. We would ordinarily take for this purpose a group of interactions of as high an order as possible, pooling sufficient terms to obtain if practicable 20 or 30 degrees of freedom for residual. In an experiment with three factors, it would be possible to use the three-factor interaction to estimate the residual standard deviation. However, three-factor interactions are sometimes of physical importance, and so one would preferably use them to estimate error only when experience with the particular application suggests that there are unlikely to be appreciable interactions of this order present.

An interesting graphical method of estimating the residual standard deviation in factorial experiments has been proposed by Daniel (1956).

Very similar methods apply when response surfaces are fitted to the results of factorial experiments with quantitative factors. The residual standard deviation is best estimated, as in a nonfactorial experiment, by a comparison of observations on different units receiving the same treatment. If, however, only one observation is made on each treatment we estimate the residual standard deviation from the dispersion of the observed points about the fitted surface. The details, which will be standard to a statistician, will not be gone into here.

A final problem of error estimation connected with factorial experiments concerns experiments with at least one sampled qualitative factor. The point at issue is best explained by an example.

Example 6.7. Consider, to take a very simple case, an experiment with two factors, the first specific qualitative, its two levels being a standard routine analytical procedure and a new quick alternative technique. Suppose that the second factor, the sampled qualitative one, represents the different technicians who use the methods, six being chosen to take part in the experiment. We may assume that a large number of equivalent specimens are available for analysis and that each technician analyzes a number of specimens, half by one method and half by the other, the allocation of specimens to technicians and to methods being completely random.

The summarized results of such an experiment might be as in Table 6.9.

TABLE 6.9

MEAN OBSERVATION ON STANDARD METHOD
– MEAN OBSERVATION ON NEW METHOD

Technician		
Technician 1	2.1	
2	1.3	
3	−0.4	The standard error of each of these differences is known to be 0.5.
4	3.2	
5	−1.1	
6	1.1	
Mean	1.03	

The standard error, 0.5, is derived from the residual standard deviation, by multiplying by $\sqrt{\{2/(\text{number of observations per technician-method combination})\}}$. The residual standard deviation is obtained in the usual way from the dispersion of the observations within the different technician-method combinations; thus the figure 0.5 measures the precision of the differences 2.1, 1.3, . . . , considered as estimates of the difference between the two methods for the particular technician concerned. The conclusion from the above figures is that there is a real difference between methods for some technicians and that the difference is not the same for all technicians. In some applications this would rule out the use of the new method, at any rate until further work had isolated and removed the cause of the systematic errors. Sometimes, however, we would want to know whether the average difference over technicians is zero, i.e., whether there is evidence that averaged over technicians the systematic error is in one direction rather than another.

When we look at the differences separately for each technician we are effectively looking on technicians as a specific qualitative factor, but when we examine the average over technicians, new problems connected with sampled qualitative factors arise.

First we may be interested in the average over the particular technicians taking part in the experiment. That is, for each technician in the experiment there is a true difference between methods. We want to make an inference about the average of these six true differences. This is just like examining any other main effect and we average the individual mean differences to give 1.03, with a standard error of $0.5/\sqrt{6} = 0.20$, the $\sqrt{6}$ occurring since we have averaged six differences. Since the estimated difference is five times its standard error there is very strong evidence that the true mean difference is positive.

Next, suppose that we want to know what the difference would have been, averaged not just for the six technicians taking part in the experiment, but for the whole population of technicians who may at some time use the methods. We usually assume that the population contains many more technicians (levels) than occur in the experiment and that the technicians in the experiment can be regarded as a random sample from the population. This last assumption has to be viewed critically unless some objective sampling procedure has been used to select technicians for the experiment. We have to ask ourselves from what population the given technicians can reasonably be regarded as sampled; this will not usually be the same as the population to which we want to apply the conclusions.

We now argue that the six mean differences given above are a random sample from a population of differences, so that the variation from technician to technician is to be regarded as additional random variation. We estimate the mean difference in the same way as before by the mean value 1.03, but now the precision of this value must be assessed from the variation among the six values 2.1, 1.3, . . . , 1.1. The estimated standard deviation of these values is 1.58 so that the estimated standard error of the mean is $1.58/\sqrt{6} = 0.65$. Thus the estimate is $1.03/0.65 = 1.58$ times its estimated standard error and reference to the tables of the t distribution with five degrees of freedom shows that as great or greater difference would occur just by chance in more than 10% of trials. Hence we cannot claim that this experiment gives definite evidence of an average bias in a positive direction.

This analysis expresses quantitatively the physically obvious point that if we wish to extend our conclusions to apply to a larger set of technicians some increase in error will in general be involved.

The variation that is used to estimate error in the second case is the extent to which the estimated methods-effect varies from technician to technician, i.e., what we have called previously the interaction methods × technicians. This provides the form for the general rule for such cases. Suppose that we have a factorial experiment with any number of factors, including one sampled qualitative factor S. Then if we want to make an inference about the main effect or interaction of other factors, averaged over an infinite population of levels of S, the estimate of error must be based on the interaction of the contrast of interest with S.

Occasionally we require to extend the conclusions not to an infinite population of levels but to a finite population. For example the six technicians might have been a sample from twelve at this particular laboratory who might have occasion to use the methods. The error in this case is intermediate between that in the two cases discussed here; formulas are given, for example, by Bennett and Franklin (1954, Chapter 7).

In the example just described it would normally be advisable not to do the whole experiment on nominally identical specimens, but to include as a third factor a number of representative types of specimen. Also one would, in practice, be interested not only in the mean values for the different methods but in the dispersions.

SUMMARY

A factorial experiment is one in which the treatments consist of all combinations of different *levels* of a set of *factors*. The advantages of a factorial experiment, as compared with a series of separate experiments on the separate factors, are economy and the possibility of investigating interactions between the different factors, i.e., of examining the extent to which the effect of one factor is different for different levels of another factor. Complicated factorial experiments with many factors, although sometimes very valuable, should however be viewed with caution.

Factors that may arise for consideration can be classified in various ways. First there is a distinction between factors representing treatments applied to the experimental units, and factors which represent simply a classification, outside the experimenter's control, of the experimental units into different types. Secondly there is a distinction between

(*a*) factors, such as temperature, pressure, and amount of reactant for which different levels correspond to well-defined values of a quantitative *carrier variable*. We call these quantitative factors;

(*b*) factors such as varieties of wheat and different experimental methods for which the different levels represent qualitatively different treatments of intrinsic interest. We call these specific qualitative factors;

(*c*) factors, such as consignments of raw material in an industrial process for which the different levels are not of intrinsic interest but are to be regarded as a sample form a population of levels. We call these sampled qualitative factors.

The interpretation of the results of factorial experiments raises various problems. With quantitative factors we usually think in terms of a response curve or response surface relating the true treatment effect to the carrier variables defining the factor levels; we can investigate the form of the surface qualitatively by inspecting treatment means or by fitting a suitable mathematical equation. The existence of a ridge system in the surface suggests that the factors affect the observation only through some particular combination of the carrier variables.

With specific qualitative factors it is often helpful to work with main effects and interactions. The main effect of a factor A gives us the differences in mean observation between the different levels of A averaged over all levels of the other factors. The two-factor interaction between say A and B examines whether, averaged over all levels of the remaining factors, the difference between levels of A is the same level of B, or vice versa. Similarly a three-factor interaction between A, B, and C examines whether, averaged over all levels of the remaining factors, the two-factor interaction between A and B has the same pattern for all levels of C, or, equivalently, whether the two-factor interaction between B and C has the same pattern for all levels of A, etc. Four and more factor interactions can be defined in a similar way but are very rarely of direct practical importance.

The absence of interaction between two quantitative factors means that the response surface is of such a form and oriented in such a way that the effect of changing one factor from one level to another is the same for all levels of the second factor.

We need in practice to have an estimate of the precision of apparent effects estimated from the observations. This usually requires the estimation of the residual standard deviation and this can be done either by the same methods as for nonfactorial experiments or, if only one replicate of the experiment is run, by the use of high-order interactions. If a sampled qualitative factor is involved, the interaction of another contrast with it determines the error, when the other contrast is to be estimated for a whole infinite population of levels of the sampled factor. In the majority of cases, however, intelligent tabulation and plotting of the results is the most important step in the analysis.

REFERENCES

Bennett, C. A., and N. L. Franklin. (1954). *Statistical analysis in chemistry and the chemical industry*. New York: Wiley.

Box, G. E. P. (1954). The exploration and exploitation of response surfaces—some general considerations and examples. *Biometrics*, **10**, 16.

——— and P. V. Youle. (1955). The exploration and exploitation of response surfaces: an example of the link between the fitted surface and the basic mechanism of the system. *Biometrics*, **11**, 287.

Campbell, R. C., and J. Edwards. (1954). Semen diluents in the artificial insemination of cattle. *Nature*, **173**, 637.

Cochran, W. G., and G. M. Cox. (1957). *Experimental designs*. 2nd ed. New York: Wiley.

Daniel, C. (1956). Fractional replication in industrial research. *Proc. 3rd Berkeley Symp. on Math. Statist. and Prob.*, **5**, 87. Berkeley: University of California Press.

Davies, O. L. (editor) (1954). *The design and analysis of industrial experiments*. Edinburgh: Oliver and Boyd.

Goulden, C. H. (1952). *Methods of statistical analysis*. 2nd ed. New York: Wiley.

Kempthorne, O. (1952). *The design and analysis of experiments*. New York: Wiley.

Mann, H. B. (1949). *The analysis and design of experiments*. New York: Dover.

Moore, P. G., and J. W. Tukey. (1954). Answer to query 112. *Biometrics*, **10**, 562.

Tukey, J. W. (1949). A degree of freedom for non-additivity. *Biometrics*, **5**, 232.

——— (1955). Answer to query 113. *Biometrics*, **11**, 111.

Van Rest, E. D. (1937). Examples of statistical methods in forest products research. *J. R. Statist. Soc. Suppl.*, **4**, 184.

CHAPTER 7

Design of Simple Factorial Experiments

7.1 INTRODUCTION

In the previous chapter we discussed the general nature and advantages of factorial experiments and some of the ideas involved in interpreting the results from such experiments. We now consider the design of these experiments. Four questions are usually involved:

(*a*) What factors shall be included?
(*b*) At what levels shall these factors be taken?
(*c*) How many experimental units shall be used?
(*d*) What measures shall be adopted to reduce the effect of uncontrolled variation?

The third of these questions will be discussed at length in Chapter 8 and so will be ignored for the time being. The other three questions will be discussed in turn in the following sections.

7.2 THE CHOICE OF FACTORS

In deciding what factors can be included in an experiment we have of course to be guided in part by considerations of economy and simplicity. However, in the initial consideration of a particular experiment it is usually best to make a fairly comprehensive list of the factors likely to be relevant, even if the study of some of them has to be postponed to later experiments. In some investigations, particularly in preliminary work, the number of factors of potential importance may be much larger than the number that can be dealt with. In such cases a less ambitious definition of the issues at stake must be adopted before drawing up the list of possible factors for inclusion.

The factors will be of two types, treatment factors and classification factors, and we discuss these separately. The treatment factors, i.e., the modifications to the system to be made during the experiment, may, as in a bioassay, be clearly defined by the nature of the problem, but more commonly there are some factors of prime importance and others whose inclusion, while perhaps desirable, is by no means necessary. We can distinguish between

(*a*) factors of direct interest;

(*b*) factors which may modify the action of the main factors or may throw light on how the main factors work;

(*c*) factors connected with the experimental technique.

Examples to make this distinction clear will be given in a little while. Classification factors fall into two rough types, namely

(*d*) groupings representing inevitable, but physically important, variations in the experimental material (e.g., age and sex differences of patients);

(*e*) deliberately inserted variations of the experimental units, designed to examine interactions and to extend the range of validity of the conclusions concerning the main factors.

In other words, in deciding on what factors to list for possible inclusion, we should ask the following questions:

(*a*) What treatments should it be the direct object of the experiment to investigate?

(*b*) What additional factors can be included whose interaction with the main factors in (*a*) may be enlightening?

(*c*) Does anything connected with the experimental technique suggest additional factors?

(*d*) Can the experimental units available for the work be divided naturally into groups in such a way that the main treatment effects are quite possibly different for the different groups?

(*e*) Is it desirable to choose deliberately experimental units of different types, so that (*d*) applies?

These principles will now be illustrated briefly by some examples.

Example 7.1. Consider the type of experiment in parasitology in which mice are stimulated with an inoculation of larvae and, after a suitable delay, presented with a challenging dose of say 200 larvae. After a further period the mice are killed and an autopsy carried out to determine the number of larvae present, i.e., to measure the degree of immunity to the challenging dose.

The answers to the questions (*a*)–(*e*) might be as follows:

(*a*) Two factors are of interest; the effect of the amount of vitamin A in the diet and the effect of including and omitting the initial inoculation. That is, we are interested in say whether absence of vitamin A lowers immunity to the challenging dose, in whether the preliminary injection increases immunity, and in any interaction between these factors.

(*b*) It is desired to keep the experiment simple and no important additional factors suggest themselves.

(*c*) Three elements of the experimental technique are the length of time between inoculation and the presentation of the challenging dose, the amount of the challenging dose, and the time interval between this and the autopsy. If it is thought that these may affect the comparisons in (*a*) to an important extent, one or more new factors may be inserted, different levels representing, for example, differing amounts of time between inoculation and the presentation of the challenging dose.

(*d*) The mice can be divided into males and females and possibly into different strains. It is also possible to make a ranked qualitative factor out of some quantitative property such as body weight by, for example, classifying the mice as light, average, or heavy.

(*e*) If all the mice initially available for the experiment happen say to be of one age, it may be worth considering having mice of a number of age groups.

This gives us a list of factors for possible inclusion in the experiment.

Example 7.2. This example is based on an experiment connected with the production of penicillin (Davies and Hay, 1950). The experiment was concerned with the first two stages of the process, namely the production of inoculum and fermentation. The answers to questions (*a*)–(*e*) might then be as follows:

(*a*) Three factors of interest in the first stage of the process are the concentration of corn steep liquor, the amount of glucose, and the quality of glucose. In the second stage the factors of principal interest are the concentration of corn steep liquor and the choice of a fermenter.

(*b*) It is possible that various operating details, temperatures, etc., connected with the processes could usefully be included as additional factors, but in the situation discussed by Davies and Hay the number of factors under (*a*) was already somewhat large for the resources it was desired to devote to the experiment.

(*c*) Suppose that no obvious factors arise here.

(*d*) and (*e*) The raw material is the corn steep liquor and if several consignments have to be used, "consignments" forms a natural factor. There is also the possibility of deliberately inserting differences between consignments.

Example 7.3. Consider an experiment to investigate the properties of varieties of sugar-beet bred for resistance to bolting. The answers to our standard questions might be as follows:

(*a*) The main factor is varieties, the different levels being both the experimentally bred strains and commercial varieties serving as controls.

(*b*) The comparison of varieties is likely to depend on the date of sowing and possibly on soil, weather, and other differences. The latter can be investigated if it is practicable to do the experiment in several centers, possibly for several years. That is we have at least two supplementary factors, namely time of sowing and centers, and possibly more.

(*c*), (*d*), and (*e*) may not raise any interesting points here.

Example 7.4. Edwards (1941) has described an investigation of the effect of initial attitude on the power to remember. Subjects were divided into three groups, favorable, neutral, and unfavorable, on the basis of their attitude to a certain political issue. They then listened to a speech containing equal numbers of statements for and against the issue and were tested immediately, and after some weeks' delay, for their recollection of the pro- and anti-statements.

This example has been included to show that the considerations of this section can be applied to investigations common, for example, in the social sciences, which are not experiments in the sense of this book. For the object here is to compare levels of a classification factor, namely different groups of subjects, not to assess the effect of a treatment imposed by the experimenter. As was explained in § 6.2 such comparisons are subject to an essential additional uncertainty. For, even if the three groups of subjects in this experiment had been drawn randomly from well-defined populations, which is hardly practicable in the present case, there is no guarantee that differences in political attitude are the only or even the principal difference between the groups.

The answers to (*a*)–(*e*) might be as follows:

(*a*) As just discussed, the main factor is a classification factor, the initial attitude of the subject.

(*b*) An important supplementary factor is the length of time between hearing the speech and receiving the memory test. The object of including this factor is not primarily to examine the average decrease of recollection with time (the main effect of time), but to see whether the differences between subject groups change with time, i.e., to see whether a possible tendency to remember better statements favorable to one's own point of view changes with time. That is why we consider the time interval as a factor under heading (*b*) rather than under heading (*a*).

(*c*) Connected with the experimental technique, we may have a factor representing the difference between questions favorable to and opposed to the basic issue.

(*d*) and (*e*) It may be practicable and desirable to group the subjects in ways additional to that in (*a*), for example, by age, sex, or education.

7.3 THE CHOICE OF LEVELS

(i) Qualitative Factors

We now assume that we have decided on a list of potential factors and so turn to the choice of levels for each factor. Here again we begin by considering what we would like to do, with not too much thought for the limitations on the resources available for the particular experiment.

Various types of qualitative factors will be considered separately.

(*a*) With specific qualitative factors of direct importance it will usually be clear from the nature of the problem what the levels should be, that is, what specific contrasts it should be the object of the experiment to investigate.

(*b*) When such factors are considered as supplementary factors it may be sufficient to use a small number of levels, typical of rather extreme conditions of the factor. This is particularly so if it is suspected that in fact the supplementary factor has no effect on the primary comparisons. For example in an agricultural fertilizer trial on wheat, "varieties" may be inserted as a supplementary factor. If it is thought that the fertilizer comparisons will probably be the same for all varieties, it may be adequate, in a smallish experiment, to take the factor "varieties" at two levels, using two varieties typical of two main very different types of wheat. On the other hand, if it is likely that there are interesting interactions between varieties and fertilizers, it would be worth-while trying to work in more varieties.

(*c*) The number of levels for factors connected with the experimental technique or with groupings of the experimental units will usually be kept as small as possible.

(*d*) The number of levels of a sampled qualitative factor should not be too small if effective inferences about the population of levels are to be drawn. If possible it should be arranged that there are at least six to ten degrees of freedom for estimating the appropriate error; essentially statistical considerations are involved here (see Chapter 8). For example, suppose that the difference between two industrial processes varies appreciably from consignment to consignment. Then the precision of the estimate of the average over a population of consignments can only be assessed if sufficient consignments are examined to provide enough independent comparisons to give a good idea of the amount of this variation from consignment to consignment.

(ii) Quantitative Factors: Choice of Extreme Levels

In many ways the most interesting case is when the factor is quantitative. Here there are three questions. What should be the extreme factor levels investigated? How many factor levels should be investigated? How should the levels be distributed and how many observations should be allotted to each level?

In some problems the extreme factor levels may be determined fairly definitely by practical considerations. For example, in an experiment in which the relative humidity in a textile spinning shed is varied, it may be that 50 and 70 per cent relative humidity represent the minimum and maximum levels at which the humidifying plant works satisfactorily. Again, in an agricultural experiment in which the quantitative factor is the amount of a certain type of fertilizer, the lowest level would be absence of the fertilizer or, less commonly, a certain minimum dressing considered essential. The maximum dressing to be used would be more difficult to determine, although the following general comments can be put forward. If the object of the experiment is the prediction of the economically optimum dressing, the best thing would usually be to arrange that the suspected position of the optimum lies near the center of the factor range investigated. On the other hand if the object is less immediately practical, it will probably be advisable to cover a somewhat wider range of levels, since the behavior of a system under rather extreme conditions is sometimes revealing scientifically.

This is an instance of an important class of situations where the range of factor levels of interest depends on the response observed. As another example, suppose that the factor level is the dose of a drug and that the observation is the proportion of a small group of animals that are killed. At high doses all animals die, at low doses none does. Usually, the range of doses of interest straddles the dose giving 50 per cent kill and might, for example, correspond to a range of true kills from 5 to 95 per cent. In the absence of appreciable prior information, the best thing will usually be to proceed in two or more stages, first determining a range of doses for detailed study. If this cannot be done, a wide range of doses must be covered and this is obviously wasteful.

(iii) Choice of Number and Position of Levels

Having settled on the range of levels to be investigated, we consider the number and position of the levels to be used. In the majority of cases it is, in practice, best to use equally spaced levels of the carrier variable, with an equal number of observations at each level. Thus in many biological experiments the carrier variable is log dose, so that the doses are equally spaced on a log scale, i.e., distributed in a geometrical progression of dose. The main exception to equal spacing is when the response curve is expected to rise to a limiting value, rather than be described by one of the simple polynomial curves of Chapter 6. In such cases it may be desirable to have a number of closely spaced levels on the ascending part of the response curve and a number of more widely spaced levels where the curve flattens out. This presupposes some prior knowledge of the general shape and position of the response curve. We have already had an instance in Example 3.2, where the levels of application of potash to a cotton crop were 36, 54, 72, 108, 144 lb K_2O per acre. Quite generally, if one range of factor values is more interesting than another the factor levels should, of course, be relatively closer together in the more interesting range.

More usually the levels will be equally spaced with an equal number of observations at each level. There are then the following cases. First we may have just two levels, i.e., we divide our units equally between the lower and upper extreme levels. This will give us an estimate of the average increase in the observation over the range investigated, but no information whatever about the shape of the response curve. Therefore two levels should be used only in preliminary experiments, or in experiments where qualitative conclusions about directions of effects are enough.

For in a reasonably comprehensive investigation we would almost certainly wish to have some idea of the shape of the response curve and would not be willing, for example, to assume it to be linear. The most important case, and the simplest one that allows the shape of the response curve to be examined, is when there are three levels. Table 7.1 helps us to understand the advantages and disadvantages of using various numbers of levels. It refers to the situation in which both the extreme factor levels and the total number of units in the experiment have been fixed; in this case the standard errors of the estimates of slope and curvature are of the form

$$\text{numerical constant} \times \frac{\text{residual standard deviation}}{\sqrt{\text{total number of units}}}.$$

Table 7.1 gives the numerical constants for various designs with equally spaced levels.

TABLE 7.1

STANDARD ERRORS OF ESTIMATES OF SLOPE AND CURVATURE USING EQUALLY SPACED LEVELS WITH FIXED TOP AND BOTTOM LEVELS

(*a*) *Equal Number of Units per Level*

Number of Levels	For Estimate of Slope	For Estimate of Curvature
2	1	—
3	1.225	2.121
4	1.342	2.250
5	1.414	2.390

(*b*) *Three Levels; Various Numbers of Observations per Level*

Proportion of Observations at Center Level	For Estimate of Slope	For Estimate of Curvature
0	1	—
1/5	1.118	2.500
1/3	1.225	2.121
1/2	1.414	2.000
3/5	1.581	2.041

In all cases the standard error of the estimate is

$$(\text{tabulated numerical constant}) \times \frac{\text{residual standard deviation}}{\sqrt{\text{total number of units}}}.$$

For example if 16 units are allocated 4 to each of 4 equally spaced levels, the standard error of the estimate of curvature is $2.250/\sqrt{16} = 0.5625$ times the residual standard deviation. The following conclusions can be drawn.

(*a*) If it is desired solely to estimate the slope, the use of two levels is best. This is most often the case in preliminary experiments in which the main object is to see qualitatively whether or not a treatment effect is present and in which direction it is. In such experiments the best thing is to use just the two extreme levels, choosing them, if possible, so that they are sufficiently far apart for any treatment effect to be substantial, but accepting the risk that substantial curvature of the response curve within the limits may vitiate the interpretation of the results.

(*b*) Both slope and curvature are more precisely estimated from three equally spaced levels than from four or more equally spaced levels with the same extreme points.

(*c*) The second part of Table 7.1 shows that for an experiment at three levels the curvature is most precisely estimated when half the observations are at the center level and the remainder equally divided between the extreme levels. However, with an equal division of observations between the three levels, i.e., with one-third of the observations at the center level, the standard error of the estimate of curvature is little increased, and the standard error of slope appreciably reduced. Hence, the equal division of units seems a reasonable system to adopt even when the curvature is of prime importance.

(*d*) If we could specify quantitatively the relative importance of slope and curvature, we could determine mathematically the optimum arrangement. However, it is unlikely that such a specification can often be made.

(*e*) If the object of the experiment is to estimate the position of a maximum or minimum value on the response curve, expected to lie somewhere in the center of the factor range, the use of equal numbers of observations at three levels can be shown to be reasonably efficient.

(*f*) The use of four levels lowers the precision of slope and curvature, but enables both the consistency of the data with a parabolic response curve to be checked and, if necessary, a more complicated cubic response curve to be estimated.

These remarks need some modification if the amount of uncontrolled variation is known to be different for different factor levels. The discussion has also assumed that the response curve is smooth: if not many levels will be needed. An example is if the factor is temperature and a physical or chemical property of a compound is under study, it being suspected that a molecular change may occur within the range investigated. A sudden, almost discontinuous, jump in the response curve is then likely and the previous discussion is not relevant.

We can sum up as follows. Use two levels when the object is primarily to examine whether or not the factor has an effect and in which direction that effect is. Use three levels whenever a description of the response curve by its slope and curvature is likely to be adequate; this should cover most cases. Use four levels if further examination of the shape of the response curve is important. Use more than four levels when it is required to estimate the detailed shape of the response curve, or when the curve is expected to rise to an asymptotic value, or in general to show features not adequately described by slope and curvature. Except in these last cases it is generally satisfactory to use equally spaced levels with equal numbers of observations per level.

The application of these remarks depends to a considerable extent on personal judgement and on the relative importance to the experimenter of different aspects of the investigation. This makes it difficult to give examples that are convincing to the general reader.

In Example 7.1 all factors could, in a preliminary experiment, be taken at two levels; for example one level of the diet factor vitamin A would be complete absence, the other level would be presence in sufficient amount to satisfy fully the requirements for the vitamin. In later work, if it is required to study the response curve and its interaction with the factor "inoculation," four or more levels might be appropriate, since the response curve is likely to be a fairly complicated one.

In the experiments on which Examples 7.2 and 7.4 are based, all factors were run at two levels. In Example 7.3 varieties are a specific qualitative factor and would normally take on a considerable number of levels, depending on the number of experimental varieties sufficiently promising to merit inclusion and on the number of different types of control variety. Two dates of sowing would probably be all that is practicable, and the number of centers at which the experiment can be done would often be outside the experimenter's control. We shall discuss the choice of levels again in connection with further examples introduced below.

7.4 THE CONTROL OF ERROR AND THE SPLIT UNIT PRINCIPLE

Suppose for the moment that we have decided on the factors for inclusion in the experiment and on the number of levels of each. Then if we are going to take all combinations of factor levels the same number of times, we know how many different treatments we wish to accommodate in the experiment. In fact that number is the product of the number of levels of each factor. That is, in an experiment with four factors at five, three, two, and two levels, the number of different treatments is $5 \times 3 \times 2 \times 2 = 60$.

We now have the same problem that we discussed for simpler experiments in Chapters 3 through 5, namely of ensuring that the uncontrolled variation in the experimental material disturbs the conclusions as little as possible. In the simpler cases our previous discussion applies without change; that is we can reduce the effect of certain sources of variation by grouping the experimental units in randomized blocks or Latin squares, or by use of adjustments based on a concomitant variable, and we can convert the remaining variation into effectively random variation by randomization.

Example 7.5. Suppose that in the situation of Example 7.1 it is decided to do a 2 (diets) × 2 (inoculated vs. not inoculated) × 2 (age groups) experiment and that 32 mice are available. There are eight treatment combinations so that an arrangement of four randomized blocks of eight mice in each is a natural one. The eight mice in each block are chosen to be as similar as possible. The assignment of treatments to mice is, of course, randomized independently within each block.

Example 7.6. Agricultural field trials in the form of factorial experiments arranged in randomized blocks or Latin squares are common. (Detailed accounts of many such experiments are contained in the Annual Reports of the Rothamsted Experimental Station for the years immediately preceding 1939.) The general experience, however, is that the number of treatments must not be too large if these designs are to be effective: often the number of treatments needs to be less than 20 for randomized blocks and 10 for Latin squares, and sometimes the limit is lower than this. With more than this number of treatments, the blocks or rows and columns tend to become too heterogeneous, resulting in a high residual standard deviation. The choice between Latin squares and randomized blocks, and the maximum size of experiment it is suitable to set out in these designs depends, of course, on the particular experimental area, on the crop, and on the size and shape of plots that are used.

Similarly in other experiments in which only a fairly small number of different treatment combinations are involved, the principles of Chapters 3–5 apply without essential modification. However, difficulties appear as soon as the number of treatment combinations exceeds the number that can be accommodated in a block of reasonably homogeneous units. For example, we discussed in Chapter 3 industrial experiments in which up to, say, four runs could be made on one day, so that with not more than four treatments the effect of variations between days can be eliminated by the use of randomized blocks. However, even a very modest factorial experiment could well have many more than four treatments, so that fresh techniques are called for if the effect of variation between days is to be eliminated in such experiments.

There is one natural method that can be used, particularly when one (or more) factors are classification factors or are supplementary factors included primarily under heading (*b*) of § 7.2. This is to have this factor (or factors) at the same level on all the units in any one block. We can best see how this works by examples.

Example 7.7. Consider the following biochemical experiment whose main features could, however, be paralleled from many fields. It is required to assess the effect on the amount of a substance S in the blood of mice of the injection of two experimental preparations A and B. Suppose that the two sexes may respond somewhat differently to A and B. Then the simplest form of experiment we could use would be a $2 \times 2 \times 2$ factorial system, with the factors being (*a*) A first absent and then present in a basic amount, (*b*) similarly for B, and (*c*) sex.

Now suppose that when the full details of the experimental method are considered, including the measurement of S which may well require delicate techniques, it is found that not more than four mice can conveniently be dealt with on one day. It is very desirable in this sort of experiment to eliminate differences between days, because there are often sources of error, particularly associated with the experimental technique, that may introduce appreciable systematic differences between determinations on different days. If it were practicable to test eight mice on one day there would be no difficulty; we could use the randomized block design in its simple form. To deal with the present situation we shall assume that the classification factor, sex, is included primarily in order to examine its possible interaction with the treatment factors A and B, i.e., the main effect of sex is assumed not to be of major interest.

Under these circumstances the following procedure is likely to be a good one.

(*a*) Suppose that 32 mice, 16 of each sex, are available for the experiment and that this number is likely to give about the precision required. Divide the mice of each sex into four groups of four, the four mice in each group to be of the same strain and, as far as possible, of similar body weights. Number the groups from one to eight.

(*b*) Independently for each group, randomize the allocation of mice to the four treatments (A absent, B absent), A_0B_0; (A present, B absent), A_1B_0; (A absent, B present), A_0B_1; (A present, B present), A_1B_1.

(*c*) Randomize the order of groups in time.

A possible arrangement and set of observations from such an experiment are shown in Table 7.2.

General inspection of the results suggests that there are substantial systematic variations from day to day and that these are not completely associated with sex differences. This means that the experimental arrangement has been a successful one in that the main comparisons, being made within days, are more precise than they would have been if an ordinary randomized block design had been used, with a block equal to two days' results.

The first step in the analysis is to calculate the tables of mean values shown in Table 7.2(*b*). First we have, both separately for each sex and also averaged over sexes, the two-way tables for the factors A and B. For example, in the last of these tables, 5.01 is the average of the eight observations receiving the treatment combination A_0B_1, each day contributing one observation and each sex four. The final table shows the mean values day by day, grouped into the two sexes. All conclusions from the experiment are based on these tables of mean values, the purpose of more refined statistical analysis, including analysis of variance,

TABLE 7.2

EXPERIMENT ON THE EFFECT OF A AND B ON AMOUNT OF SUBSTANCE S IN THE BLOOD OF MICE

(*a*) *Observations and Treatment Arrangement*

Day 1: Male —A_0B_1, 4.8; A_1B_1, 6.8; A_0B_0, 4.4; A_1B_0, 2.8
Day 2: Male —A_0B_0, 5.3; A_1B_0, 3.3; A_0B_1, 1.9; A_1B_1, 8.7
Day 3: Female—A_1B_1, 7.2; A_0B_1, 4.3; A_0B_0, 5.3; A_1B_0, 7.0
Day 4: Male —A_0B_0, 1.8; A_1B_1, 4.8; A_1B_0, 2.6; A_0B_1, 3.1
Day 5: Female—A_1B_1, 5.1; A_0B_0, 3.7; A_1B_0, 5.9; A_0B_1, 6.2
Day 6: Female—A_1B_0, 5.4; A_0B_1, 5.7; A_1B_1, 6.7; A_0B_0, 6.5
Day 7: Male —A_0B_1, 6.2; A_1B_1, 9.3; A_0B_0, 5.4; A_1B_0, 6.9
Day 8: Female—A_0B_0, 5.2; A_1B_1, 7.9; A_1B_0, 6.8; A_0B_1, 7.9

(*b*) *Tables of Mean Values*

Males

		B Absent	*B* Present	
A	Absent	4.22	4.00	4.11
	Present	3.90	7.40	5.65
		4.06	5.70	

Females

		B Absent	*B* Present	
A	Absent	5.18	6.02	5.60
	Present	6.28	6.72	6.50
		5.72	6.38	

Both Sexes Combined

		B Absent	*B* Present	
A	Absent	4.70	5.01	4.86
	Present	5.09	7.06	6.08
		4.89	6.04	

	A Absent	*A* Present	
Male	4.11	5.65	4.88
Female	5.60	6.50	6.05
	4.86	6.08	

	B Absent	*B* Present	
Male	4.06	5.70	4.88
Female	5.72	6.38	6.05
	4.89	6.04	

Day Means

Male	Female
4.70	5.95
4.80	5.22
3.08	6.08
6.95	6.95
4.88	6.05

being to assess the precision of the comparisons that can be made from the tables. In accordance with the general policy of this book the details of the more advanced statistical methodology will be omitted. The analysis of a similar experiment is described in detail by Cochran and Cox (1957, § 7.15).

The following general points can be made about the conclusions from this particular experiment and about the general implications for this type of design.

(i) The two-way tables for A and B appear to have a different form for the two sexes. For males the suggestion is that there is an increase in the observation when both A and B are present, but none otherwise. For females the data appear consistent with an absence of interaction. Statistical analysis confirms that it is unlikely that the true pattern is the same for both sexes.

(ii) In (i), if we compare any mean in a two-way table with another mean in the same table, systematic differences between days have no effect. For example if we compare, in the two-way table for males, the mean 4.22 with the mean 4.00, we are comparing two quantities each built up from one observation on each of days 1, 2, 4, and 7. Therefore if the observations on one day are, let us say, all high, the two means are affected equally and the difference between them is unaffected. Further if we compare a difference from the table for males with a difference from the table for females, this too is unaffected by systematic changes in the observation from one day to another.

(iii) But if we compare a mean value in the table for males with a mean value in the table for females, we are comparing observations on days 1, 2, 4, and 7 with observations on days 3, 5, 6, and 8. Therefore variations between days will contribute to the error in this case. In particular, if we wish to make a comparison of the overall observation on males with that on females (see the last table of means), the variation between days is important. Thus this last table shows that there is a very appreciable variation from one day to another; the apparent average difference between male and female could well be a consequence solely of the random variation in the experiment. In fact the variation between days is so large that there has, in effect, been almost complete sacrifice of information about the main effect of sexes and a substantial increase in the precision of the other comparisons.

The essence of the method just described is that a lower precision is accepted for one comparison, the difference between sexes, in order that the precision of the more interesting comparisons, namely of the treatment factors and their interaction with sex, shall be increased. Therefore the procedure should be used only when a suitable factor is available for sacrifice, or when it is, for some practical reason, convenient to arrange the experiment with a particular factor constant within each block. This last thing quite often occurs in agricultural and industrial experiments and some examples will now be given.

Example 7.8. It may happen in an agricultural experiment that one factor represents a treatment that requires large plots for its application, whereas the other factor or factors can most conveniently and accurately be tested on small plots. An example is when the first factor represents different systems of grazing; the area to which each grazing treatment is applied needs to be large enough to provide for at least one, if not several, animals and this would usually be appreciably greater than the area suitable for testing varieties or fertilizer treatments. Again, in the experiment on sugar beet described in Example 7.3, consider the two factors, date of sowing and varieties. If the sowing is done mechanically there is a great advantage in making, say, an early sowing of a number of varieties over several continuous strips of ground, and then a late sowing over the remaining continuous strips. The alternative is to make at any one time individual sowings on small plots haphazardly distributed over the experimental area, and this is troublesome.

TABLE 7.3

Two Arrangements of an Experiment on Sugar Beet

(*a*) *Randomized Block Design*

E	L	L	E	E	E	L	E	L	L	E	L
V_5	V_1	V_4	V_2	V_6	V_3	V_3	V_1	V_6	V_5	V_4	V_2

Block I

(*b*) *Modified* (*Split Plot*) *Design*

E	E	E	E	E	E	L	L	L	L	L	L
V_4	V_1	V_6	V_5	V_3	V_2	V_2	V_3	V_6	V_5	V_1	V_4

Block I

E: early sowing; L: late sowing; $V_1, \ldots, V_6$: six varieties.
One block only shown

Another example arises in work with fruit trees; for certain types of treatment, such as pruning, individual trees may receive different treatments, but with other sorts of treatment, such as mass spraying, it is convenient to treat whole sets of adjacent trees in the same way. If both types of treatment are included in the same experiment, we have the general situation that we are discussing.

Consider for definiteness the experiment on sugar beet with, for simplicity, two dates of sowing and six varieties, taking the arrangement at one center. With variety trials it is frequently convenient to have long thin plots, a few drills wide. With 12 treatments a simple randomized block design would take the form shown in Table 7.3(*a*), the 12 treatments being randomized within each block. As explained above, the objection to this is the inconvenient dispersion over the area of the plots to be sown at one time. The modified arrangement is shown in Table 7.3(*b*). It should be compared with Example 7.7. There sex is constant over groups of units close together in time. In the present example time of sowing is constant over groups of units (plots) close together in space.

In the agricultural applications the following language is frequently used. The treatment which is applied to large areas is called the *whole plot treatment*, and the treatment which is applied to small areas is called the *subplot treatment*. The experimental units to which the whole plot and subplot treatments are applied we called *whole plots* and *subplots* respectively. The whole plots are said to be split into subplots and the whole arrangement is called a *split plot experiment*. Thus in the present example there were two whole plots in each block and six subplots in each whole plot.

The arguments put forward in the discussion of Example 7.7 show that the error of a comparison of varieties or of the interaction of varieties with time of sowing* is determined by the amount of uncontrolled variation in the observations from subplot to subplot within whole plots. On the other hand, the error of the comparison of whole plot treatments is determined by the variation between whole plots within blocks. To take an extreme case, suppose that the six subplots within each whole plot have very uniform properties but that there are substantial variations between different whole plots. Then the comparison of varieties will be made precisely, because for each observation on, say, V_1 there is a corresponding, directly comparable, observation on V_2 in the same whole plot, and the difference between these observations is almost unaffected by the uncontrolled variation. Hence the overall difference between V_1 and V_2, which is the average of the individual differences, is likewise almost unaffected by uncontrolled variation. But the comparison of dates of sowing is based on the total observations on the whole plots, summed over the six varieties, and the difference between these large areas is subject to appreciable uncontrolled variation. The situation encountered in practice, although less extreme than the one just described, will tend to make the determination of the whole plot main effects less precise than that of the other comparisons.

The experiment of Example 7.7 can be described in the above language as follows. Each whole plot consists of one day's work, four mice, and the factor sex is randomly distributed over whole plots. Finally, each whole plot is split into four subplots, the four mice, and the four treatments arising from the 2×2 system of treatments, assigned randomly to them. The main difference between this and the arrangement of Example 7.8 is that in the latter, the whole plots are grouped into randomized blocks rather than completely randomized. The decision as to whether the whole plots should be completely randomized, arranged in randomized blocks, or even in Latin squares, is made by the methods of Chapter 3, exactly as if the subplots were absent.

In general, the whole plot treatments need not be restricted to a single factor but may consist of all combinations of the levels of several factors.

An important industrial application of the split plot principle is to

* That is, whether the difference between two particular varieties is greater or less for early sowing than it is for late sowing.

experiments on processes in which there are several stages. It may then be convenient to work with large batches of material at the first stage, dividing into smaller batches for the application of the treatment at the second stage.

Example 7.9. Drawing and spinning are successive stages in the production of worsted yarn. On a mill scale both these processes are carried through with large quantities of wool, but in experimental work, in which it is usually required to process small lots, the minimum lot size is appreciably greater in drawing than in spinning.

Suppose that it is required to examine the effect on the final properties of the yarn of changing the relative humidities in the two processes. This gives us two main treatment factors. Suppose that a supplementary factor, the type of wool, is included in order to examine whether the conclusions about relative humidity are dependent on the type of wool, i.e., in order to examine possible interactions.

This gives us three factors; the next step is to determine the number of levels of each. The two main factors, the relative humidities at the two stages, are quantitative, so that in accordance with the general principles set out in § 7.3 we decide first on the extreme levels. These might be 50 and 70 per cent relative humidity. Over this range it is reasonable to expect that the main yarn properties (strength, irregularity, etc.) will have response curves that are roughly parabolic, i.e., adequately described by an average slope and curvature. Hence, three equally spaced levels, 50 per cent, 60 per cent, and 70 per cent are indicated for both factors. The supplementary factor is, in a fairly small experiment, probably best taken at two levels, say a representative coarse wool and a representative fine wool denoted by C and F.

As explained above, the minimum lot size in drawing is greater than in spinning, and in fact suitable weights are obtained if a lot, after passing through the drawing process at a particular relative humidity, is split into three equal parts for subsequent spinning at the three different relative humidities. There are six whole plot treatments, namely the six combinations of three drawing relative humidities with two types of wool. Suppose that it is decided that two whole replicates will give sufficient precision. A suitable design, after randomization, is shown in Table 7.4.

An explanation of the table and further details of the experiment are as follows. For the first replicate sufficient raw material (tops) of each type of wool is taken, thoroughly mixed to make the material of each type as homogeneous as possible, and is then divided into three equal lots of C and three equal lots of F. These are numbered in random order.

Lot number 1 of coarse wool is then drawn at a relative humidity of 60 per cent. The product of this process (roving) is thoroughly mixed to make it as homogeneous as possible and then divided into three equal lots and numbered in random order. The first lot is spun at a relative humidity of 70 per cent, followed by the second lot at 60 per cent and the third lot at 50 per cent. In the meantime, the first lot of F is drawn at a relative humidity of 70 per cent, the product being divided into three equal lots for spinning, and so on. When the first replicate is complete, the second replicate is dealt with according to a similar but independently randomized scheme.

The experimental arrangement has been described on the assumption that the drawing and spinning of different lots can be carried on at the same time at different relative humidities. If this is not so, i.e., if the drawing and spinning sheds have a common humidifying plant, it may still be possible to use the plan given above, but careful organization will be needed to get a proper spacing in time of the different operations.

TABLE 7.4

DESIGN OF A TEXTILE EXPERIMENT

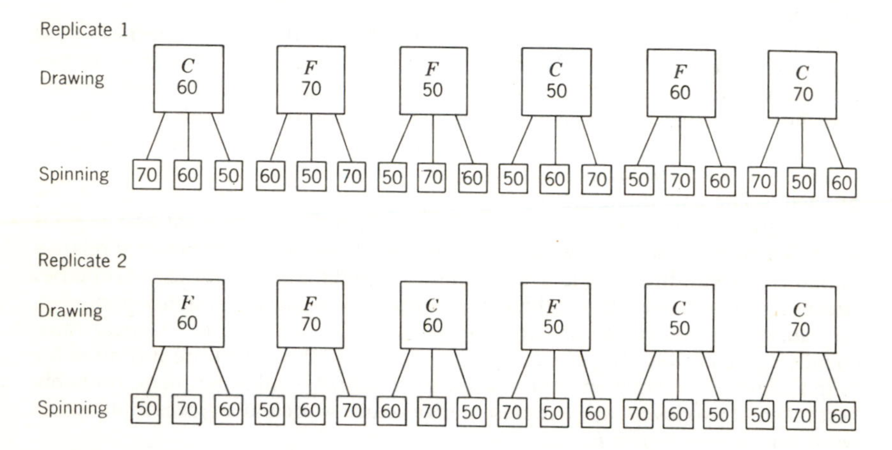

The correspondence with the split plot design can be seen if we think of each spinning lot as a subplot, receiving a different subplot treatment (spinning relative humidity). Each set of three subplots forms a whole plot and forms an experimental unit for the application of the whole plot treatments (drawing relative humidity and quality of wool).

In this experiment the experimental units are split for practical convenience. A consequence of the splitting, is that the whole plot treatments are likely to be compared less precisely than the subplot treatments; in the earlier examples this was, at least in part, one of the reasons for using the split plot design, but here this is not generally so.

If there had been a third stage of processing it might have been profitable to subdivide the spinning lots into parts for the final stage. This would have led to a split-split plot experiment.

We can sum up the points made in this section as follows. If the total number of treatment combinations is not too large, we may be able to use without modification the methods for increasing precision explained in Chapter 3, namely the grouping of experimental units into similar sets leading to the use of a randomized block or Latin square arrangement. If the number of factor combinations is too large for this method to work effectively, one possibility is that a special factor or special set of factors should be selected to be held constant within each block, so that the number of treatment combinations to be accommodated within a block is reduced to a manageable size. This device is called splitting the experimental units, since the effective experimental unit for the special factor is the block, which is split into parts to form the experimental units for the other factors. The effect is to decrease the precision of the estimates of effects connected only with the special factors, and to increase the precision of comparisons connected with the remaining factors and with the interaction of the remaining factors with the special factors. There are therefore two situations in which the method is useful. First, a particular main effect may be of little importance, the corresponding factor being included to give information about interactions. Such a factor is clearly suitable to be taken as the special factor. Secondly, there may be technical reasons why the experimental unit for some factors should be larger than the experimental unit for other factors. If neither of these reasons holds, the device of split units should not be used. In such cases it may be possible to use a more subtle technique called *confounding*. The general idea here is identical to that of the split plot experiment, except that instead of accepting lower precision for a main effect, we arrange that it is high-order interactions on which information is sacrificed. Details and discussion of this device are deferred to § 12.3.

7.5 THE FINAL CHOICE OF A DESIGN

We have now discussed three of the four key elements in the design of a factorial experiment, namely the choice of factors that it would be desirable to include, the choice of levels, and the reduction of the effect of uncontrolled variation. The fourth element is the decision about the total number of experimental units to be used. This is discussed fully in the next chapter. Sometimes we have a preassigned limited number of units available for the experiment and it is required to see whether the resulting precision of the estimates is likely to be sufficient to make the experiment worth doing at all. On other occasions we may start with some idea of the precision that we require and work out from this an approximate number of experimental units to use. In the majority of factorial experiments we decide to do a certain number of replicates of the complete factorial system of treatments, i.e., each combination of factor levels is to occur the same number of times. An exception to this is when less than one full replicate will give the precision required. In such cases the important device of *fractional replication* (§ 12.2) can sometimes be used.

When these four elements have been considered, we are in a position to reach a final decision about the design to use. This involves general judgement, common sense, and detailed knowledge of the purpose and probable outcome of the investigation, and they are difficult to formalize or exemplify. Often we find that in order to keep the experiment sufficiently simple, or within the limits of the resources available, it is necessary either to omit factors that we should like to include or to reduce the number of levels of certain factors. This means that we are faced with questions such as: is it worth having fewer levels and higher precision, or more levels and lower precision; is it better to have a simple experiment with a smaller range of validity for the conclusions; and so on? There are no easy answers to these questions. One relevant consideration, for example, is the time that the experiment will take to complete. If this time is comparatively short, and also if the experiment is inexpensive, it will probably be preferred to proceed by a series of comparatively simple experiments. On the other hand if the experiment takes a long time to complete, it will be more important to get the maximum possible information from it, and the case for a relatively complicated design is stronger, although it will nearly always be unwise to embark on a large experiment at the beginning of an investigation in a new field. These are mere vague generalities; there is no substitute for experience in the use of factorial experiments in the particular field under study, and an investigator should begin with simple experiments and gradually work up to the more complicated cases if these seem to be called for.

SUMMARY

In designing a factorial experiment we have to consider:

(i) what factors to include;
(ii) at what levels these factors should be taken;
(iii) how many experimental units should be used in all;
(iv) what measures should be taken to reduce the effect of uncontrolled variation.

We start by making a preliminary list of factors for possible inclusion. These usually fall into five types;

(*a*) factors of direct interest;
(*b*) treatment factors included primarily to elucidate the behavior of factors under heading (*a*);
(*c*) factors connected with the experimental technique;
(*d*) classification factors, suggested by natural groupings of the experimental units;
(*e*) deliberately inserted variations in the experimental units.

We then consider at how many levels we should like each factor to appear. The case calling for most discussion is that of quantitative factors. Here the use of just two levels is advisable in experiments in which the object is to see whether the factor has an appreciable effect and in which direction the effect is. Otherwise at least three levels should be used in order to give some estimate of the shape of the response curve.

The question of the total number of units is discussed in the next Chapter.

In a small factorial experiment randomized blocks and Latin squares can be used to reduce the effect of uncontrolled variation in exactly the way described in Chapter 3. In larger experiments it may be possible to obtain a satisfactory design by the device of split units, in which precision is sacrificed on one main effect in order to obtain higher precision for other comparisons of more interest. The device is also sometimes useful for reasons connected with the experimental technique. In other cases it may be possible to use the more subtle technique of confounding in which information is sacrificed about high-order interactions in order to increase the precision of the estimates of main effects and low-order interactions.

When the above considerations have been explored, it remains to reach a final decision about the best design to use. This usually involves such difficult questions as whether in the interests of simplicity and economy, certain factors should be omitted, or the number of levels of other factors reduced.

REFERENCES

Cochran, W. G., and G. M. Cox. (1957). *Experimental designs*. 2nd ed. New York: Wiley.

Davies, O. L., and W. A. Hay. (1950). The construction and uses of fractional designs in industrial research. *Biometrics*, **6**, 233.

Edwards, A. L. (1941). Political frames of reference as a factor influencing recognition. *J. Abnorm. Soc. Psychol.*, **36**, 34.

CHAPTER 8

The Choice of the Number of Observations

8.1 INTRODUCTION

We now turn to a matter more specifically statistical, namely the relation between the number of experimental units and the precision of the estimates of the treatment effects. There are two aspects to this. First the scale of effort that can be devoted to the experiment may be fixed by circumstances outside the experimenter's control. In this case it is nearly always helpful to have, before doing the experiment, some rough estimate of the precision that is likely to result. This rough estimate may, for example, show that the final estimates will probably be subject to such large errors that no effective conclusions are likely to result, thus suggesting that the experiment is not worth doing until more resources can be assembled. Or it may appear that adequate precision can be obtained with less than the full number of experimental units.

The second aspect is more positive. If the number of units is to an important extent under the experimenter's control, we may work out the precision corresponding to a range of values of the number of units. Hence reasonable compromise may be reached between, on the one hand, having too few units and low precision, and on the other, wasting time and experimental material in attaining unnecessary precision.

In the general discussion of § 1.2(ii) we noted that the final precision of the estimated treatment effects depends on

(*a*) the intrinsic variability of the experimental material and the accuracy of the experimental work;

(*b*) the number of experimental units (and the number of repeat observations per experimental unit);

(*c*) the design of the experiment (and on the method of analysis if this is not fully efficient).

In the present chapter we are concerned solely with (*b*) and so we shall assume that all practicable steps have been taken to increase precision by methods (*a*) and (*c*). We also assume that all important sources of systematic error have been removed, for example by randomization, and that the treatment comparisons are therefore subject only to random errors.

One most important point concerns the definition of an experimental unit for the purposes of the following calculations. Two observations on the same treatment are considered to come from different units only if the design of the experiment is such that the experimental material corresponding to the two observations might have received different treatments, and moreover if the corresponding material has been dealt with independently at all stages at which important variation may enter. For example, imagine that a consignment of material is divided into eight parts, four to receive each of two treatments. Let these eight parts be dealt with separately in the experimental procedure, and at the end let triplicate observations be made on each part, for example by sampling each part three times. There are then twelve observations for each treatment, but only four units. It would be legitimate to regard the twelve observations as twelve units only in the unlikely event that we may assume that negligible variation enters the experiment prior to the last stage, the taking of the observations. There is a further discussion in § 8.3(iv).

The details that follow are inevitably somewhat statistical, and the reader who wishes primarily to learn the general nature and scope of the subject may omit the following sections. He should be aware, however, of the importance of making, before an experiment is started, some estimate of the precision that is likely to be obtained.

8.2 THE MEASUREMENT OF PRECISION

We begin by discussing the way in which precision can be measured. Suppose that we have the situation of § 2.2, so that we are interested in comparisons, or *contrasts*, of the treatment constants $a_1, \ldots, a_t$. Contrasts can take various forms, for example:

(*a*) the difference, $a_1 - a_2$, between the effects of two particular treatments, say the first and the second, T_1 and T_2. This is usually estimated by the mean observation on units receiving treatment T_1 minus the corresponding mean for T_2;

(*b*) the mean difference between one group of treatments and another treatment or group of treatments. For example, in a nutritional experiment, T_1 might represent a basic diet and the remaining treatments various forms of supplemented diet. One contrast of interest might then be the average of the *a*'s for all the supplemented diets minus a_1. The contrast would usually be estimated by the corresponding difference of the observed treatment means;

(*c*) if the treatments correspond to different levels of one or more quantitative carrier variables, we may be interested in the particular combinations of the *a*'s that measure, for example, the slope and curvature of the response curve. Methods of estimating these contrasts have been considered in Chapter 6.

From the observations we construct an estimate of the particular contrast* that interests us. In general the estimate from the observations will not equal the true value of the contrast calculated from the treatment constants $a_1, \ldots, a_t$, and it is with the magnitude of the difference between the true and estimated values that we are concerned. We call this difference the *error* in the estimated contrast. Of course in any particular instance the true treatment constants $a_1, \ldots, a_t$ are unknown and so is the error in the estimated contrast. We have to work with a probability distribution of errors derived from the fact that the treatment arrangement for use has been selected from a set of possible arrangements in a random way. It can be shown that the average error is zero; this is another way of expressing the elimination of systematic error achieved by randomization. The general size of the errors for a particular contrast is usually best measured by the *standard error*, which is defined formally to be equal to the square root of the average of the squared errors. The interpretation of the standard error given in § 1.2 depends somewhat on the form of the frequency distribution of the uncontrolled variation, but has a more direct practical meaning. This interpretation is as follows. In about one occasion out of three, the randomization will lead to a design in which the error is more than plus 0.97 times standard error or less than minus 0.97 times standard error, i.e., is in absolute magnitude more than about one standard error. In about one occasion out of twenty, the randomization will lead to a design in which the error is in absolute magnitude more than 1.96 times the standard error (for practical purposes 1.96 may be replaced by 2); in about one occasion out of a hundred, the randomization will lead to a design in which the error is in absolute magnitude more than 2.58 times the standard error.

In view of these facts, the standard error is a measure of precision with a direct practical interpretation. The multiple of the standard error corresponding to other frequencies of errors can be obtained from tables of the normal distribution given in books on statistical methods; see also Table 8.1.

* The common mathematical feature of these contrasts is that they are linear combinations of the *a*'s with the sum of the coefficients zero.

It would be expected on general grounds that the standard error of a particular contrast would depend in part on the form of the contrast, in part on the numbers of observations involved, and in part on the amount of uncontrolled variation. This is confirmed by mathematical calculation which shows that, for example, the standard error of any estimate formed by taking the mean of one set of observations minus the mean of a different set, is

$$\sqrt{\left\{\frac{1}{\begin{pmatrix}\text{no. of observations}\\ \text{in first set}\end{pmatrix}} + \frac{1}{\begin{pmatrix}\text{no. of observations}\\ \text{in second set}\end{pmatrix}}\right\}} \times \begin{pmatrix}\text{residual}\\ \text{standard}\\ \text{deviation}\end{pmatrix}, \quad (1)$$

where the residual standard deviation is a measure of the amount of that part of the uncontrolled variation which affects the error of the treatment contrasts. More precisely imagine that we could obtain observations from the experimental units in the absence of true treatment effects. We then remove that part of the variation in these observations that can be accounted for by the blocks in a randomized block design or by the rows and columns in a Latin square design, i.e., we take residuals eliminating blocks in the randomized block design and eliminating rows and columns in the Latin square. The residual standard deviation measures the amount of variation in the residuals, in a way analogous to that in which the standard error measures the error in an estimated contrast, i.e., we can think of the standard error and the standard deviation as similar quantities, the first referring to estimated effects, the second to individual observations. As we saw in the discussion of split plot designs in Chapter 7, different contrasts may have different residual standard deviations.

Equation (1) gives the standard error of the difference between two means and the formulas for slopes and curvatures were given in Chapter 6. A fairly nonmathematical discussion for a general contrast* is given by Cochran and Cox (1957, § 3.5). For the particular case when the two sets contain equal numbers of observations, (1) becomes

$$\sqrt{\left\{\frac{2}{\begin{pmatrix}\text{no. of observations}\\ \text{per set}\end{pmatrix}}\right\}} \times \begin{pmatrix}\text{residual}\\ \text{standard deviation}\end{pmatrix}. \quad (2)$$

It follows that, if we can determine or estimate the standard deviation,

* The general formula, from which all those given here and in Chapter 6 follow as special cases, is that the standard error of $l_1\bar{x}_1 + \ldots + l_k\bar{x}_k$ is $\sigma\sqrt{\{l_1^2/n_1 + \ldots + l_k^2/n_k\}}$, where $\bar{x}_1$ is the mean of n_1 observations, etc., no two $\bar{x}$'s having observations in common, and where σ is the residual standard deviation.

we are able to find the standard error for any particular contrast and hence can

(*a*) determine from the observations limits within which the true value of the contrast lies, at any given level of probability. Thus, the true value lies within the estimated value plus or minus two standard errors, with a probability of 95%;

(*b*) determine, before the experiment is done, the width of the interval of uncertainty at any given level of probability.

In this chapter we are mainly concerned with (*b*), but (*a*) will be briefly illustrated by an example.

Example 8.1. Suppose that in comparing the growth rates of animals receiving two diets, T_1 and T_2, it is known from previous experience of similar experiments that the residual standard deviation is likely to be about 2.5 units. Let ten animals be devoted to each diet. Then the standard error of the estimated difference between the growth rates for the two diets is, by formula (2), $\sqrt{(2/10)} \times 2.5 = 1.12$ units.

If now the observed mean observation on T_1 is 6.10 units more than that on T_2, we can calculate limits for the true difference between the diets as follows: with a chance of 2/3, the true difference lies between $6.10 - 0.97 \times 1.12$ and $6.10 + 0.97 \times 1.12$, i.e., between 5.01 and 7.19; with a chance of 19/20, the true difference lies between $6.10 - 1.96 \times 1.12$ and $6.10 + 1.96 \times 1.12$, i.e., between 3.90 and 8.30. With a chance of 99/100, the true difference lies between $6.10 - 2.58 \times 1.12$ and $6.10 + 2.58 \times 1.12$, i.e., between 3.21 and 8.99.

These statements enable us to form an objective picture of what can be inferred about the true difference from the results of the experiment.*

Quite often we are interested not just in estimating a particular contrast but also in examining its *statistical significance*. This idea needs careful explanation.

Suppose for definiteness that we are interested in the relative effect of two particular treatments T_1 and T_2, i.e., in the true contrast $a_1 - a_2$. Now imagine that using the results of a particular experiment we find that the estimated difference is roughly equal in magnitude to the standard error. Then the frequency interpretation of the standard error tells us that even if there were no real difference between the treatments, a difference as large or larger than the one observed would occur by chance about once in every three times. That is, the difference is just such as would be expected to occur if the true treatment effect were zero. The consequence of this is not that the true difference is asserted to be zero, but that *on the basis of the results under analysis* we would not be justified in claiming that there is a real difference between the treatments, or, in other words, that the data are consistent with a zero true treatment difference.

* The precise interpretation of these probability statements is explained in textbooks on statistics and needs careful qualification.

To put this slightly differently, the data do not, at an interesting level of significance, establish the sign of the true treatment effect. For the positive estimated difference is reasonably consistent with a zero or negative true difference. Significance tests, from this point of view, measure the adequacy of the data to support the qualitative conclusion that there is a true effect in the direction of the apparent difference.

Imagine next that the estimated contrast is just over twice its standard error. An apparent treatment difference as great or greater than this would occur by chance less than one time in twenty and we say that the difference is statistically significant at the 5 per cent (1 in 20) level. Similarly, if the estimated contrast is more than about 2.6 times its standard error, an apparent difference as great or greater than the observed one would occur by chance less than one time in one hundred, and we say that the difference is statistically significant at the 1 per cent level. The level of significance corresponding to other values of the estimated contrast is shown in Table 8.1. Any estimate statistically significant at the 1 per cent level is automatically statistically significant at the 2 per cent, 5 per cent, etc. levels.

The level of statistical significance attained measures the uncertainty involved in taking say an apparent difference between two treatments to be real. For example, if high statistical significance is attained, e.g., the 0.1 per cent level, the statistical uncertainty involved in treating the apparent difference as real is very slight. The following is a general guide to the practical meaning of the various levels:

not statistically significant at 10% level	data are consistent with a zero true contrast.
statistically significant at or near the 5% level but not near the 1% level	data give good evidence that the true contrast is not zero.
statistically significant at or near the 1% level	data give strong evidence that the true contrast is not zero.

Example 8.2 illustrates some of these ideas.

Example 8.2. Consider again the experiment described in Example 8.1. The estimated difference is 6.10 units and the standard error is 1.12 units. The ratio of these is $6.10/1.12 = 5.45$, and this is considerably larger than the largest value in Table 8.1. Therefore the difference is very highly significant statistically, and negligible uncertainty is involved in taking there to be a real difference between the two sets of observations in the direction indicated.

If the estimated difference had been 1.50 units the ratio to the standard error would have been $1.50/1.12 = 1.34$, and from Table 8.1 there is a probability of rather less than 20 per cent of obtaining as great or greater a difference just by chance, the true difference being zero. Since this probability is quite appreciable, we may consider the data as consistent with the absence of a true treatment difference. The 95 per cent limits of error for the true effect are 1.50 plus and minus 1.96 × 1.12, i.e., (−0.70, 3.70). Note first that this includes negative values, so that we cannot, at this level of probability, infer that the true difference is positive. Note also that it depends entirely on the circumstances of the application whether the data are consistent with the existence of practically important true treatment differences.

As a final example, note that if the estimated difference had been 2.38 units, it would have been statistically significant at the 5 per cent but not at the 2 per cent level. Usually this would be taken as moderately good evidence that the true treatment difference is positive.

TABLE 8.1

LIMITS OF STATISTICAL SIGNIFICANCE FOR AN ESTIMATED CONTRAST WITH THE STANDARD ERROR KNOWN

Ratio of Estimated Contrast to its Standard Error	Level of Statistical Significance
1.28	20%
1.64	10%
1.96	5%
2.33	2%
2.58	1%
2.81	0.5%
3.09	0.2%
3.29	0.1%

If the ratio exceeds the value in the left-hand column, it is statistically significant at the level given in the right-hand column.

Derived, by permission of the Biometrika trust and the authors, from Table 1 of *Biometrika Tables for Statisticians* by E. S. Pearson and H. O. Hartley, Cambridge University Press, 1954.

The following further points should be noted.

(*a*) The significance test is concerned with what the data under analysis tell us. If further data become available, or if we have relevant information about the contrast from general experience or theoretical considerations, our overall conclusions about the contrast may be changed.

(*b*) If the contrast is statistically significant at say the 1 per cent level, we are in little doubt that there is a nonzero true contrast. This is only part of the matter. It will be necessary to consider also the magnitude of the true contrast, not just whether or not it is zero, and we do this by the method indicated at the beginning of this section, that is, by working out the estimated contrast plus and minus appropriate multiples of the standard error, in order to give limits between which the true contrast lies at assigned levels of probability.

(*c*) It can happen that although a contrast is statistically significant, the limits, at a reasonable probability level, for the true value correspond to differences of no practical importance. That is, we may sometimes conclude that although two treatments differ, the difference between them is of no importance. Statistical significance is not the same as technical importance.

(*d*) On the other hand, if the estimated contrast is consistent with a zero true contrast, it is nevertheless still possible that an important true contrast may exist. For instance in the case discussed above in which the estimated difference is equal to the standard error, and equal, say, to one unit, the limits for the true difference at the 5 per cent probability level are

$$\text{estimated contrast} \pm 1.96 \times \text{standard error},$$

i.e., are very nearly -1 and 3. That is, there is a 95 per cent chance that the true contrast lies between the limits worked out in this way. Now depending entirely on what magnitude of difference we regard as of practical importance, this range from -1 to 3 may or may not include differences of practical concern to us. All that the absence of statistical significance tells us is that we cannot reasonably claim that these data show that T_1 gives a higher observation than T_2, because the range -1 to 3 includes negative as well as positive differences. Further data may, or may not, show that a practically important difference exists, unless we can say from practical knowledge that differences in the range -1 to 3 are of no interest. It follows that we should ordinarily consider the limits of error for a true contrast, even when the estimated difference is not statistically significant. Significance tests fulfil an important, but limited, role in the analysis of data.

This last point (*d*) suggests the need to consider the sensitivity or *power* of a significance test. The whole idea of statistical significance hinges around the desire to protect ourselves against claiming that our data show a treatment contrast in a particular direction, when, in fact, the true contrast is zero (or in the opposite direction). But it is also important to arrange if possible that if the true contrast is sufficiently different from zero to be of practical importance, then the estimated contrast should stand a good chance of being judged statistically significant. This suggests that we should consider the probability that for a given value of the true contrast, the estimated contrast should be statistically significant at some particular level, for example the 5 or 1 per cent levels. This gives what is called the power of the significance test, i.e., it measures the chance of detecting a certain true contrast at a specified level of significance. Power is important in choosing between alternative methods of analysing data and in deciding on an appropriate size of experiment. It is quite irrelevant in the actual analysis of data.

Table 8.2 shows the results of such calculations. We illustrate the meaning of the Table on the same situation that has been used for Examples 8.1 and 8.2.

TABLE 8.2

POWER OF THE SIGNIFICANCE TEST FOR A CONTRAST WHERE STANDARD ERROR IS KNOWN

Magnitude of True Contrast/Standard Error	Probability that Estimated Contrast will be Positive and Statistically Significant at the Following Level		
	10 per cent	5 per cent	1 per cent
0	5 per cent	$2\frac{1}{2}$ per cent	$\frac{1}{2}$ per cent
0.5	13	7	2
1.0	26	17	6
1.5	44	32	14
2.0	64	52	28
3.0	91	85	66
4.0	99	98	92

Example 8.3. The standard error in the example under discussion is 1.12 units. Therefore the third line of Table 8.2 tells us that if a true difference of this magnitude existed, with say T_1 giving a greater observation than T_2, there is a 26 per cent chance that the estimated difference will show the mean for T_1 to be greater than the mean for T_2, the difference being statistically significant at the 10 per cent level. Similarly, from the next to the last line, if the true difference is $3 \times 1.12 = 3.36$ units, there is a 91 per cent chance that the estimated difference will be statistically significant at the 10 per cent level, etc.

We can express these quantitative statements roughly by saying that if the true difference is equal to one standard error, there is not a high chance that the sample difference will be statistically significant at a useful level, but that if the true difference is equal to three times the standard error, there is a reasonably high chance that statistical significance will be attained.

Suppose now that in the example the number of experimental units for each diet is increased from 10 to 40, the residual standard deviation being unchanged. This halves the standard error of the difference between diets and so the true difference of $3 \times 0.56 = 1.68$ units has the same probability of leading to statistically significant differences as a true difference of 3.36 units had before, and so on.

We can sum up the discussion so far as follows. The standard error of an estimated contrast is a measure of the difference that is likely to arise between the estimate of the contrast and its true value. If we know, before doing the experiment, what the standard error will be, we can predict the resulting width of the interval of uncertainty for the true contrast and can also work out the power of the test of the statistical significance of the estimated contrast. The standard error depends on the form of the contrast, the number of experimental units involved, and the residual standard deviation, which measures the relevant portion of the uncontrolled variation of the observations.

It has been assumed throughout that the amount of uncontrolled variation is constant. If, for example, some treatments lead to more uncontrolled variation than others, the formulas are changed. The possibility that the standard deviation is different in different sections of the experiment has quite often to be allowed for in complicated statistical analyses, but only affects the planning of the experiment, when we know beforehand roughly what variations in standard deviation will occur. The general effect is that we should take relatively more observations where the variability is expected to be high.

The immediate object of most experiments of the type we are considering is the estimation of the magnitude of certain contrasts among the treatments, and often also the examination of the statistical significance of the resulting estimates. We usually require, therefore, that the standard errors of our estimated contrasts should not be too large. Very occasionally, however, the idea that we require to estimate the magnitude of a particular contrast, say the difference between two treatments, is misleading. We may be interested solely in deciding which of the two treatments gives the higher observation, or in picking out from a number of treatments a small set with particular properties. In such cases we want to end with a simple recommendation that, for example, T_1 gives a higher observation than T_2. Although we want assurance that this is in some sense the proper decision to reach, we do not necessarily ask for a measure of the uncertainty of the final decision or for an estimate of the magnitude of the difference between the treatments. The effect of this on the design of the experiment can be seen from the following example.

Consider an experiment to determine whether a proposed new medical treatment effects a higher proportion of cures than a standard treatment. Suppose that on the basis of the results of the experiment we propose to reach one or other of two possible decisions, namely to use in future either always the standard treatment or always the new treatment. Suppose also that observations become available sequentially in time, as suitable patients present themselves.

Now if one treatment is markedly superior to the other, this may become apparent very soon in the experiment. Then, provided that the evidence is statistically convincing, there are compelling reasons for discontinuing the experiment and using the better treatment on all future patients. On the other hand, if the difference between the treatments is slight, many observations will be usually required to reach a decision. In the first case the estimate of the magnitude of the difference between the treatments may, owing to the small number of observations, be very imprecise; this is the price that has to be paid for carrying out the experiment with just the choice between the two particular decisions in mind. In the medical application just described, it would often be very reasonable to regard the problem as purely one of reaching a decision between two (or more generally a small number) of alternative courses of action. However experiments that can be profitably regarded in this way are not so common as might be thought. Although many experiments, particularly in technology, are done primarily to determine some course of action, for example to decide which of a number of industrial processes or experimental methods to use, it does seem to be the case that we nearly always need a reasonably precise estimate of the differences involved. There are various reasons for this, such as the following:

(*a*) Decisions are rarely as simple as in the case outlined above, in that they may depend on several types of observation and also on the relative expensiveness of the alternative treatments or processes. The final decision has often to be made by an act of judgement, weighting these different factors in a rather intuitive way. Estimates of the magnitude of the treatment effect for each type of observation are needed for this process to be at all satisfactory.

(*b*) Even in experiments with an immediate practical aim, it is usually advisable to try to reach some understanding of the system under investigation in addition to the decision of immediate concern. For this, quantitative estimation of the magnitudes of treatment effects is usually desirable.

(*c*) It often happens that the results of an experiment are useful in a somewhat unexpected way, for example in helping to settle a question different from that for which the investigation was first set up. If the results are obtained in a form bearing solely on the immediate point at issue, much of this potential usefulness may be lost.

To sum up this discussion, an estimate of the relevant treatment contrasts is nearly always required in experiments designed to add to fundamental knowledge. In experiments intended to decide between alternative courses of action, it is important to consider in designing the experiment exactly what the possible decisions are and how they are related to the observations to be made. The experiment should, within reason, be designed to give just information relevant to the decision, and considerable economy is sometimes achieved by determining the total number of units in the light of the initial results of the experiment.* Usually, however, it will again be necessary to estimate the magnitudes of the relevant contrasts.

* The statistical technique for doing this is called *sequential sampling* and is described briefly in § 8.5.

8.3 THE ESTIMATION OF PRECISION

In the preceding section we saw that the precision with which a contrast is estimated is measured by the standard error. This depends in a known way on the number of observations and the form of the contrast and on the residual standard deviation, which measures the amount of the relevant part of the uncontrolled variation. Therefore the numerical determination of the standard error in any particular case depends largely on finding or estimating the residual standard deviation, and this we now consider.

We have to consider the estimation of the residual standard deviation both in the analysis of the final results and in the preliminary calculations to determine the appropriate number of experimental units. There are essentially five methods of determining the residual standard deviation:

(i) by the observed dispersion, in the experiment itself, of the observations on different units receiving the same treatment. This can be used only in the final analysis and not in preliminary calculations;

(ii) from the magnitude of high-order interactions in factorial experiments;

(iii) by theoretical considerations;

(iv) from within-unit sampling variation. This will be described later;

(v) from past experience of similar experiments.

These methods will be considered in turn.

(i) Use of Observed Variation between Experimental Units

This is the most frequently used method and has already been described in Chapter 3 in connection with randomized block and Latin square designs. An experimental unit was defined to correspond to the smallest subdivision of the experimental material, such that it is possible for different units to receive different treatments. By considering the variation between different units receiving the same treatment, we have a direct measure of the reproducibility of the observations obtained in independent repetitions of the experiment.

The requirements for this method to give a correct estimate of the residual standard deviation are

(*a*) that the different units should respond independently of one another (see § 2.4); and

(*b*) that any source of uncontrolled variation balanced out in the design of the experiment should also be removed before calculating the standard deviation (see § 3.3).

The reader should re-read §§ 3.3, 3.4 for an account of the method by which the estimate of the residual standard deviation can be calculated.

The estimate of standard deviation can now be used to obtain an *estimated standard error* for any particular contrast. Thus for the difference between two treatment means, both based on the same number of observations, the estimated standard error is

$$\sqrt{\left\{\frac{2}{\begin{matrix}\text{no. of observations}\\ \text{per treatment}\end{matrix}}\right\} \times \begin{pmatrix}\text{estimate of residual}\\ \text{standard deviation}\end{pmatrix}}. \tag{3}$$

This formula is obtained from formula (2) for the true standard error by replacing the standard deviation by our estimate of it.

In the discussion in § 8.2 of the meaning and use of the standard error, it has been assumed that the true standard deviation is known. Thus in Example 8.1, the interpretation of the limits given is correct only if the standard deviation of 2.5 units is not itself subject to random errors. If the standard deviation is only an estimate, it is intuitively clear that the limits for the true contrast, at a given level of probability, must be pushed further apart to allow for the additional uncertainty in the system.

The additional allowance in the uncertainty that has to be made depends on the *degrees of freedom* for the residual which measure, roughly speaking, the number of independent pieces of information available to estimate the standard deviation. There is some further discussion in Example 3.2. The residual degrees of freedom depend to a considerable extent on the total number of units in the experiment and to a lesser extent on the particular design adopted. The most important cases will be put down here for reference:

For a completely randomized experiment in which t treatments are tested on N experimental units, not necessarily with the same number of units for each treatment, the residual degrees of freedom are $(N - t)$.

In a randomized block experiment in which t treatments are tested on N experimental units arranged in k randomized blocks with N/k units in each, the residual degrees of freedom are $N - t - k + 1$. In particular if each block contains each treatment just once, $N = tk$ and the residual degrees of freedom are $(k - 1)(t - 1)$.

In a single $t \times t$ Latin square experiment in which t treatments are compared on t^2 experimental units, the residual degrees of freedom are $(t - 1)(t - 2)$.

In a composite Latin square design with r separate $t \times t$ squares, the residual degrees of freedom are $(t - 1)(rt - r - 1)$.

In a composite Latin square design with r squares each of size $t \times t$ and in which the rows, say, are intermixed, the residual degrees of freedom are $(rt - 2)(t - 1)$.

The derivation of these formulas need not concern us, although the reader who has followed the discussion in Example 3.2 should be able to work them out.

The effect of errors of estimation in the standard deviation is illustrated in Table 8.3.

TABLE 8.3

MULTIPLIERS TO OBTAIN LIMITS OF ERROR AT ASSIGNED LEVELS OF PROBABILITY WHEN THE STANDARD DEVIATION HAS TO BE ESTIMATED

Degrees of Freedom for Residual	Level of Probability 90 per cent	95 per cent	99 per cent
5	2.02	2.57	4.03
10	1.81	2.23	3.17
15	1.75	2.13	2.95
20	1.72	2.09	2.85
25	1.71	2.06	2.80
30	1.70	2.04	2.75
Standard deviation known	1.64	1.96	2.58

The precise interpretation of these limits depends on an assumption about the form of the frequency distribution of the uncontrolled variation.

Extracted, by permission of the Biometrika trust and the authors, from Table 12, *Biometrika Tables for Statisticians* by E. S. Pearson and H. O. Hartley, Cambridge University Press, 1954.

Thus, in Example 8.1, with a known standard error of 1.12, and an estimated difference of 6.10 there is a chance of 19/20 that the true difference lies between $6.10 - 1.96 \times 1.12$ and $6.10 + 1.96 \times 1.12$, i.e., between 3.90 and 8.30. If the standard error had been obtained from an estimated standard deviation with 20 degrees of freedom and had happened again to have the value 1.12, the limits would have been $6.10 - 2.09 \times 1.12$ and $6.10 + 2.09 \times 1.12$, that is, 3.76 and 8.44. Similarly with 10 degrees of freedom the limits are 3.60 and 8.60. Although Table 8.3 could be extended down to a single residual degree of freedom, general experience suggests that standard deviations based on less than about five degrees of freedom should not be used for the estimation of standard errors.

In the analysis of experiments, Table 8.3 is used for calculating statistical significance and finding limits of error. If it is required, during the design stage of the experiment, to estimate the precision to be expected, the following rough rule is useful. The effect on the estimated precision of contrasts of having to estimate the residual standard deviation is approximately to multiply the standard deviation by

$$1 + \frac{1}{\text{(residual degrees of freedom)}}. \tag{4}$$

This rule tends to underestimate the effect when the residual degrees of freedom are small and, as noted above, the degrees of freedom should not, if possible, fall below five.

The increase in error arises from errors in the estimation of the residual standard deviation. If we were solely concerned with obtaining estimated contrasts as close as possible to the true values, and not with estimating the precision of our conclusions, the residual degrees of freedom would be irrelevant. That is, the factor in formula (4) applies to the estimated precision and not to the true precision.

As an example of the use of the rule, suppose that we have 5 treatments and 20 experimental units and wish to choose between a completely randomized experiment and a design in 4 randomized blocks. In the first design the standard deviation is effectively multiplied by $1 + 1/15 = 1.067$, and in the second by $1 + 1/12 = 1.083$; the degrees of freedom 15 and 12 have been obtained from the general formulas given above. The ratio of these factors is 1.015, so that the completely randomized design is the more accurate unless at least a $1\frac{1}{2}$ per cent reduction in the residual standard deviation is attained by blocking. Usually skilful use of the randomized block design would produce an appreciably greater reduction in standard deviation than this. In general it is clear that little information is lost by having to estimate the residual standard deviation provided that the residual degrees of freedom exceed 15 or 20.

In a more complex design, such as a split plot experiment, there are two or more residual standard deviations and these have to be estimated separately, each having its appropriate degrees of freedom.

To sum up, we can estimate the residual standard deviation directly from the results of the experiment, by considering the dispersion of the observations on different units receiving the same treatment. Any portion of the uncontrolled variation whose effect has been eliminated in the design of the experiment must likewise be eliminated before calculating the standard deviation. If we are designing an experiment in which the residual standard deviation is to be estimated in this way, the ultimate precision to be expected is lower than if the standard deviation had been known exactly; an allowance for this can be made. The advantage of this method of determining precision is that it makes the interpretation of the experiment selfcontained, in that the standard deviation is determined under the actual conditions of the experiment, and is not dependent on any assumption that, for example, the standard deviation is the same as in previous similar experiments.

(ii) Use of High-Order Interactions in a Factorial Experiment

This has been discussed in § 6.11. In a complicated factorial experiment we may attain sufficient precision from one replicate, and in this case all experimental units receive different treatments, so that method (i) is inapplicable. We can however estimate the residual standard deviation, if it can be assumed that the true values of certain high-order interactions are negligible. Once this assumption has been made, an estimate can be obtained with a certain number of degrees of freedom, and the discussion in (i) applies. If the assumption about the high-order interactions is false, the true residual standard deviation will be overestimated.

(iii) From Theoretical Considerations

It is sometimes possible to calculate theoretically what the residual standard deviation should be under idealized conditions. Such a calculation is useful

(*a*) to estimate the residual standard deviation in the analysis of small experiments, in which very few degrees of freedom are available for residual;

(*b*) to provide, in the planning of the experiment, an estimate of the precision that is likely to be attained;

(*c*) to use in the interpretation of an estimate of standard deviation obtained by methods (i) or (ii). It is often instructive to compare the observed standard deviation with a theoretical value. If the theoretical value is too small there are important sources of variation present not accounted for in the theoretical analysis.

The calculation of theoretical standard deviations is, of course, a matter of statistical technique and will not be gone into in detail here. The following are the most important cases.

First the observation on each experimental unit may be that out of, say, N individuals, r have a certain property and the remaining $N - r$ do not. For example on each plot of an agricultural field trial, we may examine 100 randomly selected plants and count the number diseased. In this case N is 100 and r is the number with the disease actually counted on the plot. Suppose that the only source of uncontrolled variation present arises from sampling the plot rather than counting all plants, i.e., in general, arises from examining N randomly selected individuals rather than an indefinitely large number. Then it can be shown mathematically that the residual standard deviation of the observed proportion with the property is equal to

$$\sqrt{\left\{\frac{\text{true proportion with property} \times \text{true proportion without property}}{N}\right\}}. \quad (5)$$

For instance, if the true proportion of diseased plants was 0.3 in the above example, the standard deviation would be $\sqrt{(0.3 \times 0.7/100)} = 0.046$.

The second case is when the observation made on each experimental unit is the rate of occurrence of a randomly occurring event. Examples are counts of radioactive particles, or accidents, or breakdowns of a machine, or mutations of genetic material. In all these cases we have events occurring in a haphazard way in time, and the observation on each experimental unit is that a certain number, n, of events occur in a period of observation T. The rate of occurrence is thus n/T. Suppose that the only source of uncontrolled variation arises from having observed each unit only for a time T, rather than for a much longer time,* and that on each unit events occur completely randomly, the occurrence of one event being entirely independent of the occurrence of all other events. Then it can be shown mathematically that the residual standard deviation of the rate of occurrence is equal to

$$\sqrt{\left\{\frac{\text{true rate of occurrence}}{T}\right\}} = \frac{\sqrt{\begin{pmatrix}\text{expected number of}\\ \text{events observed}\end{pmatrix}}}{\text{period of observation, } T}. \quad (6)$$

For example, suppose that each experimental unit is a batch of wool and that the observation made on each unit is the end breakage rate in spinning. If the true end breakage rate is expected to be about 10 per 1000 spindle hours, and the period of observation is 3000 spindle hours, the standard deviation will, under the above assumptions, be $\sqrt{(10/3)} = 1.83$ if the unit of time is taken to be 1000 spindle hours. Similarly with a period of observation of 1000 spindle hours per experimental unit, the residual standard deviation would be $\sqrt{10} = 3.16$. Other applications of this formula are to counting problems in bacteriology and serology.

A third situation is when the observation for each experimental unit is a measure of dispersion. For example the treatments may be different experimental methods and one object of the experiment may be to compare

* That is, we assume that all units receiving the same treatment would give effectively the same rate of occurrence, if observed for a sufficiently long time.

the reproducibilities of the different methods. For one batch of material several observations are made by a particular method, and the dispersion of these observations measured, for example by the standard deviation. We now have for each experimental unit a standard deviation and we treat these as the "observations" for analysis. These "observations" have a residual standard deviation, i.e., we have to consider the standard deviation of a standard deviation. It can be shown that if the frequency distribution of the readings on any one unit approximates to a special mathematical form, called the normal or Gaussian distribution, the residual standard deviation is approximately equal to

$$\frac{\text{true value of the standard deviation}}{\sqrt{(2 \times \text{number of readings per determination of st. dev.})}}. \qquad (7)$$

These are not the only cases where a theoretical calculation of the residual standard deviation can be made; whenever the observation under analysis can be considered as originating from a probability model, a theoretical calculation of standard deviation may be possible. The disadvantage of using the theoretical standard deviation is that the theoretical model assumed, for example the sampling of a completely random series of events with no other sources of variability, may be quite inaccurate as a representation of the uncontrolled variation actually occurring. If it can be obtained, an estimate of the residual standard deviation calculated from the observed variation between units is to be preferred for the direct assessment of the precision of treatment effects. In analyzing data on proportions, counts, and variabilities it is frequently desirable to work with mathematically transformed values. The theoretical standard deviation is different after transformation but can always be found.

(iv) From Within-Unit Sampling Variation

It frequently happens that the observation of main interest for any one unit is the mean of independent readings obtained from randomly selected portions of the unit. Some examples should make this clear.

In an agricultural field trial, it may be required to analyze the total yield of product per plot. If each plot is large, the yield may be estimated by selecting a number of small areas within each plot and weighing the product only from these. From the total yield of the sampled areas, we can estimate the yield of the whole plot.

In many types of industrial experiment, we are interested, among other things, in comparing the mean strengths of articles produced by alternative processes. Each experimental unit consists of a batch of articles, processed at one time by one method, and the mean strength will usually be estimated by testing a relatively small number of articles, randomly selected from the batch. For example in a textile experiment each batch of yarn, i.e., each experimental unit, would probably consist of many miles of yarn, the mean strength being estimated from tests on say 100 1-ft lengths randomly selected from the batch.

More generally, it is a common characteristic of methods of chemical and biological analysis that duplicate or triplicate independent determinations are made on samples of material from the same experimental unit, the final observation for the unit being the average of the separate determinations. In complex methods of chemical analysis there may be several stages of sampling, corresponding to the different stages of the analysis.

In this type of situation we can distinguish several components of uncontrolled variation, as shown by the following example.

Example 8.4. Consider as a typical example a simplified form of an experiment for comparing two methods S_1 and S_2 of spinning wool yarn. There will be several experimental units, each being a batch of raw material from which yarn is to be spun. Suppose that the batches are processed in random order, half by process S_1 and half by process S_2. Finally from each batch a number of lengths are randomly selected and tested for strength. This gives us a collection of observations of the following general type:

Unit 1	Unit 2	Unit 3	. . .
S_1	S_2	S_2	. . .
—	—	—	
—	—	—	
—	—	—	
.	.	.	
.	.	.	
.	.	.	

Quite generally think of a situation in which the primary observation on each unit is the average of several readings.

Now it would in principle be possible to make a large number of observations on each experimental unit. From the variation of the observations on one unit, we could then obtain a measure of the within-unit variability. If we measure variation by the standard deviation we thus obtain the *within-unit standard deviation*. In the present example this is a measure of the variation of strength within a batch of yarn spun in one lot; it takes no account of variation in mean strength from batch to batch. Next if we had the true mean strength for each experimental unit, we could define the *between-unit standard deviation* to measure the uncontrolled variation between units receiving the same treatment in the true mean strength for each unit. This would measure the effect of variations between batches of raw material and of nonconstancy in the conditions of processing. Notice that the between-unit standard deviation is unaffected by variations of strength within a batch, since it refers to the mean of a very large number of observations per batch.

In a practical situation we usually have only a small or moderate number of observations on each unit, and it can be seen that the comparison of the mean strength for the two processes is then subject to errors arising from both sources. The effective residual standard deviation for the comparison of mean strengths can be shown to be

$$\sqrt{\left\{(\text{between-unit st. dev.})^2 + \frac{(\text{within-unit st. dev.})^2}{\text{no. of obs. per unit}}\right\}}.$$

There is a discussion of this formula with numerical applications in Example 8.9; for the present, note that if we can estimate both component standard deviations, we can predict the residual standard deviation corresponding to any number of observations per experimental unit.

The two components of standard deviation can be estimated from the results of an experiment in which there are at least two observations per experimental unit; the analysis of variance for doing this is described in textbooks on statistical methods (Goulden, 1952, p. 67). The estimation of the separate components is, however, only necessary in order either to examine the nature of the uncontrolled variation or to predict what the standard deviation would have been with a different number of observations per unit. If all that is required is to estimate the precision of the process comparisons in the experiment as performed, it is enough to analyze the mean strengths per batch as if they were single observations.

Now consider the type of experiment, which is sometimes done, in which there is only one experimental unit for each treatment. That is, one batch of material is processed by S_1 and one by S_2, and several measurements of strength are made for each batch. Clearly no fully satisfactory estimate of precision can be obtained from such an experiment, because there is no way of estimating the between-unit standard deviation. The most that can be shown is that the two units have different mean strengths; whether or not this difference is due to the processes or is just random between-unit variation cannot be determined from the observations themselves. An essential condition for a selfcontained analysis of the observations and for the correct estimation of precision is that for each treatment there should be several experimental units, run independently. Nevertheless, in cases where it is impracticable to have more than one, or a small number, of experimental units for each treatment, and in which prior knowledge suggests that the between-unit component of variation is relatively unimportant, the estimation of precision from the within-unit standard deviation is permissible, i.e., we in effect assume that the between-unit standard deviation is zero. This is not a good procedure, however, and should be avoided wherever possible, by running enough independent experimental units for each treatment to provide a satisfactory estimate of the residual standard deviation by method (i).

In more complicated cases, with several stages of sampling, there will be several components of standard deviation but the general principles involved remain the same.

The use of within-unit sampling variation to measure the precision of treatment contrasts is therefore in general undesirable. However, in experiments with a very small number of units, so that no effective estimate of the correct residual standard deviation can be made, the within-unit may, if used with caution, be useful in giving the minimum error to which the treatment contrasts are subject. More generally the magnitudes of the within-unit standard deviation and the between-unit standard deviation give information about the importance of the different sources of uncontrolled variation, and also determine, for future experiments, what is a suitable value for the number of readings per unit.

Notice that the use of within-unit variation is analogous to that of certain theoretical values for the standard deviation, in that both are obtained assuming that some sources of variability are negligible.

(v) From the Results of Previous Similar Experiments

The last method of estimating the residual standard deviation is from the statistical analysis of the results of previous similar experiments. Particularly in routine laboratory work, large bodies of previous data may be available for such an analysis from which an estimate based on a large number of degrees of freedom may be obtained.

Such an estimate is particularly useful in the determination of an appropriate size for an experiment being designed. It is also useful in the analysis of the results of an experiment

(*a*) to estimate the residual standard deviation when few degrees of freedom for residual are available in the experiment;

(*b*) to compare with a residual standard deviation obtained from the experiment itself. It is frequently a good check on the experimental work to see how the standard deviation compares with that in previous similar experiments.

If a reasonably accurate estimate of the residual standard deviation can be obtained from the experiment itself, we would normally use this for the calculation of standard errors rather than the estimate from prior work, even though the latter is nominally more accurate. We thereby avoid the assumption that the amount of uncontrolled variation is the same as in previous work, make the interpretation of the experiment more selfcontained, and, other things being equal, the conclusions more cogent. A possible exception to the use of the observed residual standard deviation is when it is appreciably less than the value from prior work, and yet it is fairly certain from knowledge of the system that no real increase in precision can have occurred.

(vi) Summing Up

We have seen that methods (i), (ii), and (iv) of estimating the standard deviation are applicable only in the analysis of the observations, not in the design of the experiment. In order to obtain an estimate of precision prior to the performance of the experiment we must use methods (iii) or (v), theoretical calculation or the analysis of the results of previous similar experiments. Occasionally, as for example when the experiment is the first of its type, neither of these methods can be used; in such a case the size of the experiment must either be settled by general judgement or, alternatively, if a prior calculation of precision is very desirable, the experiment must be done in two or more stages, the observations from the first stage being used to determine the appropriate size of the second stage. This technique is discussed briefly in § 8.5.

8.4 SOME STANDARD FORMULAS

We can now give some formulas for deciding on an appropriate number of observations to take. First determine the residual standard deviation as well as possible, by one or another of the methods of the previous section.

If we have a set of treatments, and the comparisons of all pairs are of equal importance, we devote an equal number of units to each treatment.* Then the standard error of the estimate of the difference between any two treatments is

$$\sqrt{\left\{\frac{2}{\text{no. of units per treatment}}\right\}} \times \begin{array}{c}\text{residual standard}\\ \text{deviation}\end{array}$$

and therefore the number of units per treatment leading to a preassigned standard error is

$$2 \times \left\{\frac{\text{residual standard deviation}}{\text{required standard error}}\right\}^2. \tag{8}$$

If we can now decide what standard error we require, either by considering the width of the limits of error for the true contrast, or by considering the power of the associated significance test, the appropriate number of units per treatment is determined. Similar calculations may be made for contrasts other than simple differences between pairs of treatments.

Example 8.5. In a certain type of agricultural field trial it may be known that the residual standard deviation is about 10 per cent of the mean yield. Suppose that we require to make the limits of uncertainty for a true difference at the 95 per cent level of probability extend 5 per cent on each side of the estimated difference. This implies a standard error of $2\frac{1}{2}$ per cent, and hence from (8) the appropriate number of plots per treatment is $2 \times (10/2\frac{1}{2})^2 = 32$.

With even a moderate number of treatments this represents a large experiment, and it might well be decided that our requirements on precision have to be

* An exception to this would be if it were expected that observations on different treatments would have different amounts of uncontrolled variation. This possibility is noted briefly in § 8.2. It would then be reasonable to take more observations on those treatments for which the variability is expected to be high.

weakened. The next step would then be to work out the extent of the interval of uncertainty at the 95 per cent level for various sizes of experiment. We get

Number of Plots per Treatment	Approximate 95 Per Cent Limits, plus and minus
32	5 per cent
25	5.7 per cent
16	7.1 per cent
9	9.4 per cent

It is now usually a matter of intuitive judgement to decide what to do. The additional expense of an increase in the size of the experiment has to be balanced against the resulting gain in precision in the conclusions. We shall give an example below in which these considerations can be weighed quantitatively, but this is rather unusual.

Alternatively it may seem better to think in terms of the power of the significance test of the difference between two treatments. For example suppose that a true difference between two treatments of 10 per cent is considered of appreciable practical importance. Then it will be desirable that if a true difference of this magnitude exists there should be a good chance that a reasonable degree of statistical significance should be attained by the observed values. Now Table 8.2 shows that if the true difference is three times the standard error, there is 91 per cent chance of attaining statistical significance at the 10 per cent level, an 85 per cent chance of attaining statistical significance at the 5 per cent level, and so on. If the ratio to the standard error is much less than three the chance of attaining significance is appreciably reduced. Therefore it would be reasonable to arrange that the true difference, 10 per cent, is three times the standard error, i.e., to arrange that the standard error is 10/3 per cent. If we substitute this value into formula (8), we get for the number of plots per treatment $2 \times (10 \times 3/10)^2 = 18$. Again, if this calculation makes the total number of plots in the experiment intolerably large, the effect of weakening the requirements can be investigated.

If it seems that an experiment with a small number of experimental units will be adequate, the residual degrees of freedom in the design will be small, and this, as we have seen, increases the effective standard deviation. Usually, however, the allowance for this is relatively small compared with the general uncertainty involved in the whole calculation of the appropriate number of units. An exception is when the size of the experiment as determined from the first calculation would leave five or fewer degrees of freedom for residual; it would then be impracticable to estimate the standard deviation from the observed dispersion of the observations, and it may be desirable to increase the number of units solely in order to get enough degrees of freedom for residual. This is especially the case when methods of estimating the standard deviation other than from the observed dispersion of the observations are unreliable.

It must be stressed, however, that the condition that there should be enough degrees of freedom for residual is not to be used as a general criterion for determining the size of experiments. The main consideration is the standard error of the contrasts, with the degrees of freedom for residual a subsidiary matter.

Similar methods apply when the residual standard deviation is calculated theoretically, or when the contrast of interest is not a simple difference.

Example 8.6. Suppose that in an investigation in nuclear physics it is desired to examine the frequency with which certain conditions which can be set up experimentally lead to specified types of transition. Imagine that the mechanism of the transition is unknown and that to find out something about it, the effect on the frequency of transitions of various modifications to the experimental conditions is to be investigated.

Suppose that initially the transition occurs in about 20 per cent of occasions and that it is thought desirable that if a certain treatment increases this true proportion to 30 per cent, there should be a good chance of attaining statistical significance. Let each experimental unit consist of n trials, the proportion of these leading to transitions being observed. If we calculate on the basis of a 25 per cent transition rate, the residual standard deviation is, from (5), $\sqrt{(0.25 \times 0.75/n)} = \sqrt{(0.1875/n)}$. If r units are tested with each treatment, the standard error of the estimated difference between the treatments will be $\sqrt{(2/r)} \times$ standard deviation, which equals $\sqrt{[0.375/(rn)]}$. Arguments similar to those used for the previous example suggest arranging that the true difference of interest 30–20 per cent, i.e., 0.1, is three times the standard error. This gives the equation

$$3 \times \sqrt{\left\{\frac{0.375}{rn}\right\}} = 0.1, \qquad \text{that is } rn \simeq 340.$$

Thus we need about 340 trials for each treatment. Tables and a nomogram for the appropriate number of units, calculated by a more refined method, are available (Eisenhart et al., 1947, p. 247).

Now the calculation has given just the total number of trials that should be carried out for each treatment; from the point of view of the calculation it makes no difference whether we have for each treatment one unit with 340 trials, two units with 170 each, and so on. This is because formula (5) for the standard deviation is based on the assumption that there are no sources of uncontrolled variation to make two trials on different units receiving the same treatment any less alike than two trials on the same unit. In practice this would be at best a good approximation and it would be preferable to have as many different units as practicable, in order to attain the best possible sampling of other sources of uncontrolled variation that may be present. In the present example a good arrangement would possibly be to have, for each treatment, seven experimental units, each unit consisting of, say, 50 trials made as far as possible under identical conditions.

Example 8.7. Suppose that we are particularly interested in the slope of the response curve for a certain quantitative factor and that the factor is investigated at three equally spaced levels, with equal numbers of units at each level. Table 7.1 shows that the standard error of the slope is

$$1.225 \times \frac{\text{residual standard deviation}}{\sqrt{(\text{total number of experimental units at the three levels})}}.$$

If we can estimate the residual standard deviation and decide on the standard error that we require, we can determine the total number of experimental units as before.

In determining the number of experimental units, we are compromising between, on the one hand, having high precision and an expensive experiment and, on the other, having an economical experiment giving low precision. Usually this compromise has to be reached in a somewhat intuitive way, but if it is possible to measure in the same units, for example of money, the cost of the experiment and the loss caused when the conclusions are inaccurate, it will be possible to calculate explicitly the appropriate number of units. Yates (1952) has provided an interesting discussion of this.

Example 8.8. Yates's example is of the determination of the optimum dressing of nitrogen for sugar beet. The optimum dressing will be such that the cost of a small additional dressing just equals the value of the additional yield produced; the determination of this optimum will be subject to random errors of experimentation and it is possible to express the average "loss" arising from the use of an incorrect dressing in terms of the square of the standard error of the estimated optimum, and of the total area of crop to which the conclusions are to be applied, etc. The standard error will depend in part on the size, and hence on the cost, of the experiment, and if the cost of an experiment of given size is known, the average "loss" from an incorrect recommendation plus the cost of experimenting can be minimized, and so the most economical size of experiment calculated.

To carry through the calculation it is necessary to know approximately the cost of experimenting, the loss per unit of experimental material due to a given departure from optimum conditions, the residual standard deviation, and the quantity of material to which the conclusions are to be applied.

There are several things that may complicate such a calculation. It may be advisable to determine optimum treatments separately for various portions of the experimental material. Again, it often happens that the conditions under which the experimental work is done are not fully representative of conditions under which the results are to be applied, i.e., there may be a bias. Effort spent in removing this bias rather than in increasing the size of the experiment is often worth-while.

The main discussion at the beginning of this section has been of the case where the comparisons of all pairs of treatments are of equal importance, so that we arrange that each treatment occurs the same number of times. It may happen, however, that some comparisons are of more interest than others. There are two main possibilities.

We may have one control treatment and a number, m, of alternative treatments. Sometimes the most interesting thing is to compare the alternative treatments individually with the control, comparisons of the alternative treatments among themselves being of secondary importance.

It can be shown that if all observations have the same precision the best procedure is to arrange that for each unit receiving a particular alternative treatment there are approximately $\sqrt{m}$ units receiving the control treatment. A second case is when the main interest is in the difference between the control and the average of the other treatments. Then we should have m observations on the control for each observation on a particular alternative treatment.

For example, in a nutritional experiment we may compare a control diet deficient in a certain constituent with, say, three other diets, all containing substantial amounts of the constituent, but differing in the form in which it is presented. Since the nearest whole number to $\sqrt{3}$ is 2, the recommended arrangement is to have two observations on the control to one on each of the other three treatments, whenever the main interest is in comparing an individual supplemented diet with the control. If the main interest is in comparing the average of the three supplemented diets with the control, the recommendation would be three observations on the control to one on each of the other three treatments. The experiment could be set out in randomized blocks with five units per block in the first case and six in the second case. The total number of blocks would probably be determined so as to reduce the standard error of the principal comparison to an acceptable level.

A different situation arises when the treatments can be divided into two groups, one relatively more important than the other. Here trial and error combined with the use of formula (1) will usually indicate a suitable arrangement. For example if the total number of available units is severely limited, we may prefer to attain a specified precision for comparisons within the important group, accepting whatever precision can be obtained from the remaining units for the remaining comparisons. Or we may decide to have the standard errors for comparisons within the more important group and within the less important group to be in the ratio of say 1 : 2. This would be achieved by having the corresponding ratio for the number of observations per treatment be 4 : 1. In fact, by formula (1), if there are $4n$ observations on each of one group of treatments and n observations on each of the second group, the following standard errors for estimated differences are obtained:

for two treatments in first group, $\sqrt{[2/(4n)]}$ × standard deviation = $\sqrt{[1/(2n)]}$ × standard deviation;

for two treatments in second group, $\sqrt{(2/n)}$ × standard deviation;

for a treatment in one group compared with a treatment in another group, $\sqrt{[1/n + 1/(4n)]}$ × standard deviation = $\sqrt{[5/(4n)]}$ × standard deviation.

Again the number of units per second group treatment, n, can be determined if the required value of the standard error is known.

The final group of problems for consideration is connected with the within-unit sampling variation of the type illustrated in Example 8.4. Here we have to decide not only on the total number of experimental units but also on the number of repeat observations per experimental unit.

The general principle involved is the obvious one that if the main expense and time are in taking the observations, so that repeat observations on the same unit cost as much as the testing of the same number of new units, the best procedure is to have as many units as are necessary, making either one observation on each unit or two if the variation within units is of intrinsic interest. On the other hand, if, as is commonly the case, the main expense is in the provision and testing of the experimental units, it will be best to use a small number of experimental units, making a relatively large number of observations on each unit. For instance, in Example 8.4, an increase in the number of experimental units would involve the processing of fresh batches of wool and would be expensive, whereas an increase in the number of observations per unit merely involves selecting further lengths for test and carrying out the strength-testing, and the expense of this is slight.

A statistician should be consulted for details on how to proceed in such cases. Two components of standard deviation are involved (see Example 8.4). The within-unit standard deviation measures the variation that would be obtained if a large number of observations were made all on one unit. The between-unit standard deviation measures the variation, in the absence of treatment effects, when the average of a large number of observations on each unit is analyzed. The effective residual standard deviation, for the comparison of the treatment means, is

$$\sqrt{\left\{(\text{between-unit st. dev.})^2 + \frac{(\text{within-unit st. dev.})^2}{\text{no. of repeat obs. per unit}}\right\}}, \qquad (9)$$

and once approximate values for the two components of standard deviation can be found, we can determine as before the standard error that will result from any given number of units and number of observations per unit.

Example 8.9. Suppose that in an experiment such as that of Example 8.4, it is known from previous work that the between-unit and within-unit standard deviations are, respectively, about 1 and 2 units. The standard error of the estimated difference between two treatments is therefore

$$\sqrt{\left\{\frac{2}{\text{no. of units per treatment}}\left(1 + \frac{4}{\text{no. of repeat obs. per unit}}\right)\right\}}.$$

The numerical values of this are shown in Table 8.4.

TABLE 8.4

STANDARD ERROR OF DIFFERENCE BETWEEN TWO TREATMENTS IN A HYPOTHETICAL EXPERIMENT

No. of Repeat Obs. per Unit	1	2	4	8	16	32	Infinity
No. of Units per Treatment							
2	2.24	1.73	1.41	1.22	1.12	1.06	1.00
4	1.58	1.22	1.00	0.87	0.79	0.75	0.71
6	1.29	1.00	0.82	0.71	0.65	0.61	0.58
8	1.12	0.87	0.71	0.61	0.56	0.53	0.50
10	1.00	0.77	0.63	0.55	0.50	0.47	0.45
12	0.91	0.71	0.58	0.50	0.46	0.43	0.41

The following conclusions can be drawn from Table 8.4 and can be paralleled in a more general case.

(*a*) Not much decrease in the standard error is produced by increasing the number of repeat observations per unit beyond about 16 (the corresponding number in the general case is about four times the square of the ratio of the within-unit standard deviation to the between-unit standard deviation).

(*b*) A particular standard error can be produced in various ways. For example a standard error of 0.71 can be obtained with 20 units per treatment and 1 observation per unit (not shown), with 12 units per treatment and 2 observations per unit, with 8 units per treatment and 4 observations per unit, with 6 units per treatment and 8 observations per unit, and also with 4 units per treatment and a very large number of observations per unit. This is the smallest number of units per treatment that will give the required standard error; no allowance has been made for the effective loss of precision consequent on a reduction in the degrees of freedom for residual.

(*c*) If it is possible to assess the relative costs of a unit and of an observation, the most economical combination to produce a given standard error can be found, either mathematically or by direct examination of a table such as Table 8.4.

In more complicated cases with several stages of sampling, there will be several components of standard deviation. The general remarks above are applicable, but details are too complicated to go into here.

8.5 SOME SEQUENTIAL TECHNIQUES

In the methods described in § 8.4, a single calculation is made of the number of observations to be made to attain given precision. It is sometimes useful to fix the number of observations not in one step at the beginning of the experiment, but in several steps, that is, to make the number of observations depend on the actual outcome of the experiment. An experiment set up in this way is called *sequential*.

The general idea of working in stages, deciding what to do at one stage only after examination of all the results obtained up to that point, is of course widely used; here we are concerned with a specialized aspect of it in which the results already obtained determine, not the objective of further work, but simply how many more observations shall be taken. There are four situations in which these methods may be useful:

(*a*) when initially no reliable estimate of the residual standard deviation is available;

(*b*) when the residual standard deviation depends in a known way on the quantity to be estimated;

(*c*) when a clear-cut decision is required between a small number of courses of action;

(*d*) when an estimate is required with a precision depending on the value of the contrast under estimation.

Some experiments lend themselves naturally to a sequential approach. For example if experimental units have to be dealt with singly, so that observations become available at intervals in time, a sequential determination of the number of observations is often perfectly feasible. In other situations, notably in agricultural field trials, the experimental work has to be planned and started at one time and the results become available together much later. A sequential method is not then practicable for an individual experiment, although it may well be applicable to a series of similar experiments, for example in the repetition for several years of an important variety trial. The general point that the following discussion applies mainly to the first type of experiment should be borne in mind throughout.

In any experiment in which the total number of observations is influenced by the values of the observations care is needed in applying the conventional statistical methods of analysis (Anscombe, 1954), since the practical interpretation of statistical significance limits, etc., requires that the total number of observations is chosen without regard to the outcome of the experiment. If, for example, units are tested in small groups and statistical significance is calculated after each step, the experiment being stopped as soon as, say, the 1 per cent level of statistical significance is reached, the true statistical significance of the conclusions is usually considerably exaggerated. This consideration means that ideally the appropriate method of statistical analysis has to be worked out theoretically for each method of determining the number of observations; in practice the point is usually of importance for the decision problem, (*c*), and sometimes for the fourth type of problem, but not in the other cases.

Consider first the type of problem that would be dealt with by a simple calculation like that of Example 8.5 were a sufficiently reliable estimate of the residual standard deviation available. If there is no such estimate, a common-sense procedure is to do a preliminary experiment using as many units as possible, subject to the proviso that the final required precision is unlikely to be attained. From the observations, the residual standard deviation and the standard error of the interesting contrasts are estimated in the usual way. If the standard error is already as small or smaller than the required value, the experiment is complete; if the standard error is too large, the residual standard deviation is used, as in Example 8.5, to calculate the total number of observations required, and then the appropriate number of additional units is taken. Such a two-stage procedure is called *double sampling;* it is the simplest form of sequential technique.

Example 8.10. In an experiment to compare three methods of measuring the percentage of red cells in blood, it was required to estimate the true mean difference between any two methods with a standard error of about $\frac{1}{2}$ per cent. For each subject the percentage is measured, in random order, by all methods, i.e., we use a randomized block design. Suppose that from the results of a preliminary test of fifteen subjects, the residual standard deviation is estimated to be 1.85 per cent.* The standard error of the difference between two treatments is $1.85\sqrt{(2/15)} = 0.68$. This is somewhat greater than the standard error originally required, and to calculate how many more subjects should be tested to attain the required precision we argue as follows. If the residual standard deviation is approximately the same for future subjects, the standard error for any particular total number of subjects is about

$$1.85\sqrt{\left(\frac{2}{\text{no. of subjects}}\right)}$$

and if this is to equal $\frac{1}{2}$, we have the equation

$$1.85\sqrt{\left(\frac{2}{\text{no. of subjects}}\right)} = \frac{1}{2},$$

or

$$1.85^2\left(\frac{2}{\text{no. of subjects}}\right) = \frac{1}{4},$$

i.e., number of subjects $= 8 \times 1.85^2$, which is approximately 27. This is the total number of subjects; we have 15 already and so need to test about another 12.

The observations on the 27 subjects are analyzed as a whole by the ordinary methods of analysis for a randomized block design and, provided that the second set of observations are not markedly different from the first, approximately the required precision will result.†

* The methods did not depend on the direct counting of cells, so that the theoretical formula (5) of § 8.3(iii) was not applicable.

† There is an approximation involved in applying ordinary methods of analysis to the results of a sequential experiment, but this is unlikely to be important. The common-sense procedure here is a modification of one due to Stein (1945), who, at the cost of some loss of information, arranged that the precision should exactly equal the required value.

Double sampling is a simple procedure in that only one intermediate stage of calculation is involved, and the experiment falls into just two parts. In some situations, particularly when observations become available singly at say weekly intervals, it may be worth elaborating the scheme somewhat. This can be done by calculating, at frequent intervals, the estimated standard error of the important contrasts. So long as this is greater than the required standard error the experiment is continued, but as soon as it becomes less than or equal to the required value, the experiment is stopped and a full analysis of the results is made by conventional statistical methods. There is an approximation involved in using conventional methods of analysis in a sequential problem, but the effect here is small. The disadvantage of the fully sequential procedure, as compared with double sampling, is that it involves more intermediate calculation; the main advantage is that the rather arbitrary choice of an initial sample size is avoided.

The second type of problem referred to above arises mostly in connection with the special theoretical formulas (5) and (6) of § 8.3(iii).

Example 8.11. In some types of work the observation to be made on each unit is a proportion, for example the proportion of a particular subject's blood cells showing a certain abnormality. The question arises of determining not the number of subjects (units), but the number of observations per subject. A reasonable requirement is often to obtain a certain *fractional* precision in the estimate for each subject. If this were say 20 per cent, we require that if the true proportion abnormal is 5 per cent, the standard error of our estimate should be 1 per cent, whereas if the true proportion abnormal is 1 per cent, the standard error should be 0.2 per cent, and so on.

Now from formula (5) it may be shown that to attain this standard error the number of cells to be counted depends markedly on the true value of the proportion to be estimated. Thus if the true proportion is 5 per cent, the standard error is, from (5), $\sqrt{(0.05 \times 0.95/\text{number of cells per subject})}$. This is equal to the required value, 0.01, when the number of cells counted per subject is 475. Similarly, if the true proportion is 1 per cent, the number to be counted should be slightly less than 2500.

Now if we have to begin with a fairly good idea of the value of the proportion to be estimated, this calculation will determine the number of cells that should be counted. But if all we know is that the proportion lies somewhere between, say, 5 and 1 per cent, the calculation is not helpful, because the appropriate number of cells is so critically dependent on the unknown proportion to be estimated.

Therefore some sequential method seems called for. Double sampling is one possibility. Another method was proposed by Haldane (1945) and is called inverse sampling. The idea here is that instead of fixing the total number of cells to be examined and recording the number of abnormals, counting should continue until a certain predetermined number of abnormals have been obtained. This number is the reciprocal of the square of the fractional error required, and in the above case is $1/(0.2)^2 = 25$. That is, counting should continue until 25 abnormals have been obtained. The resulting proportion of abnormals is an estimate with approximately the required precision.

This case has been described as a simple illustration of a technique for adjusting the number of observations to produce an estimate with the desired precision. The same type of method can be used in more complicated cases; a statistician should be consulted for details.

There is a review of the statistical literature on methods of this type by Anscombe (1953) and a discussion of double sampling methods for these problems by Cox (1952).

The third type of problem, in which a decision is required between two, or sometimes more, courses of action, raises fresh points. There are two approaches; first we may attempt a direct balancing of the monetary loss due to reaching the wrong decision and the cost of experimenting. This is analogous to the approach of Example 8.8. A double-sampling scheme for choosing between two alternative decisions, based on these considerations, has been given by Grundy et al. (1954). The general idea is that a choice has to be made between say two alternative processes; if an initial experiment indicates a substantial superiority for one process, the appropriate decision is reached. If not, further units are tested, the number of new units being chosen to minimize the sum of the average loss arising from reaching the wrong decision and the cost of testing the further units. For these results to be applied a reasonably accurate economic analysis of the problem has to be possible and this is frequently not so. In that case different and more intuitive methods have to be used; the following is an example.

Example 8.12. Kilpatrick and Oldham (1954) have described an experiment to decide between two methods of relieving bronchial spasm in patients with a chronic pulmonary disease. The treatments were the inhalation of either adrenaline, the customary treatment, or calcium chloride, which had been suggested as a possible substitute with certain advantages. The assessment of the drugs was made in terms of an objective measure, the expiratory flow rate (e.f.r.).

The experimental procedure was of the paired comparison type, as follows. In the morning the e.f.r. of a subject was determined and then a 15-minute inhalation of one or other substance was given. Neither patient nor observer knew which, the decision having been reached by randomization. The e.f.r. was again determined on completion of inhalation, and on the evening of the same day the procedure was repeated using the other substance. The difference between the gain of e.f.r. resulting from calcium chloride inhalation and that resulting from adrenaline inhalation was worked out as each subject's results were obtained.

Suitable subjects were expected to become available for the experiment only infrequently, so that it was of some importance to reach the appropriate decision in the most economical way. The problem is of the same general type as that discussed immediately above, but any economic analysis of the consequences of a wrong decision is out of the question. Instead the following considerations were formulated, after careful thought:

(*a*) If calcium chloride caused subjects to gain, in the long-run average, 10 liters per min of e.f.r. more than they gained on adrenaline, it was desired to reach the decision to prefer calcium chloride:

(*b*) If the gain of e.f.r. with calcium chloride was no greater than with adrenaline, in the long-run average, it was desired to reach the decision to prefer adrenaline, in view of its well-established virtues.

(*c*) The chance that, in the situations described in (*a*) and (*b*), we reach the wrong decision as a result of chance fluctuations in the observations is to be only 1 per cent.

For any scheme that is set up, there will be an operating characteristic of the

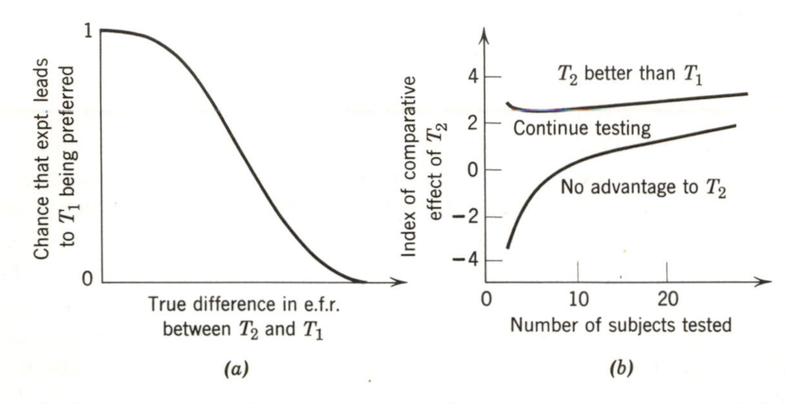

Fig. 8.1. A sequential scheme for comparing two treatments. (a) Typical operating characteristic curve; (b) Boundaries defining the test.

general shape shown in Fig. 8.1(*a*); what we are doing in (*a*)–(*c*) is to find two points on this curve, one towards each end. It is implicit that, for example, if the true difference in gains exceeds 10 per cent in favor of calcium chloride, the probability of deciding to prefer adrenaline is even less than 1 per cent.

When the requirements (*a*)–(*c*) have been formulated, it is a statistical problem to determine a rule for conducting the experiment, so that the requirements (*a*)–(*c*) are satisfied. The rule will say, after each subject's results become available, either

(i) that the experiment should be stopped and adrenaline preferred; or
(ii) that the experiment should be stopped and calcium chloride preferred; or
(iii) that a further subject should be tested.

That is, the experiment is allowed to continue until sufficient results are available to justify a decision.

The most convenient form for representing the sampling rule is on a diagram, such as Fig. 8.1(*b*). As explained above, for each subject the difference is calculated between the gains in e.f.r. on calcium chloride and on adrenaline. After each subject is tested, the cumulative total of the differences is worked out and divided by the square root of the cumulative total of the squares of the differences. Thus if the first three differences are 2.6, 7.3, and −1.4, the quantity calculated is $(2.6 + 7.3 - 1.4)/\sqrt{(2.6^2 + 7.3^2 + 1.4^2)} = 1.080$. This is an index of the comparative effect of calcium chloride. Its use is mathematically equivalent to that of the mean difference divided by its standard error. The index is large and positive if calcium chloride is much the better treatment, large and negative if adrenaline is much the better treatment, and zero if the average of the observed differences is zero. This index is plotted step by step on Fig. 8.1(*b*). On this diagram are two boundaries; as soon as one or other of them is crossed, the experiment stops and the corresponding decision is reached. So long as the plotted point remains between the boundaries, the experiment is continued. If calcium chloride is much the superior treatment, it is very probable that the index will rapidly cross the upper boundary, leading to a speedy decision; similarly if adrenaline is much superior, the lower boundary is likely to be crossed after the testing of only a small number of subjects. If, however, the situation is less clear-cut, it is likely that the index will remain between the boundaries until an appreciable number of subjects has been tested. That is, the boundaries arrange that the experiment is speedily ended in the clear-cut cases, and is continued in the doubtful cases. The formulas for determining the boundaries are given by Rushton (1950).

The use of a sequential scheme of this type, which suits the number of observations to the requirements of choosing between two alternatives with preassigned chances of error, leads on the average to substantial economies in the number of subjects needed to reach a decision. This is particularly so when there is a big difference between the treatments.

In the experiment described by Kilpatrick and Oldham, the observations were all substantially negative, suggesting adrenaline to be the superior substance, and after only four subjects had been tested, the lower boundary was crossed, see Fig. 8.1(*b*), and the experiment was ended with the decision to prefer adrenaline.

Now in a strict statistical sense, this is the only conclusion that can be given formal justification, i.e., that in accordance with the criteria (*a*), (*b*), and (*c*) the appropriate decision is to prefer adrenaline. Yet the fact that all results were negative suggests not only that adrenaline is to be preferred, but also that it is actually superior; in accordance with (*b*), adrenaline would have been preferred even if there had been no difference between the substances. It is natural to try to estimate, with limits of error, the amount of the difference between the substances, but this cannot be done, at any rate until there has been further research into the statistical problems involved. Nor can we measure the statistical significance of the observed difference, beyond saying that there is a significant departure from (*a*) at the 1 per cent level. Thus, the economy of the sequential scheme has been achieved at the cost of restricting the conclusions to the choice between two decisions. Kilpatrick and Oldham point out that the experiment might better have been designed to choose between three decisions: adrenaline superior, no difference, and calcium chloride superior. However, even then the statistical conclusions are severely limited, and no estimation of limits for the amount of the true difference is at present possible.*

* This raises interesting issues of general statistical theory. It is not clear whether an essential defect in conventional theory is involved, or an essential property of the design, or merely a mathematical difficulty in working out details of appropriate techniques.

This example has been discussed in some detail because it illustrates the advantages and disadvantages of these sequential decision procedures. The formulas for determining the boundaries in the sampling diagram have been worked out for many standard statistical situations (Wald, 1947); this was done with the application to industrial inspection problems in mind, where a clear-cut decision between accepting and rejecting a batch has to be made. There have not been many applications of the methods in research work. An excellent account of the methods with a view to their application in medicine has been given by Armitage (1954), who has also given (Armitage, 1957) some very interesting new types of

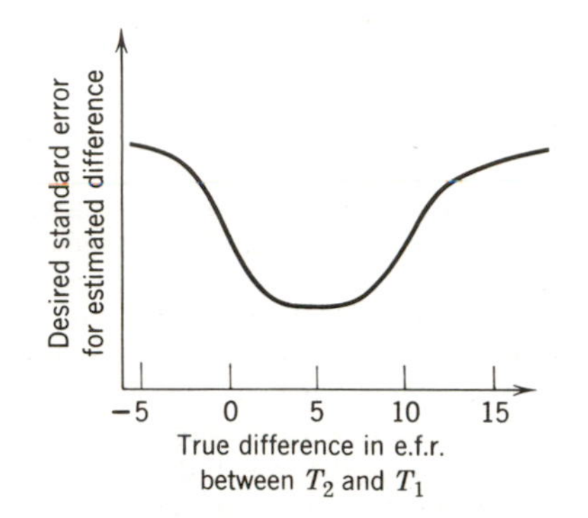

Fig. 8.2. Desired standard error in a situation where the precision required depends on the true value being estimated.

sequential scheme that may be more suited to some sorts of experimental work than Wald's methods.

To sum up, sequential decision procedures merit serious consideration whenever a choice between two or three (or any small number) of courses of action is required, when experimental units are tested in order and it is practicable to do a certain amount, usually small, of calculation after the testing of each unit or group of units, and when requirements analogous to (*a*), (*b*), and (*c*) of Example 8.12 can be formulated. The disadvantage is that, at present, the statements at the end of the experiment that can be given statistical justification are rather limited.

This last point means that the sequential decision procedures just described are inappropriate whenever it is part of the object of the experiment to estimate, with limits of error, the magnitude of contrasts. A natural method to use in cases where such an estimate is required, even though the main objective is the decision, is to try to set up a scheme for estimating, say, the difference between treatments with a standard error depending on the true value of the difference. This is the fourth type of sequential scheme. For example, in the situation of Example 8.12 it might have been reasonable to require an estimate of the true difference in e.f.r. between substances with a standard error of the general form in Fig. 8.2. If the true difference lies in the "doubtful" range (0, 10), we require a low standard error, so that a suitably precise significance test can be made for reaching the appropriate decision; if the difference lies outside this range, we still require an estimate of the difference, but are content with a much higher standard error. Double sampling schemes for achieving these objectives have been discussed theoretically by Cox (1952), but no examples of their use in practice are known.

SUMMARY

It is very often desirable either to make a preliminary calculation of the precision to be expected from an experiment of the size contemplated, or, preferably, to determine the size so as to just attain a desired precision. Precision can, for most purposes, be measured by the *standard error* of the contrasts of interest, for example by the standard error of the estimated difference between two treatments.

The value of the standard error depends on the contrast to be estimated, on the number of units, and on the amount of uncontrolled variation, which is itself measured by the residual *standard deviation*. This can be estimated

(*a*) from the observed dispersion of the observations between units receiving the same treatment, eliminating any part of the variation that is balanced out by the design of the experiment;

(*b*) from the magnitudes of high-order interactions in a factorial system;

(*c*) from theoretical considerations, as for example when the observation is the count of a number of occurrences of a randomly occurring event;

(*d*) from the magnitude of the within-unit sampling variation, when the main observation on each unit is the mean of several readings;

(*e*) from the results of previous similar experiments.

Method (*a*) is usually the best for the analysis of data, although the comparison of measures of residual variation from several sources is frequently instructive. Methods (*c*) and (*e*) are the ones applicable for the preliminary estimation of precision.

Once an approximate value for the standard deviation has been obtained, the standard error of any particular contrast can be worked out corresponding to any given number of observations, or, alternatively, the number of observations to achieve a given standard error can be predicted. If some comparisons are more important than others, the number of units should not be the same for each treatment. In more complicated cases both the number of experimental units and the number of observations per unit have to be determined.

Sequential methods, of which the simplest is double sampling, are sometimes useful, particularly when the experiment can conveniently be done in stages with some intermediate calculation between stages. The idea here is that the number of units should be determined in the light of the observations actually obtained and not settled definitely in advance. This technique is especially worth considering when (*a*) the standard deviation is initially completely unknown, or is appreciably dependent on the quantities being estimated or when, (*b*) a clear-cut decision is required between two or more courses of action or when, (*c*) an estimate of a contrast is required with a precision depending markedly on the value of the contrast.

REFERENCES

Anscombe, F. J. (1953). Sequential estimation. *J. R. Statist. Soc.* B, **15,** 1.

——— (1954). Fixed-sample-size analysis of sequential observations. *Biometrics*, **10,** 89.

Armitage, P. (1954). Sequential tests in prophylactic and therapeutic trials. *Q. J. of Medicine*, **23,** 255.

——— (1957). Restricted sequential procedures. *Biometrika*, **44,** 9.

Cochran, W. G., and G. M. Cox. (1957). *Experimental designs*. 2nd ed. New York: Wiley.

Cox, D. R. (1952). Estimation by double sampling. *Biometrika*, **39,** 217.

Eisenhart, C., M. W. Hastay, and W. A. Wallis. (1947). *Selected techniques of statistical analysis*. New York: McGraw-Hill.

Goulden, C. H. (1952). *Methods of statistical analysis*. 2nd ed. New York: Wiley.

Grundy, P. M., D. H. Rees, and M. J. R. Healy. (1954). Decision between two alternatives—how many experiments? *Biometrics*, **10,** 317.

Haldane, J. B. S. (1945). On a method of estimating frequencies. *Biometrika*, **33,** 222.

Kilpatrick, G. S., and P. D. Oldham. (1954). Calcium chloride and adrenaline as bronchial dilators compared by sequential analysis. *Brit. Med. J.*, part ii, 1388.

Rushton, S. (1950). On a sequential *t*-test. *Biometrika*, **37,** 326.

Stein, C. (1945). A two-sample test for a linear hypothesis whose power is independent of the variance. *Ann. Math. Statist.*, **16,** 243.

Wald, A. (1947). *Sequential analysis*. New York: Wiley.

Yates, F. (1952). Principles governing the amount of experimentation in developmental work. *Nature*, **170,** 138.

CHAPTER 9

Choice of Units, Treatments, and Observations

9.1 INTRODUCTION

The central techniques of the subject have been described in the preceding chapters. Ideally, randomization is used to achieve the absence of systematic error; the devices of blocking and of adjusting for concomitant observations lead to increased precision; and the idea of factorial experiments allows both the effective precision to be increased and also the range of validity of the conclusions to be extended. Finally, the methods outlined in the preceding chapter enable a suitable size of experiment to be set up. In this way the criterion for a satisfactorily designed experiment, set out in § 1.2, are in principle satisfied.

All this, however, assumes that three matters have been settled: that we have decided on the treatments to be compared, the types of observation to be made, and the nature of the experimental units to be used. These issues are clearly of central importance and it might fairly be claimed that they are the essence of experimental design. They are, however, generally regarded as being technical questions specific to the subject matter of the experiment and so are not considered in statistical studies of experimental design. In the following, a few general points will nevertheless be made.

9.2 CHOICE OF EXPERIMENTAL UNITS

An experimental unit was defined in § 1.1 to correspond to the smallest subdivision of the experimental material such that different units may receive different treatments. We consider to be included in the definition of the unit all those aspects of the experimental set-up not involved in the treatments, i.e., those that are independent of the particular assignment of treatments adopted. Thus in an industrial investigation an experimental unit might be defined as a particular batch of raw material processed at a certain time, tested on a particular apparatus by a certain observer, etc., a treatment being a particular method of processing.

In setting up experimental units, the following are among the questions to be settled:

(i) an appropriate size for the experimental units;

(ii) whether it is important that the conditions investigated should be representative of "practical" conditions, and whether a wide range of validity for the conclusions is desirable. In particular, it may be desirable to introduce more variation than is present among the experimental units initially available, or to introduce special treatments to represent the effect of variations in external conditions;

(iii) whether one physical object can profitably be used as a unit several times and whether this, or any other aspect of the experiment, introduces important lack of independence in the responses of different units.

(i) Size of Units

In some experiments a suitable choice for the amount of material to be included in each unit is important. An example is the selection of plot size (and shape) in an agricultural field trial.

Technical considerations enter into such choices to a considerable extent. Sometimes, however, the following situation arises. A certain total amount of material is available and, within certain limits, can be divided into any number of experimental units. The number of repeat observations to be made on each unit is also at our disposal. What is the optimum procedure? Alternatively, how much experimental material is needed to attain a specified precision in the most economical way? This problem is closely related to that of § 8.3, where the idea of components of dispersion between and within units was introduced.

Suppose, to take a specific case, material is available for 100 hours' production, with four treatments for comparison, the minimum run on one process being say 5 hours. One possibility is to have just four units, i.e., to run on one process for the first 25 hours, and so on. Numerous observations of each type are taken within each unit; for example if the production flow is continuous, single observations might be taken at hourly intervals, to give finally 25 observations on each process. This is a bad design, since a proper estimate of error, based on the comparison of different whole units receiving the same treatment, is not available. To evaluate the precision of comparisons between treatments, quite specific assumptions have to be introduced about the form of the haphazard variation. Not only are such assumptions best avoided, but also the resulting precision is likely to be rather low. Hence we should use this arrangement only if there are strong practical reasons for keeping the number of treatment changes to a minimum.

At the other extreme we could have 20 experimental units, each corresponding to 5 hours' production, and these might, for instance, be arranged in 5 randomized blocks, grouping into blocks on the basis of order in time and still taking hourly observations. Similarly, there are various intermediate possibilities. According to the formula of § 8.4, the standard error of the estimated difference between two treatments is

$$\sqrt{\left[\frac{2}{\begin{pmatrix}\text{total no. of}\\ \text{obs. per treatment}\end{pmatrix}}\left\{\begin{pmatrix}\text{st. dev.}\\ \text{within units}\end{pmatrix}^2 + \begin{pmatrix}\text{no. of obs.}\\ \text{per unit}\end{pmatrix}\times\begin{pmatrix}\text{st. dev.}\\ \text{between units}\end{pmatrix}^2\right\}\right]},$$

the total number of observations per treatment being the product of the number of experimental units per treatment and the number of repeat observations per unit.

Note first that if the standard deviation between units is zero, the precision depends only on the total number of observations per treatment and not on the number of units per treatment. Second, the precision corresponding to any desired distribution of effort can always be worked out, provided that we can specify how the two standard deviations in the formula depend on unit size. As a first approximation, the standard deviation within units may be treated as constant and the standard deviation between units as varying according to the law

$$\text{standard deviation} = A \times (\text{size of unit})^{-B},$$

where B is a constant usually between 0 and $\frac{1}{2}$; for the application of this to agricultural experiments, see Fairfield Smith (1938). Both A and B and the standard deviation between units are characteristics of the experimental material to be estimated by the analysis of suitable data from previous experiments. Once they have been determined, the precision of any set-up can be determined. For a given total number of observations and a given total quantity of material, maximum precision will be attained with minimum block size, but this is not usually the relevant comparison. A more realistic analysis is to assume that the total cost of the experiment is approximately

$$\begin{aligned} & C_1 \times \text{total amount of material used} \\ + \; & C_2 \times \text{total number of experimental units used} \\ + \; & C_3 \times \text{total number of observations made,} \end{aligned}$$

where C_1, C_2, and C_3 are constants expressing in the same units the costs of a unit of material, of an experimental unit (e.g., of changing from one process to another), and of an observation.

If all the quantities mentioned above can be determined approximately, the information is thus available from which to decide on the most efficient arrangement of resources. It is clear that to do this with any precision calls for quite detailed knowledge of the system.

(ii) Representative Nature of Units

In many technological investigations, particularly in the final stages immediately preceding practical application, it is important that the conditions investigated should be as representative as possible of the conditions under which the results are to be applied. This can have an important influence on the design of the experiment, especially if the practical conditions are very variable or if the treatment effects are likely to be rather sensitive to changes in the external conditions. In any case proper control treatments should of course be included to check that an apparent treatment effect is not solely a consequence of, for example, increased attention paid to the units during the experiment. The point at issue now is that a treatment effect which may be perfectly genuine under experimental conditions may be quite changed under working conditions.

There are various steps which may be taken. Thus, apart from direct efforts to reproduce practical conditions as closely as possible, a factorial experiment may be set up in which one or more factor levels correspond to artificially severe forms of complications likely to arise in practice, that is, we may insert a separate treatment "conditions" and examine the interaction of this with the treatment effects of direct concern.

If the main variation connected with the experimental units lies in the experimental material itself rather than in the external conditions, it will be important to consider where the units used come from. For instance, if in an animal feeding experiment it is required to apply the conclusions to pigs of a certain breed, then ideally the pigs in the experiment should be a sample of pigs of this breed chosen by a sound statistical sampling procedure. Again, in an industrial experiment, if it is suspected that the treatment effects depend somewhat on the particular consignment of raw material used, then the material used in the experiment should be chosen appropriately from the whole set or population of consignments to which the conclusions are to apply. This is a counsel of perfection which is probably practicable only very rarely. It is, however, desirable in such cases to check so far as possible that the units used do not differ in an obvious respect from the population to which it is required to extend the conclusions. Further, it is almost always advantageous to arrange the experiment so that any variation in the treatment effects from unit to unit can be detected and its nature analyzed, for if the treatment effects may be shown to be effectively constant over the experiment, confidence in extrapolation of the conclusions is much greater.

In scientific work, on the other hand, the representative nature of the experimental units is often not of great interest. The choice of experimental material so that the treatment effects can be observed in a simple and illuminating form will be of considerable importance; however, the emphasis is, to begin with, usually on the deliberate choice of the most suitable material. Even here it may be desirable to include a range of experimental units, both in order to get conclusions with a broader basis and possibly also to provide a link with earlier work.

In both sorts of experiment it may, therefore, be desirable to include deliberately additional variations between experimental units with the related objects of examining nonconstancy of the treatment effects and of extending the range of validity of the conclusions. These additional variations, such as those between sexes or between units of two very different types, are best introduced as a classification factor in a factorial experiment (Chapter 6). With four main treatments, T_1, T_2, T_3, and T_4 and two levels of the classification factor, M and F, the two main types of design are recalled in Table 9.1. In the first type

TABLE 9.1

TWO DESIGNS FOR THE INCLUSION OF A SECOND FACTOR

(*a*) *Simple Factorial Arrangement in Randomized Blocks*

Block 1:	MT_4;	MT_1;	FT_2;	MT_3;	FT_1;	MT_2;	FT_4;	FT_3
Block 2:	FT_1;	FT_3;	MT_4;	FT_2;	MT_1;	MT_3;	FT_4;	MT_2

(*b*) *Split Unit Arrangement*

Whole Unit 1:	F;	T_2	T_1	T_3	T_4
Whole Unit 2:	M;	T_4	T_3	T_1	T_2
Whole Unit 3:	M;	T_1	T_3	T_4	T_2
Whole Unit 4:	F;	T_2	T_4	T_3	T_1

a simple factorial arrangement is used, with units of the two types mixed together in each block. In the second, we have in effect a randomized block design for the treatments $T_1, \ldots, T_4$, with each block consisting of units of one type, blocks of different types being randomly intermixed. (In the language of § 7.4, this is a split unit experiment with the treatment of direct interest considered as the subunit treatment.)

The second arrangement is to be preferred if the units of the two types are most conveniently dealt with separately, or if the smaller block size of the second design is likely to lead to an increase of precision in the estimation of the important effects.

(iii) Independence of Different Units

It is very desirable that the different experimental units should respond independently of one another, in the senses that there should be no way in which the treatment applied to one unit can affect the observation obtained on another unit, and that the occurrence of, say, an unusually high or low observation on one unit should have no effect on what is likely to occur on another unit. The first requirement is necessary to allow the effects of the different treatments to be sorted out from one another; the second ensures that a proper estimate of error is obtained from the comparison of observations on units receiving the same treatment.

The precautions to be taken depend on the nature of the experiment, but they usually consist in physical isolation of the different units and, in particular, of the units receiving the same treatment. This needs to be done at all stages of the experiment at which important variations may be introduced. Thus if appreciable variation is likely to occur in obtaining the observations, e.g., in testing an industrial product, it will be desirable to deal with different units in an order involving some randomization. This will ensure that systematic errors arising in the testing procedure do not bias the comparisons and will also tend to minimize any subjective tendency to make observations on the same treatment more alike.

The main situation in which observations on one unit may be affected by the treatment applied to a different unit is when the same physical object (subject, animal, etc.) is used as a unit several times. A considerable increase in precision may often be obtained by doing this, because of the elimination of the effect of differences between subjects, but even if special precautions are taken, it may not be possible to avoid a carry-over of treatment effects from one unit to another. Designs that allow for carry-over effects are discussed in Chapter 13. A further difficulty with this sort of design is that it may involve comparing the treatments under conditions rather different from those that apply in practice.

9.3 CHOICE OF TREATMENTS

So far in this book, we have been discussing how to plan experiments so that reliable and precise comparisons of treatments can be made when uncontrolled variation is present. The selection of treatments to be compared falls almost entirely outside the discussion as being partly a technical question specific to the field under study and partly a question of general scientific procedure.

Thus the investigation of a rather complicated phenomenon might take the form first of a general survey of the effect on the system of a variety of changes; a factorial experiment may be very appropriate here. Then one or more ideas about how the system, or part of it, "really works" are tested as rigorously as possible by suitable modification of the system (treatments). Nearly always a series of experiments is necessary, the initial ideas being modified at each stage, wherever necessary. In so far as we have to disentangle the treatment effects from irrelevant variations, the methods we have been discussing are, of course, applicable; treatments are chosen so as to give as direct an indication as possible of the underlying mechanism. In suitable cases many different types of modified systems may be used, possibly involving quite distinct experimental techniques, and possibly using systems quite different from the initial one.

Even in experiments with a direct practical aim, it may be possible to include special treatments intended to give fundamental knowledge about the process. Thus in an industrial experiment to compare new and standard processes, various modified forms of the new process might be used, at any rate in a small-scale trial, even though there might be no intention of putting these additional processes to direct practical use.

In many experiments the inclusion of a control treatment is essential; the correct specification of the control can be very important too. It should consist in applying to the experimental units a procedure identical to that received by the "treated" units in all respects except that which it is desired to test. To take a simple example, in assessing a new drug we would not usually wish to consider as part of the treatment effect, the improvement that normally results from receiving treatment with pharmacologically inactive substances. Therefore, the control treatment should be a placebo, indistinguishable to the subject from the new treatment* (see Chapter 5); it may be profitable to include an untreated control group as well. Again, if it is required to assess the effect of the excision of a portion of an experimental animal, the control group of animals should be subjected to as much as possible of the procedure applied to the experimental group; here again a group of untreated animals can be included too, but they should not be regarded as the main control.

Example 2.5, concerned with the comparison of drugs for the relief of

* Rutstein (1957) has described consequences, in some experiments of this type, of failing to include proper control groups.

headaches, illustrates the importance of controls in a rather different way. There the control treatment was in effect used to divide the subjects into two types and hence to show a nonuniformity in response. Had this experiment been done without a control group, it is extremely likely that misleading conclusions would have been obtained. Example 1.8 is another illustration of the importance of controls.

Where the treatments differ qualitatively, it is usually desirable that the treatments differ in single specifically identified ways. Otherwise the implications of any treatment differences found will not be clear. Thus, suppose that we compare a standard industrial process A with a process B, modified in several respects. The resulting comparison may be of immediate practical interest, but is unlikely to add much to understanding of the process, since the cause of any change that is observed cannot be identified. In scientific experiments it is very desirable that any treatment differences found should have as far as possible unique interpretations. It often happens that an experiment establishes clear-cut treatment effects, but that on consideration it is seen that these have two or more quite different interpretations. Further treatments ought to have been included to discriminate between the different explanations. The initial choice of treatments so that ambiguities of this sort are avoided is one of the most important and difficult steps in experimental design. It will often be an advantage to set up a factorial structure for the treatments under investigation.

The discussion and examples of § 7.2 concern the choice of factors for inclusion in factorial experiments, and this section should be re-read at this point. The selection of factor levels when the factors are quantitative has been discussed in § 7.3; see also § 14.3.

A final general remark is that one of the best checks of the general reliability of an experiment is to show agreement with previously established results in that field. Hence it is quite often worth including a treatment solely with this object.

9.4 CHOICE OF OBSERVATIONS

The preceding chapter dealt with the number of repeat observations of a particular type that should be made on each unit, and with the number of units. In most experiments, however, observations of several different types are made on each unit and we now discuss briefly the selection of quantities for observation. They can be classified first according to the purpose for which they are made and second according to their mathematical nature, for example whether they are quantitative or whether they amount solely to an ordering of different objects.

Observations may be classed roughly into the following types.

(i) Primary Observations

These either measure properties for which the treatment comparisons are of direct interest, or are needed in the calculation of such quantities. Yield of product is an obvious example; another is the set of observations that are needed on each plot of a sugar-beet experiment in order to calculate the yield of sugar in cwt/acre.

(ii) Substitute Primary Observations

If a particular primary observation is difficult to obtain, it may be profitable to use a substitute observation that is easier to get. For example, the primary properties in a textile experiment may be the wearing properties, handle, and appearance of the woven fabric. All these in principle can be measured by consumer trials, but it is much easier to measure physical properties of the fabric, such as life in a laboratory wearing test, flexural rigidity, and yarn irregularity, which are thought to be closely related to the more nebulous qualities judged by the consumer. Another textile example is that in a spinning experiment, a primary quantity might well be the behavior of the yarn in the next process, weaving, as measured, say, by the number of yarn breaks per loom running hour. A substitute observation for this is the yarn strength as measured in a laboratory winding or strength test. In rather small experiments of this sort we would probably rely entirely in such substitute observations.

We normally regard an observation as a substitute one only if there is no theoretical or empirical relation converting the observation taken into the primary one, and more particularly if there is far from complete correlation between the two. The principle of measuring one thing by observing another closely dependent on it is of course the basis of many methods of measurement.

(iii) Explanatory Observations

These are taken to attempt an explanation of any treatment effects found on the primary observations. Thus suppose that in the experiment discussed briefly in (ii), the primary observations that we are directly interested in are taken, and are subjective in nature. Then we shall want to explain treatment differences found for these observations in terms of the physical or chemical behavior during processing, and to do this additional observations need to be made; some of these may, in fact, be those considered as substitute observations under (ii).

(iv) Supplementary Observations for Increasing Precision

In § 4.3, a method was explained for reducing the effect of variation between experimental units by a process of *adjustment*. In this, the mean value, for each treatment, of the main observation is adjusted to what it would have been had the units had the same value of a secondary or concomitant observation. In an alternative simpler, but in general less efficient, procedure an index of response is used. Thus in a psychological experiment the main observation might be the score obtained by a subject after exposure to a treatment, the concomitant observation being the score obtained in a similar test administered before applying the treatment. A natural index of response is the difference between the two scores for the subject.

The condition for the validity of this is that the observation used as a basis for adjustment should be unaffected by the treatments. This is so if the observation is made before the treatments are applied. For the procedure to be useful there should, of course, be high correlation between the two types of observation. If it is suspected that the main source of uncontrolled variation lies in the individual nature of the experimental units, rather than in the variation between natural groups of units, the skilful choice of concomitant observations can lead to an appreciable increase in precision.

(v) Supplementary Observations for Detecting Interactions

A further important use of the type of observation discussed in (iv) is to examine whether treatment effects vary systematically between units, for example whether subjects with a high initial score tend to respond differently to the treatments from those with low initial scores. The general importance of examining whether treatment effects are constant has been discussed in Chapter 2.

(vi) Observations for Checking the Application of the Treatments

For example if the treatments correspond to different temperatures, it will be natural to check independently that the appropriate temperature has in fact been achieved. If there is a serious discrepancy, and corrective action is impossible, it may be worth adjusting for the error in the statistical analysis of the results.

(vii) Observations to Check on External Conditions

Routine observations are often necessary to check that no unexpected gross change in external conditions occurs, and that no mishap, irrelevant to the treatment comparisons, occurs to any of the units.

A systematic consideration of these types might often be a good thing. Of course it remains true that the best experiments are often simple in conception and that many of the types of observation may not be required.

In some situations there may be no doubt how a particular property should be measured, a generally accepted objective and quantitative scale of measurement being available. This is the case with such things as yield of product, etc. Once what is to be considered as effective product is clearly defined, there will usually be little difficulty, in principle, in measuring what is wanted. Similarly, classical research has established reliable methods of measuring the standard physical and chemical quantities. In new fields, however, the situation may be quite different and some discussion will now be given of some of the general issues that may need consideration.

First, problems may arise in reducing a complex response to a manageable form. This, however, usually is a question more of analysis and interpretation than of experimental design. Thus in studies of the smoothness of metallic surfaces or the irregularity of textile yarns the initial observation will be an irregular trace showing the variation of thickness. In learning experiments the observation on one animal may consist of a series of "successes" and "failures," corresponding to successive attempts at a task. During learning the proportion of successes increases to near unity, during extinction the proportion decreases again. For further study it is usually very desirable to reduce such data and two general procedures can be used for this. One is to define one or two quantities which have a direct interpretation in terms of the more complex response. For instance, the rate of learning is often measured by the number of trials necessary before, say, five consecutive successes are obtained; the irregularity traces might be summarized by the coefficient of variation of thickness. The second general procedure is to estimate the parameters in a mathematical model that is thought to represent the system. In the examples just discussed these models would be probabilistic, for example one of the various stochastic models that have been advanced to represent learning. If such models give real insight into the system, their use is of course very desirable; if not (if the model is purely empirical and if appreciable extra labor is involved in fitting the model) it will be worth considering whether some simpler method can be used. The danger of the first method is that if the response can change in various ways, the empirical indices may be misleading. For example if a learning curve should really be described by the limiting proportion of correct responses when learning is complete and by the rate at which this limit is reached, it is clear that any single measure of learning may be misleading. This sort of consideration is particularly important when it is planned to record directly the final indices of interest.

Sometimes, instead of recording observations on a scale, we may take dichotomous observations, for example, on whether the diameter of a circular cylinder is or is not greater than a critical value, whether a foodstuff is judged satisfactory or unsatisfactory, whether one foodstuff is judged preferable to another, and so on. Such observations may be considered either for convenience and simplicity, or because only qualitative judgements are possible.

An ingenious example of the first use is a technique due to Anderson (1954) for comparing the strengths of two textile yarns *A* and *B*. Two 10 in. lengths, one of *A* and one of *B*, are joined and the resulting 20 in. length pulled until it breaks. The position of the break shows which is the weaker of the two lengths. If this is repeated, the weaker specimen will be from yarn *A* in about one-half the trials if the strengths are really the same, and if a significant departure from equal proportions is observed a difference between the strengths of the yarns has been established. This gives a simple comparative test without special apparatus and without quantitative measurement. Such a method has a statistical efficiency of about 64 per cent for establishing the existence of small differences, as compared with the quantitative method in which the strength of each section is measured. That is, a certain number of dichotomous observations and 64 per cent of that number of quantitative observations give about the same precision. This may be considered quite a high efficiency if appreciable simplification is attained; the main disadvantage of the qualitative method is that if a moderate or large difference exists, the method will indicate its presence but its magnitude will be estimated only with low precision. Similar remarks apply to comparison with a standard (as when the diameter of a cylinder is compared with a critical value) although this sort of procedure is perhaps not very likely to be used in experimental work.

A natural generalization of the pairing method just discussed is the *ranking* of more than two objects in order either of scale value or of subjective preference. Again quite sensitive tests can be made of the existence of small differences, but only poor estimates are obtained of the size of large differences. One problem of design that sometimes arises is that of choosing between either ranking, say, five objects or placing in order all possible pairs of objects. In general, ranking is likely to be the more economical method provided that it is practicable to judge several objects simultaneously without loss of precision.

An alternative to the qualitative judging of pairs and to ranking is the scoring of differences on a simple scale, for instance the five-point scale,

$+2$: *A* is much preferred to *B*,
$+1$: *A* is preferred to *B*,
0: no difference,
-1: *B* is preferred to *A*,
-2: *B* is much preferred to *A*.

General instructions are given to the judges to try to standardize the use of the scale as much as possible. The advantage of this over the previous method is partly that an analysis, even of complex designs, may usually be made by standard statistical methods (Scheffé, 1952) and partly the added sensitivity that should, in theory at any rate, result from the distinction that may be drawn between strong preference and preference. No comparative studies seem to have been published of how the differences between the two methods work out in practice. A tentative recommendation is to use a five or more point scale whenever the comparison of results from different judges is not of particular importance.

Statistical techniques are available for determining from the data a system of scoring the qualitatively different responses that will give most sensitive discrimination between treatments (Fisher, 1954, p. 289).

SUMMARY

The choice of experimental units, of treatments for comparison, and of types of observation is considered briefly.

The size and representative character of the units may be important and the insertion of additional variations among the units profitable. Treatments are chosen in some cases because of their direct interest, in others because of the information they may give about the mechanism underlying the system and in others to attain wider range of validity for the conclusions about the main treatments.

The main purposes for taking observations of a particular quantity are: to get information of direct technological or scientific importance, to substitute for such primary observations, to explain treatment effects occurring with the primary observations, and to use as a supplementary variable to increase precision or to detect variations in the treatment effects.

A brief discussion is given of the forms in which observations may be obtained.

REFERENCES

Anderson, S. L. (1954). A simple method of comparing the breaking load of two yarns. *J. Text. Inst.*, **45**, T472.

Fairfield Smith, H. (1938). An empirical law describing heterogeneity in the yields of agricultural crops. *J. Agric. Sci.*, **26**, 1.

Fisher, R. A. (1954). *Statistical methods for research workers*. 12th ed. Edinburgh: Oliver and Boyd.

Rutstein, D. D. (1957). The cold-cure merry-go-round. *Atlantic Monthly*, **199**, No. 4, 63.

Scheffé, H. (1952). An analysis of variance for paired comparisons. *J. Am. Statist. Assoc.*, **47**, 381.

CHAPTER 10

More about Latin Squares

10.1 INTRODUCTION

In Chapter 3 we saw the importance of the Latin square as a device for eliminating the effect of two sources of uncontrolled variation. The general idea was to achieve this elimination by arranging the experimental units in a square array, each treatment to occur once in each row and once in each column of the square.

In the present chapter we give tables of Latin squares and related designs, and discuss various detailed points connected with their use.

10.2 A TABLE OF LATIN SQUARES

Table 10.1 gives a Latin square arrangement of each size from 3×3 to 8×8. Higher-order squares can be produced by the same method used to write down the 7×7 and 8×8 squares, i.e., each row of letters consists of the preceding row shifted one place to the left. Thus, a 10×10 Latin square can be written down by starting with $AB \ldots IJ$ as the first row, $BC \ldots JA$ as the second, $CD \ldots JAB$ as the third, and so on.

To select a treatment arrangement for use, we have to randomize the appropriate square given in Table 10.1, which is in a standardized order with the letters in alphabetical order along the first row and first column. To do this the procedure is

(*a*) for the 3×3 square to randomize the rows and the columns of the tabulated square;

(*b*) for the 4×4 square to select at random one of the four tabulated squares and then to randomize its rows and columns as in (*a*);

(*c*) for the higher-order squares, the rows and columns should be independently randomized and the treatments assigned randomly to the letters A, B, C, . . .

For the 3×3 and 4×4 squares this procedure selects a square at random from the set of all squares of the required size; for the higher-order squares it selects at random from a set of squares which, although not including all Latin squares of the particular size, is a suitable set for both practical and theoretical purposes. The theoretical justification for

TABLE 10.1

EXAMPLES OF LATIN SQUARES

3 × 3

A	*B*	*C*
B	*C*	*A*
C	*A*	*B*

4 × 4

(i)				(ii)				(iii)				(iv)			
A	*B*	*C*	*D*	*A*	*B*	*C*	*D*	*A*	*B*	*C*	*D*	*A*	*B*	*C*	*D*
B	*A*	*D*	*C*	*B*	*C*	*D*	*A*	*B*	*D*	*A*	*C*	*B*	*A*	*D*	*C*
C	*D*	*B*	*A*	*C*	*D*	*A*	*B*	*C*	*A*	*D*	*B*	*C*	*D*	*A*	*B*
D	*C*	*A*	*B*	*D*	*A*	*B*	*C*	*D*	*C*	*B*	*A*	*D*	*C*	*B*	*A*

5 × 5

A	*B*	*C*	*D*	*E*
B	*A*	*E*	*C*	*D*
C	*D*	*A*	*E*	*B*
D	*E*	*B*	*A*	*C*
E	*C*	*D*	*B*	*A*

6 × 6

A	*B*	*C*	*D*	*E*	*F*
B	*C*	*F*	*A*	*D*	*E*
C	*F*	*B*	*E*	*A*	*D*
D	*E*	*A*	*B*	*F*	*C*
E	*A*	*D*	*F*	*C*	*B*
F	*D*	*E*	*C*	*B*	*A*

7 × 7

A	*B*	*C*	*D*	*E*	*F*	*G*
B	*C*	*D*	*E*	*F*	*G*	*A*
C	*D*	*E*	*F*	*G*	*A*	*B*
D	*E*	*F*	*G*	*A*	*B*	*C*
E	*F*	*G*	*A*	*B*	*C*	*D*
F	*G*	*A*	*B*	*C*	*D*	*E*
G	*A*	*B*	*C*	*D*	*E*	*F*

8 × 8

A	*B*	*C*	*D*	*E*	*F*	*G*	*H*
B	*C*	*D*	*E*	*F*	*G*	*H*	*A*
C	*D*	*E*	*F*	*G*	*H*	*A*	*B*
D	*E*	*F*	*G*	*H*	*A*	*B*	*C*
E	*F*	*G*	*H*	*A*	*B*	*C*	*D*
F	*G*	*H*	*A*	*B*	*C*	*D*	*E*
G	*H*	*A*	*B*	*C*	*D*	*E*	*F*
H	*A*	*B*	*C*	*D*	*E*	*F*	*G*

this system of randomization is that it ensures the mathematical properties discussed briefly in § 5.6. More detailed study might show that there were theoretical advantages to different methods of randomization from those used here, but at the time of writing nothing seems to be known about this and it is doubtful if the matter is of practical importance.

Example 10.1. First let it be required to select a 4 × 4 Latin square. We use the random permutations of 1, . . . , 9 given in the Appendix. Suppose that the first three permutations for use are 2, 1, 3, 4; 3, 2, 1, 4; 3, 1, 4, 2, where the numbers 5, . . . , 9 have been omitted. The first number of the first permutation is 2 so that we start with the second tabulated square

$$\begin{matrix} A & B & C & D \\ B & C & D & A \\ C & D & A & B \\ D & A & B & C \end{matrix} \tag{1}$$

Now permute the rows in accordance with the second permutation, placing first the third row of (1), then the second row, and so on. This gives

Original Row Number				
3	*C*	*D*	*A*	*B*
2	*B*	*C*	*D*	*A*
1	*A*	*B*	*C*	*D*
4	*D*	*A*	*B*	*C*

(2)

Finally permute the columns in accordance with the third permutation, i.e., put first the third column of (2), and so on. This gives

Original Column No:	3	1	4	2
	A	*C*	*B*	*D*
	D	*B*	*A*	*C*
	C	*A*	*D*	*B*
	B	*D*	*C*	*A*

(3)

If, as is frequently the case, the final design is to consist of several such squares, the component squares must be randomized independently.

To select a larger square, say, a 7×7, three random permutations of the numbers 1, . . . , 7 are required. The first is used to permute the rows of the standard square in Table 10.1; this corresponds to the step from (1) to (2) just described for the 4×4 square. The second and third permutations are used respectively to permute the columns, and to assign the letters $A, \ldots, G$ to the seven treatments.

10.3 A TABLE OF GRAECO-LATIN SQUARES

In § 3.4 Graeco-Latin squares were described. These are developments of the Latin square in which two alphabets are used, each individually forming a Latin square and such that each letter of one alphabet occurs with each letter of the other alphabet just once.

Table 10.2 gives such squares of sides 3, 4, 5, and 7. To select an actual arrangement for use, the rows, columns, Latin letters, and Greek letters should be independently randomized.

TABLE 10.2

EXAMPLES OF GRAECO-LATIN SQUARES

3 × 3

$A\alpha$	$B\beta$	$C\gamma$
$B\gamma$	$C\alpha$	$A\beta$
$C\beta$	$A\gamma$	$B\alpha$

4 × 4

$A\alpha$	$B\beta$	$C\gamma$	$D\delta$
$B\delta$	$A\gamma$	$D\beta$	$C\alpha$
$C\beta$	$D\alpha$	$A\delta$	$B\gamma$
$D\gamma$	$C\delta$	$B\alpha$	$A\beta$

5 × 5

$A\alpha$	$B\beta$	$C\gamma$	$D\delta$	$E\epsilon$
$B\delta$	$C\epsilon$	$D\alpha$	$E\beta$	$A\gamma$
$C\beta$	$D\gamma$	$E\delta$	$A\epsilon$	$B\alpha$
$D\epsilon$	$E\alpha$	$A\beta$	$B\gamma$	$C\delta$
$E\gamma$	$A\delta$	$B\epsilon$	$C\alpha$	$D\beta$

7 × 7

$A\alpha$	$B\beta$	$C\gamma$	$D\delta$	$E\epsilon$	$F\phi$	$G\eta$
$B\delta$	$C\epsilon$	$D\phi$	$E\eta$	$F\alpha$	$G\beta$	$A\gamma$
$C\eta$	$D\alpha$	$E\beta$	$F\gamma$	$G\delta$	$A\epsilon$	$B\phi$
$D\gamma$	$E\delta$	$F\epsilon$	$G\phi$	$A\eta$	$B\alpha$	$C\beta$
$E\phi$	$F\eta$	$G\alpha$	$A\beta$	$B\gamma$	$C\delta$	$D\epsilon$
$F\beta$	$G\gamma$	$A\delta$	$B\epsilon$	$C\phi$	$D\eta$	$E\alpha$
$G\epsilon$	$A\phi$	$B\eta$	$C\alpha$	$D\beta$	$E\gamma$	$F\delta$

There is no 6 × 6 Graeco-Latin square. Graeco-Latin squares of size 8 × 8, 9 × 9, 11 × 11, and 12 × 12 exist; larger squares are unlikely to be of practical value.

Example 10.2. Let it be required to select a 5 × 5 Graeco-Latin square. We need four random permutations of the numbers 1, . . . , 5. Suppose that these are 1, 4, 3, 5, 2; 3, 1, 2, 4, 5; 4, 3, 1, 5, 2; 1, 5, 4, 3, 2. Following the method of Example 10.1, we use the first permutation to rearrange the rows of the standard square in Table 10.2. This gives

$A\alpha$	$B\beta$	$C\gamma$	$D\delta$	$E\epsilon$
$D\epsilon$	$E\alpha$	$A\beta$	$B\gamma$	$C\delta$
$C\beta$	$D\gamma$	$E\delta$	$A\epsilon$	$B\alpha$
$E\gamma$	$A\delta$	$B\epsilon$	$C\alpha$	$D\beta$
$B\delta$	$C\epsilon$	$D\alpha$	$E\beta$	$A\gamma$

Rearrangement of the columns according to the second permutation now gives

$C\gamma$	$A\alpha$	$B\beta$	$D\delta$	$E\epsilon$
$A\beta$	$D\epsilon$	$E\alpha$	$B\gamma$	$C\delta$
$E\delta$	$C\beta$	$D\gamma$	$A\epsilon$	$B\alpha$
$B\epsilon$	$E\gamma$	$A\delta$	$C\alpha$	$D\beta$
$D\alpha$	$B\delta$	$C\epsilon$	$E\beta$	$A\gamma$

Now let the treatments to be represented by Latin letters be $T_1, \ldots, T_5$ and the treatments represented by Greek letters be $S_1, \ldots, S_5$. The last two permutations are now used to assign treatments to letters in accordance with the following scheme

$$\begin{array}{ccccc} T_1 & T_2 & T_3 & T_4 & T_5 \\ D & C & A & E & B \end{array} \qquad \begin{array}{ccccc} S_1 & S_2 & S_3 & S_4 & S_5 \\ \alpha & \epsilon & \delta & \gamma & \beta \end{array}$$

The resulting square is

$$\begin{array}{ccccc} T_2S_4 & T_3S_1 & T_5S_5 & T_1S_3 & T_4S_2 \\ T_3S_5 & T_1S_2 & T_4S_1 & T_5S_4 & T_2S_3 \\ T_4S_3 & T_2S_5 & T_1S_4 & T_3S_2 & T_5S_1 \\ T_5S_2 & T_4S_4 & T_3S_3 & T_2S_1 & T_1S_5 \\ T_1S_1 & T_5S_3 & T_2S_2 & T_4S_5 & T_3S_4 \end{array}$$

It is remarkable that no 6×6 Graeco-Latin square exists;* this was conjectured in 1782 by Euler, who considered such squares in connection with various combinatorial problems, but the nonexistence has been proved only fairly recently, by systematic enumeration of all possibilities. It is generally thought that no $n \times n$ Graeco-Latin square exists whenever the size n leaves a remainder of two when divided by four, e.g., that no 10×10 Graeco-Latin square exists. However no correct mathematical proof of this has been produced, and the task of systematic enumeration is too formidable, so that it is not known for certain that no 10×10 Graeco-Latin square exists. In all other cases, i.e., when the size is not divisible by two or is divisible by four, Graeco-Latin squares can be constructed (Mann, 1949, Chapter 8).

10.4 ORTHOGONAL PARTITIONS OF LATIN SQUARES

The Graeco-Latin square arises from the need to insert into a Latin square design a further classification, or grouping of the units, having the same number of levels as there are treatments and rows and columns in the Latin square. Thus in Example 3.11 we started from a 4×4 Latin square with 4 treatments, and in which the experimental units are classified by days, (by rows of the square) and by time of day (by columns of the square). Then, as a further classification, we considered four observers, requiring that each observer measured once for each treatment and once on each day and at each time of day. These requirements are satisfied by the Graeco-Latin square.

Suppose now that there are two observers not four. We require that

* There are 9408 distinct reduced 6×6 Latin squares, i.e., squares with the letters in the first row and in the first column in alphabetical order.

each observer measures each treatment twice and observes two experimental units on each day and makes observations in all twice at each time of day. A design with these properties can easily be obtained from the Graeco-Latin square arrangement, by taking the units receiving old observers one and two to form those for the new observer I, and those originally receiving observers three and four to form those for new observer II. Table 10.3(*b*) shows the design thus formed.

TABLE 10.3

AN ORTHOGONAL PARTITION DERIVED FROM A 4 × 4 GRAECO-LATIN SQUARE

(*a*) *Arrangement of Processes and Observers with four Observers* (*see Table* 3.10*a*)

	Time 1	Time 2	Time 3	Time 4
Day 1	P_2O_3	P_4O_1	P_3O_2	P_1O_4
Day 2	P_3O_4	P_1O_2	P_2O_1	P_4O_3
Day 3	P_1O_1	P_3O_3	P_4O_4	P_2O_2
Day 4	P_4O_2	P_2O_4	P_1O_3	P_3O_1

(*b*) *Arrangement of Processes and Observers with two Observers*

	Time 1	Time 2	Time 3	Time 4
Day 1	P_2O_{II}	P_4O_I	P_3O_I	P_1O_{II}
Day 2	P_3O_{II}	P_1O_I	P_2O_I	P_4O_{II}
Day 3	P_1O_I	P_3O_{II}	P_4O_{II}	P_2O_I
Day 4	P_4O_I	P_2O_{II}	P_1O_{II}	P_3O_I

We call an arrangement such as this an *orthogonal partition* of the Latin square; orthogonal partitions can always be formed from Graeco-Latin squares by combining the letters of the Greek alphabet suitably. Thus in an 8 × 8 Graeco-Latin square we could change the 8 Greek letters into

(*a*) two, each occurring four times in each row and column and four times with each Latin letter;

(*b*) four, each occurring twice in each row, etc.;

(*c*) two, each occurring three times in each row, etc., and one occurring twice in each row, etc.

Therefore the Graeco-Latin square deals easily with this more general situation.

This method does not work for squares of side six, since, as we have seen, no Graeco-Latin square of side six exists. This nonexistence, however, does not preclude the possibility of other orthogonal partitions (Finney, 1945), and the two most interesting are shown in Table 10.4.

TABLE 10.4

ORTHOGONAL PARTITIONS OF THE 6 × 6 LATIN SQUARE

(*a*) *Second Alphabet with Two Letters, Each Occurring Three Times in Each Row, etc.*

$A\alpha$	$B\alpha$	$C\beta$	$D\beta$	$E\alpha$	$F\beta$
$B\beta$	$C\alpha$	$F\beta$	$A\alpha$	$D\beta$	$E\alpha$
$C\alpha$	$F\alpha$	$B\beta$	$E\beta$	$A\beta$	$D\alpha$
$D\alpha$	$E\beta$	$A\alpha$	$B\beta$	$F\alpha$	$C\beta$
$E\beta$	$A\beta$	$D\alpha$	$F\alpha$	$C\beta$	$B\alpha$
$F\beta$	$D\beta$	$E\alpha$	$C\alpha$	$B\alpha$	$A\beta$

(*b*) *Second Alphabet with Three Letters, Each Occurring Twice in Each Row, etc.*

$A\alpha$	$B\beta$	$C\gamma$	$D\beta$	$E\alpha$	$F\gamma$
$B\gamma$	$C\alpha$	$A\beta$	$F\alpha$	$D\gamma$	$E\beta$
$C\beta$	$A\gamma$	$B\alpha$	$E\gamma$	$F\beta$	$D\alpha$
$D\gamma$	$F\beta$	$E\alpha$	$B\gamma$	$A\alpha$	$C\beta$
$E\beta$	$D\alpha$	$F\gamma$	$C\alpha$	$B\beta$	$A\gamma$
$F\alpha$	$E\gamma$	$D\beta$	$A\beta$	$C\gamma$	$B\alpha$

The following is an example of the use of the arrangement in Table 10.4(*a*).

Example 10.3. Consider an agricultural experiment comparing six varieties and laid out in a 6 × 6 Latin square. Suppose that it is required to superimpose on the variety trial a comparison of the effect of the presence of a basic dressing of nitrogenous fertilizer. That is, we have a second factor with two levels N_0, N_1 representing absence and presence of the fertilizer.

Let this be represented by the Greek letters in Table 10.4(*a*); then this table will give us a suitable design, with each fertilizer treatment occurring the three times in each row, three times in each column, and three times with each variety.

The actual design for use is obtained by randomization of Table 10.4(*a*), in three steps; (*a*) the randomization of rows, (*b*) the randomization of columns, and (*c*) the random assignment of varieties to Latin letters and of fertilizer treatments to Greek letters. If the varieties are denoted by $V_1, \ldots, V_6$, the final design might be

V_1N_1	V_4N_0	V_6N_0	V_3N_0	V_5N_1	V_2N_1
V_6N_1	V_5N_1	V_3N_1	V_2N_0	V_1N_0	V_4N_0
V_4N_0	V_3N_1	V_5N_0	V_6N_1	V_2N_0	V_1N_1
V_5N_0	V_2N_1	V_1N_1	V_4N_1	V_6N_0	V_3N_0
V_2N_1	V_6N_0	V_4N_1	V_1N_0	V_3N_1	V_5N_0
V_3N_0	V_1N_0	V_2N_0	V_5N_1	V_4N_1	V_6N_1

10.5 SQUARES WITH MORE THAN TWO ALPHABETS

In the Latin square, the experimental units are classified by rows, by columns, and by Latin letters representing treatments. In the Graeco-Latin squares the experimental units are classified by rows, by columns, by Latin letters and by Greek letters. It is natural to consider whether further "alphabets" can be added preserving the property that each letter of the new alphabet occurs once in each row, once in each column and once in combination with each letter of the other alphabets.

Table 10.5 gives examples for 4 × 4 and 5 × 5 squares. These

TABLE 10.5

4 × 4 AND 5 × 5 SQUARES WITH SEVERAL ALPHABETS

(*a*) 4 × 4 *Square with Three Alphabets*

$A\alpha a$	$B\beta b$	$C\gamma c$	$D\delta d$
$B\gamma d$	$A\delta c$	$D\alpha b$	$C\beta a$
$C\delta b$	$D\gamma a$	$A\beta d$	$B\alpha c$
$D\beta c$	$C\alpha d$	$B\delta a$	$A\gamma b$

(*b*) 5 × 5 *Square with Four Alphabets*

$A\alpha a1$	$B\beta b2$	$C\gamma c3$	$D\delta d4$	$E\epsilon e5$
$B\gamma d5$	$C\delta e1$	$D\epsilon a2$	$E\alpha b3$	$A\beta c4$
$C\epsilon b4$	$D\alpha c5$	$E\beta d1$	$A\gamma e2$	$B\delta a3$
$D\beta e3$	$E\gamma a4$	$A\delta b5$	$B\epsilon c1$	$C\alpha d2$
$E\delta c2$	$A\epsilon d3$	$B\alpha e4$	$C\beta a5$	$D\gamma b1$

arrangements are of considerable mathematical interest and are frequently useful as a basis for deriving further designs. However they are not often used directly in practice and hence will not be discussed at length.

Thus with the 5 × 5 square, any pair of "alphabets" constitute a Graeco-Latin square; for example it may be verified that each combination of a Greek letter with a number occurs just once, and each number and each Greek letter occurs once in each row and once in each column.

The main facts about such squares are that the maximum number of alphabets is one less than the size of the square, and that this maximum number can be attained when the number of rows in the square is a prime number,* or a power of a prime number. That is, for a 3 × 3 square only two alphabets can be inserted, i.e., nothing larger than a

* A prime number is a number (e.g., 2, 3, 5, 7, . . .) with no whole number divisors other than one and itself.

Graeco-Latin square is possible. For 4×4 and 5×5 squares, the arrangements in Table 10.5 contain the maximum number of alphabets. We have already remarked that no 6×6 Graeco-Latin square (two alphabets) is possible. For 7×7, 8×8 and 9×9 squares, arrangements with the maximum number of alphabets, 6, 7, and 8, can be constructed.

Direct applications of these squares are, as stated above, rare, but the following is based on a celebrated example (Tippett, 1934).

Example 10.4. In a cotton winding process it was found that the product of one of five spindles was defective, the cause of the defect not being known. However, four component parts of the spindles could be interchanged; we denote these by the four alphabets, i.e.,

Component part I:	A,	B,	C,	D,	E
II:	α,	β,	γ,	δ,	ϵ
III:	a,	b,	c,	d,	e
IV:	1,	2,	3,	4,	5.

Thus, the combination $B\gamma a5$ would refer to a spindle built up from the second component of type I, the third of type II, the first of type III, and the fifth of type IV.

Five runs were then made on each spindle, the components being arranged in accordance with a randomized form of Table 10.5(b).

Suppose that the arrangement of components is as follows, the runs leading to defective product being marked with an asterisk.

Period	Spindle	No. 1	No. 2	No. 3	No. 4	No. 5
1		$E\alpha d2$	$C\epsilon b1$	$A\gamma e4$	$D\delta c5$	$B\beta a3*$
2		$C\gamma c3$	$A\delta a2$	$D\beta d1$	$B\alpha b4$	$E\epsilon e5$
3		$A\beta b5$	$D\alpha e3$	$B\epsilon c2*$	$E\gamma a1$	$C\delta d4$
4		$D\epsilon a4$	$B\gamma d5*$	$E\delta b3$	$C\beta e2$	$A\alpha c1$
5		$B\delta e1*$	$E\beta c4$	$C\alpha a5$	$A\epsilon d3$	$D\gamma b2$

The four runs leading to defective product all have the second component of type I, i.e., B, strongly suggesting that B is the source of the trouble. Note that the Latin square property ensures that B has occurred on each spindle and in each period, showing that the trouble is not connected with spindles or periods. Similarly, the Graeco-Latin square property ensures that each component of types II, III, and IV occurs once together with B; had this not been the case, it would have happened, say, that component γ occurred fairly frequently with B. There would then have been some possibility that the trouble was with γ. As an exercise, the reader may consider why, if the experiment had involved quantitative measurement, rather than a qualitative observation of the occurrence of a very defective product, the advantages of the Graeco-Latin square property would have been even stronger.

This is an example of what is called the fractional replication of a factorial experiment (see § 13.2). We have a system with 6 factors, each at 5 levels, and one replicate of the complete system would require $5^6 = 15{,}625$ observations. Here it was possible, with $5^2 = 25$ observations, to obtain useful information about the main effects of the factors.

10.6 MISCELLANEOUS EXAMPLES

In this section we give some further examples of the use of Latin square and related designs. These are intended primarily to indicate further fields in which the Latin square design has been found useful.

Example 10.5. Brunk and Federer (1953) have discussed some instances of carefully planned investigations in market research. One of these concerned the effect of varying practices of pricing, displaying, and packaging apples on the sale of apples.

In each experiment of this series, four merchandising practices A, B, C, and D were compared and four supermarkets took part. Since it was clearly desirable that each treatment should be used in each store, it was sensible to arrange for the experiment to continue for a multiple of four time periods. The experimental units were therefore grouped into sets of 16 and classified by stores and by periods,

TABLE 10.6

LATIN SQUARE DESIGN IN MARKET RESEARCH

First Part of Week

Day	Store 1	2	3	4
Monday	B	C	D	A
Tuesday	A	B	C	D
Wednesday	D	A	B	C
Thursday	C	D	A	B

Second Part of Week

Day	Store 1	2	3	4
Friday, a.m.	B	A	C	D
Friday, p.m.	C	D	B	A
Saturday, a.m.	A	B	D	C
Saturday, p.m.	D	C	A	B

The letters A, B, C, D refer to four alternative methods of displaying, etc. apples. The observation for each experimental unit is the sale of apples per 100 customers in the store.

i.e., by order in time. To eliminate systematic differences between stores, and between different periods, a 4 × 4 Latin square was appropriate. In fact, however, the week was divided into two parts, Monday through Thursday, and Friday and Saturday, and one 4 × 4 Latin square built up for each part of the week. This was a good move because the grocery order per customer was larger at the week-end and it was quite possible that the treatment differences would not be the same in the two parts of the experiment.

Table 10.6 shows the resulting design. This is for an experiment lasting one week and comparing four treatments. In fact a whole series of such experiments was done, the four treatments for test in any one week being chosen partly in the light of the results obtained up to that point.

A question that always arises in connection with experiments to determine some practical course of action concerns the extent to which the conditions of the experiment agree with the conditions under which the results are to be applied. One difference between experimental and practical conditions that is an inevitable consequence of the present type of design is that the merchandising practices are under frequent change. It is well known from experiments on industrial productivity that the mere occurrence of changes has an effect on behavior, although it is perhaps unlikely in this instance that the comparison of the treatments would be much affected, even if the general level of sales was. Brunk and Federer report that further investigations connected with this were made.

Example 10.6. Vickery et al. (1949) have given an ingenious application of the Latin square to the selection of uniform experimental units. In this work each experimental unit consists of five leaves and the problem is not the assignment of treatments to units, but the formation of sets of five leaves, each set to be as alike the others as possible.

There are five leaflet positions on each plant and to form a group of five units, five plants are used. The following easily remembered Latin square is used to form experimental units, each roman numeral denoting a set of leaves to form one unit;

	Plant				
Leaflet Position	1	2	3	4	5
1	I	II	III	IV	V
2	II	III	IV	V	I
3	III	IV	V	I	II
4	IV	V	I	II	III
5	V	I	II	III	IV

That is, the five leaves are removed in order from the first plant. The leaves are likewise removed from the second plant, placing the first leaf with the second leaf of the first plant, the second leaf with the third leaf of the first plant, and so on. The procedure is the same for plants 3, 4, and 5, staggering the arrangement one step each time. At the end we have five sets of leaves, each set containing one leaf from each plant and one from each position. It is therefore reasonable to expect the bulk properties of the sets of leaves to vary little from one set to another, i.e., for the sets to form a suitable collection of experimental units for the comparison of treatments.

Vickery et al. showed experimentally that this method of forming experimental units leads to less random variation between units than alternative methods. There would be no point in randomizing the Latin square in this case and indeed randomization would make the method impracticable, since for the grouping of leaves to be done reasonably speedily it is essential that the method of grouping should be memorized.

Example 10.7. Davies (1945) has used a Graeco-Latin square in the comparison of miles per gallon achieved on a particular car with seven alternative forms of gasoline. Each test involved driving the car over a fixed route of 20 miles, including various gradients. To remove possible biases connected with the driver, seven drivers were used and to remove possible effects connected with the traffic conditions the experiment was run at seven different times of the day. We therefore have, in addition to the seven treatments under comparison, three classifications of the experimental units, namely by drivers, by days, and by times of the day. A double classification of the experimental units suggests the use of a Latin square, a triple classification a Graeco-Latin square.

The following shows a possible arrangement, obtained by randomization of the plan in Table 10.2.

	Time of Day 1	2	3	4	5	6	7
Day 1	G_4D_5	G_1D_3	G_3D_6	G_2D_4	G_6D_2	G_5D_7	G_7D_1
2	G_7D_6	G_4D_1	G_2D_2	G_6D_7	G_5D_3	G_1D_5	G_3D_4
3	G_6D_1	G_2D_5	G_1D_4	G_4D_2	G_7D_7	G_3D_3	G_5D_6
4	G_2D_3	G_3D_7	G_5D_1	G_1D_6	G_4D_4	G_7D_2	G_6D_5
5	G_1D_7	G_5D_2	G_7D_5	G_3D_1	G_2D_6	G_6D_4	G_4D_3
6	G_3D_2	G_7D_4	G_6D_3	G_5D_5	G_1D_1	G_4D_6	G_2D_7
7	G_5D_4	G_6D_6	G_4D_7	G_7D_3	G_3D_5	G_2D_1	G_1D_2

$G_1, \ldots, G_7$ denote the seven gasolines, $D_1, \ldots, D_7$ the seven drivers.

Note that each gas is used once on each day, once by each driver, and once at each time of day, ensuring a balanced comparison. The observation for analysis is the gasoline consumption for the 20-mile journey.

Other methods of running trials of this type have been discussed by Menzler (1954). The design just described leads to a single set of estimates of the treatment differences averaged over drivers, days, etc. It might, however, be advantageous to include as a factor in the experiment driving "strategy," i.e., average speed to be aimed at, and to set up the experiment as a factorial experiment designed in part to look for interaction between gasoline differences and driving strategy. Also it would usually be a good thing to include several cars in the experiment, to give the conclusions a wider range of validity.

Example 10.8.* Suppose that in an experiment in prosthodontics, seven treatments are to be compared, the treatments being commercial dentures of different materials and set at different angles. To eliminate as much as possible of the variation due to differences between patients, each patient wears dentures

* I am grateful to Dr. B. G. Greenberg for a description of this example.

of one type for a month, then dentures of another type for another month, and so on, until after seven months each patient has worn each type of denture, i.e., has been subjected to each treatment.

In this procedure a time trend in the results is quite likely, and hence it is sensible to arrange that each treatment is used equally often in each time position. This requirement of balancing out two types of variation, namely between-patient and between-time variation, naturally suggests the use of a Latin square, and with seven patients, the following design might be obtained after randomization:

		Patient						
		1	2	3	4	5	6	7
Month Number	1	T_5	T_4	T_3	T_7	T_2	T_6	T_1
	2	T_7	T_6	T_4	T_1	T_3	T_5	T_2
	3	T_6	T_3	T_2	T_5	T_1	T_4	T_7
	4	T_2	T_7	T_5	T_3	T_6	T_1	T_4
	5	T_4	T_2	T_1	T_6	T_7	T_3	T_5
	6	T_3	T_1	T_7	T_4	T_5	T_2	T_6
	7	T_1	T_5	T_6	T_2	T_4	T_7	T_3

Here $T_1, \ldots, T_7$ denote the seven alternative treatments.

The seven patients are incorporated into the experiment as they become available, i.e., are the first seven suitable patients to appear after the beginning of the experiment. If it is likely from the start that more than seven patients are necessary to get the required precision, it would be desirable to use a simple multiple of seven patients. With, say, 21 patients, it would be reasonable to use three intermixed Latin squares (see Table 3.9(*b*)), i.e., three 7×7 Latin squares with the 21 columns completely randomized.

The quantity used to assess the success of each treatment is an index of masticatory efficiency based on the ability to chew peanuts under standardized conditions. This is measured immediately after the fitting of each denture and again after the lapse of one month. The initial and final values measure different aspects of the treatment and are analyzed and interpreted separately.

Two of the basic assumptions of Chapter 2 need particular discussion here; these are that the treatment differences are constant, and that there is no overlap of treatment effect from one experimental unit to another. It seems quite possible in this experiment that the treatment differences may depend in an important way on the type of patient. The main thing that can be done about this is to attempt to correlate the apparent treatment differences for individual patients with properties of the patients such as their previous experience with dentures. Also it would be advantageous to have an estimate of how much of the variation in masticatory efficiency is measurement error, i.e., to arrange that the measure at any one time is obtained as the average of duplicate determinations.

The second point, concerning the overlap of treatment effects, can be dealt with in part by elaboration of the design (Chapter 13); alternatively, inspection of the data may show that the effect is in fact unimportant or that some simple correction for it can be introduced.

SUMMARY

A Latin square is an arrangement of n letters in an $n \times n$ square such that each letter occurs once in each row and once in each column. A table of such squares is given, together with instructions on randomization.

A Graeco-Latin square is an arrangement of n letters of each of two alphabets in pairs in an $n \times n$ square in such a way that each alphabet separately forms a Latin square and that each letter of one alphabet occurs with each letter of the other alphabet just once. For values of n of practical interest, Graeco-Latin squares exist except when n is six; tables are given. For n equal to six some special arrangements having some of the properties of a Graeco-Latin square are occasionally useful.

Squares with more than two alphabets are considered briefly.

The Latin square is likely to be useful when the experimental units are cross-classified in two ways, the Graeco-Latin square when the units are cross-classified in three ways. In all cases the number of levels of each classification is to be equal to the number of treatments and to the number of units per treatment, if the designs are to be used in an unmodified form.

REFERENCES

Brunk, M. E., and W. T. Federer. (1953). Experimental designs and probability sampling in marketing research. *J. Am. Statist. Assoc.*, **48,** 440.

Davies, H. M. (1945). The application of variance analysis to some problems of petroleum technology. London: Publication of Inst. of Petroleum.

Finney, D. J. (1945). Some orthogonal properties of the 4×4 and 6×6 Latin squares. *Ann. of Eugenics*, **12,** 213.

Mann, H. B. (1950). *Analysis and design of experiments*. New York: Dover.

Menzler, F. A. A. (1954). The statistical design of experiments. *Brit. Transport Rev.*, **3,** 49.

Tippett, L. H. C. (1934). Applications of statistical methods to the control of quality in industrial production. *Manchester Statistical Society*.

Vickery, H. B., C. S. Leavenworth, and C. I. Bliss. (1949). The problem of selecting uniform samples of leaves. *Plant Physiology*, **24,** 335.

CHAPTER 11

Incomplete Nonfactorial Designs

11.1 INTRODUCTION

We shall consider in this chapter experiments in which either the treatments are not arranged in factorial form or in which the factorial structure is unimportant. That is, we suppose that we have a collection of treatments any pair of which may need comparison.

In Chapter 3, and repeatedly in subsequent chapters, we have seen the importance of the randomized block and Latin square principles as methods for increasing precision, the idea being to group the experimental units into sets expected to behave similarly. The assumption has been made throughout the discussion of these methods that the blocks, or the rows and the columns of the Latin square, contain sufficient experimental units for each treatment to occur at least once in each block, row, or column. Suppose now that this is not so, and that, for example, the number of treatments exceeds the most suitable number of experimental units per block.

An example will illustrate the type of difficulty that occurs. Suppose that it is very desirable to eliminate differences between days, that three experimental units are the most that can be dealt with on one day, and that there are five treatments. Table 11.1 gives a design that would be useful in such a case; clearly the randomized blocks design in its simple form is no use here, since we cannot arrange for each treatment to occur once in each block. Table 11.1 is in fact an example of what is called a balanced incomplete block design.

Suppose, for example, that all the results on the first day are high. Now in an ordinary randomized block experiment all the treatments would be affected equally, so that the differences between the treatment means would be unaffected. This is not so here. Treatments 4, 5, and 1 receive an increment not experienced by the other treatments; the consequence is that the estimate of treatment differences is not to be based on the treatment means, if the desired elimination of block effects is to be achieved. Each treatment mean has to be adjusted for the effect of the blocks in which the particular treatment does not occur. The problem in selecting a design in such a case is to arrange that

(*a*) the resulting estimated treatment effects have as high a precision as possible;

(*b*) the adjustments just mentioned can be made as simply as possible;

(*c*) treatment comparisons of equal practical importance are made with approximately equal precision.

TABLE 11.1

A SIMPLE BALANCED INCOMPLETE BLOCK DESIGN FOR FIVE TREATMENTS IN BLOCKS OF THREE UNITS

Day			
1	T_4	T_5	T_1
2	T_4	T_2	T_5
3	T_2	T_4	T_1
4	T_5	T_3	T_1
5	T_3	T_4	T_5
6	T_2	T_3	T_1
7	T_3	T_1	T_4
8	T_3	T_5	T_2
9	T_2	T_3	T_4
10	T_5	T_1	T_2

The order of treatments within days has been randomized.

Fortunately these requirements can be satisfied simultaneously.

In general we shall call an experimental arrangement an *incomplete block design* if the experimental units are grouped into blocks with the object of eliminating differences between blocks and if the number of units in each block is less than the number of treatments.

If in the example set out in Table 11.1 we desired also to eliminate from the error of treatment comparisons any systematic effects connected with order within the block (time of day), we should need an incomplete design analogous to a Latin square. Such designs, with two-way elimination of error, are considered in §§ 11.3 and 11.5.

The general plan of this chapter is that in § 11.2 the simplest type of incomplete block design is described fairly fully. A corresponding account of the simplest types of design with two-way control of error is given in § 11.3. Finally §§ 11.4 and 11.5 consist of brief surveys of some of the more complicated designs that are occasionally useful.

11.2 BALANCED INCOMPLETE BLOCK DESIGNS

(i) General

The simplest and most important type of incomplete block designs have the following properties:

(*a*) each block contains the same number of units;

(*b*) each treatment occurs the same number of times in all;

(*c*) if we take any two treatments and count the number of times they occur together in the same block, the same number is obtained for all pairs of treatments.

Thus in the arrangement of Table 11.1, each block contains three units and each treatment occurs six times. Further, two treatments, say, T_2 and T_5, occur together in blocks 2, 8, and 10, i.e., three times. Similarly any other two treatments will be found to occur together in just three blocks. Thus the conditions (*a*), (*b*), and (*c*) are satisfied in this case. Any incomplete block design which satisfies the conditions is called a *balanced incomplete block* arrangement.

If we contemplate using such a design in a particular situation, we shall be concerned with four quantities, namely, the number of treatments, the number of times each treatment appears (the number of replicates), the number of blocks and the number of experimental units per block. Now the total number of experimental units is equal to

$$\text{number of treatments} \times \text{number of replicates of each treatment} = t \times r, \tag{1}$$

and to

$$\text{number of blocks} \times \text{number of experimental units per block} = b \times k, \tag{2}$$

so that expressions (1) and (2) must be equal to one another; that is, only three of the above four quantities can be assigned independently, the fourth then being determined. For instance in Table 11.1, $t = 5$, $r = 6$ and $b = 10$, $k = 3$, so that $t \times r = b \times k = 30$.

The problem of choosing a design often arises in somewhat the following way: There are a certain number of treatments to be investigated and we wish to use blocks each with not more than a certain number of units, and finally there is some restriction, possibly rather vague, on the total size of the experiment. If all comparisons of pairs of treatments are of about equal importance, and if the maximum number of units per block is less than the number of treatments, the use of a balanced incomplete block design will seem natural, but it may well happen that for the first set of values of the quantities t, r, b, and k that suggest themselves, no balanced incomplete block design exists. For example, if the equation $tr = bk$, discussed above, is not satisfied, no balanced incomplete block design can exist. However, even if the equation is satisfied, there still may be no design. Thus we need a concise table to show when balanced incomplete block designs do exist, as well as tables to show the actual form of the designs. These will be discussed in the next section.

(ii) Existence and Form of Designs

Table 11.2 summarizes the values of t, b, k, r for which balanced incomplete block designs of common practical importance exist. The designs not included in the table correspond to experiments with more than 150 units and are likely to be required only in special types of application.

TABLE 11.2

INDEX OF BALANCED INCOMPLETE BLOCK DESIGNS USING NOT MORE THAN 150 EXPERIMENTAL UNITS

No. of Units per Block, k	No. of Treatments, t	No. of Blocks, b	No. of Replicates, r	Total no. of Units, $bk = rt$	Description and Notes
2	3	3	2	6	unr.*; Y†
2	4	6	3	12	unr.
2	5	10	4	20	unr.
2	any t	$\frac{1}{2}t(t-1)$	$(t-1)$	$t(t-1)$	unr.
3	4	4	3	12	unr.; Y
3	5	10	6	30	unr.
3	6	10	5	30	
3	6	20	10	60	unr.; res.‡
3	7	7	3	21	Y
3	9	12	4	36	res.
3	10	30	9	90	
3	13	26	6	78	
3	15	35	7	105	res.
4	5	5	4	20	unr.; Y
4	6	15	10	60	unr.
4	7	7	4	28	Y
4	8	14	7	56	res.
4	9	18	8	72	
4	10	15	6	60	
4	13	13	4	52	Y
4	16	20	5	80	res.

TABLE 11.2 (*Continued*)

No. of Units per Block, k	No. of Treatments, t	No. of Blocks, b	No. of Replicates, r	Total no. of Units, $bk = rt$	Description and Notes
5	6	6	5	30	unr.; Y
5	9	18	10	90	
5	10	18	9	90	
5	11	11	5	55	Y
5	21	21	5	105	Y
5	25	30	6	150	res.
6	7	7	6	42	unr.; Y
6	9	12	8	72	
6	10	15	9	90	
6	11	11	6	66	Y
6	16	16	6	96	Y
6	16	24	9	144	
7	8	8	7	56	unr.; Y
7	15	15	7	105	Y
8	9	9	8	72	unr.; Y
8	15	15	8	120	Y
9	10	10	9	90	unr.; Y
9	13	13	9	117	Y
10	11	11	10	110	unr.; Y
11	12	12	11	132	unr.; Y

* unr. means unreduced.
† Y means that a corresponding Youden square (§ 11.3) exists.
‡ res. means resolvable.

A fuller index of designs is available in Cochran and Cox (1957, § 9.6).
Consider now some examples of the use of this table.

Example 11.1. In an experiment to compare a number of detergents used for domestic dishwashing, the following procedure has been used (Pugh, 1953). In order to obtain a series of homogeneous experimental units, a pile of plates from one course in a works canteen is divided into groups. Each group of plates is then washed with water at a standard temperature and with a controlled amount of one detergent, the observation being the logarithm of the number of plates washed before the foam is reduced to a thin surface layer. The groups of plates from the one course form a block, and the washing for one block is done by one person. Thus the experimental conditions are as constant as possible within one block. Different blocks may consist of plates soiled in different ways, washed by different people.

Now the number of plates available in one block is limited and usually allows only four tests to be completed, i.e., there is a restriction to four experimental units per block. Suppose that it is required to compare eight treatments. Table 11.2 shows that a balanced incomplete block design exists with 4 units per block ($k = 4$) and 8 treatments, and requires 56 experimental units, each treatment occurring 7 times, and this is the natural design to consider first.

If this seems a reasonable size of experiment, having regard to the resources available and to the precision that is required, this would settle the design to be used. If it were desired to attain higher precision than this gives, two or more whole repetitions of the basic design could be used. If the basic design is considered too large, various possibilities can be tried:

(*a*) It sometimes happens that if we add one or more additional treatments, a more suitable design will be found. In the present case the designs with nine and ten treatments do not involve fewer units than that with eight and therefore this device does not work.

(*b*) It quite frequently happens that if one or more treatments are omitted, a design of approximately the required size exists. Thus in the present case if one treatment is omitted, a design for 7 treatments, using only 28 experimental units, can be used. In many investigations the number of treatments for inclusion is somewhat arbitrary, and a treatment excluded from one experiment can always be taken up for examination in a later part of the investigation. (In others it would be very undesirable to leave out an important element of the investigation in order to obtain a simple design.)

(*c*) We may use fewer than four units per block, the *maximum* number. There is no design in Table 11.2 with three units per block and eight treatments, although potentially suitable designs do exist for seven and for nine treatments. We shall see in the discussion of the analysis of the results from such an experiment that for a given number of treatments, the number of units per block should be as large as is consistent with holding experimental conditions uniform within the block. Therefore it is best to use four units per block if possible.

(*d*) We may abandon the idea of using a balanced incomplete block design. In an experiment where complexity of analysis is of little concern we may be able to use one of the more complicated partially balanced designs described in § 11.4. Or if it is very important to have a simply arranged experiment we may have to abandon this system of blocking, use blocks with more experimental units, and accept any resulting loss of precision.

These considerations are typical of those involved in selecting a balanced incomplete block design.

The general conclusions to be drawn from Table 11.2 are two. First, the existence of designs is very limited for the larger values of k, the number of units per block, unless we can embark on a large experiment. Second, for the smaller values of k, balanced incomplete block designs exist for most values of the number of treatments, t, but even here the possible values for the total number of units are limited.

Next we must consider the form of the designs. Those cases marked in Table 11.2 by the letters unr., standing for unreduced, are derived by taking all possible sets of k treatments from the full t, each set forming a block. Thus with $k = 3$ and $t = 5$ we can form the design by writing down systematically all ways of selecting three treatments from $T_1, \ldots, T_5$. This gives

$$
\begin{array}{lll@{\qquad}lll}
T_1 & T_2 & T_3 & T_1 & T_4 & T_5 \\
T_1 & T_2 & T_4 & T_2 & T_3 & T_4 \\
T_1 & T_2 & T_5 & T_2 & T_3 & T_5 \\
T_1 & T_3 & T_4 & T_2 & T_4 & T_5 \\
T_1 & T_3 & T_5 & T_3 & T_4 & T_5
\end{array}
$$

The last step would be to randomize this design by permuting the blocks at random, rearranging the three treatments randomly within each block, and numbering the treatments randomly. This would lead to an arrangement such as that of Table 11.1.

In other cases the methods of construction are more complicated and raise interesting mathematical problems (Mann, 1949). These need not, however, concern the practical worker who can in most cases obtain the arrangements from tables such as Table 11.3. This lists the smaller designs; a fuller table is given by Cochran and Cox (1957, Chapter 11). Before use, the arrangements given in the table should be randomized.

Some of the designs are marked in Tables 11.2 and 11.3 as *resolvable*. This means that the blocks can be grouped into sets, each treatment occurring just once within each set. This leads to an increase in precision under certain circumstances and a resolvable design should be used in preference to a nonresolvable one, whenever a choice exists.

Example 11.2. Wadley (1948) has used balanced incomplete block designs in work on cattle connected with the comparison of allergen preparations. Tuberculosis in cattle can be diagnosed by injecting the skin with an appropriate allergen and observing the thickening produced. In an experiment to compare allergens, the observation for each allergen is the log concentration required to produce a 3-mm thickening, this being estimated by interpolation from the thickenings observed at four different concentrations.

In Wadley's experiment 16 allergens were under comparison. On each cow there were 4 main regions and in each region about 16 injections could be made. This suggests using each region as a block, with 4 allergen preparations in each block, each being used at 4 concentrations, thus making 16 injections in each block. If no suitable design had been available for this case, it would have been natural to have considered whether 5 preparations could have been included in each region (20 injections).

We therefore examine the part of Table 11.2 with 4 units per block, and a

TABLE 11.3

Details of the Smaller Balanced Incomplete Block Designs that are not Unreduced and do not Correspond to Youden Squares

Design						
(*a*) $k = 3$, $t = 6$,	(1, 2, 5);	(1, 2, 6);	(1, 3, 4);	(1, 3, 6);		
$b = 10$, $r = 5$	(1, 4, 5);	(2, 3, 4);	(2, 3, 5);	(2, 4, 6);		
	(3, 5, 6);	(4, 5, 6)				
(*b*) $k = 3$, $t = 9$,	Repl. I.	(1, 2, 3);	(4, 5, 6);	(7, 8, 9)		
$b = 12$, $r = 4$	Repl. II.	(1, 4, 7);	(2, 5, 8);	(3, 6, 9)		
	Repl. III.	(1, 5, 9);	(7, 2, 6);	(4, 8, 3)		
	Repl. IV.	(1, 8, 6);	(4, 2, 9);	(7, 5, 3)		
(*c*) $k = 3$, $t = 10$,	(1, 2, 3);	(1, 2, 4);	(1, 3, 5);	(1, 4, 6)		
$b = 30$, $r = 9$	(1, 5, 7);	(1, 6, 8);	(1, 7, 9);	(1, 8, 10);		
	(1, 9, 10);	(2, 3, 6);	(2, 4, 10);	(2, 5, 8);		
	(2, 5, 9);	(2, 6, 7);	(2, 7, 9);	(2, 8, 10);		
	(3, 4, 7);	(3, 4, 8);	(3, 5, 6);	(3, 7, 10);		
	(3, 8, 9);	(3, 9, 10);	(4, 5, 9);	(4, 5, 10);		
	(4, 6, 9);	(4, 7, 8);	(5, 6, 10);	(5, 7, 8);		
	(6, 7, 10);	(6, 8, 9)				
(*d*) $k = 3$, $t = 13$,	(1, 2, 3);	(1, 4, 5);	(1, 6, 7);	(1, 8, 9);		
$b = 26$, $r = 6$	(1, 10, 11);	(1, 12, 13);	(2, 4, 9);	(2, 5, 13);		
	(2, 6, 11);	(2, 7, 10);	(2, 8, 12);	(3, 4, 8);		
	(3, 5, 10);	(3, 6, 12);	(3, 7, 9);	(3, 11, 13);		
	(4, 6, 10);	(4, 7, 13);	(4, 11, 12);	(5, 6, 8);		
	(5, 7, 12);	(5, 9, 11);	(6, 9, 13);	(7, 8, 11);		
	(8, 10, 13);	(9, 10, 12)				
(*e*) $k = 3$, $t = 15$,	Repl. I.	(1, 2, 3);	(4, 8, 12);	(5, 10, 15);	(6, 11, 13);	(7, 9, 14)
$b = 35$, $r = 7$	Repl. II.	(1, 4, 5);	(2, 8, 10);	(3, 13, 14);	(6, 9, 15);	(7, 11, 12)
	Repl. III.	(1, 6, 7);	(2, 9, 11);	(3, 12, 15);	(4, 10, 14);	(5, 8, 13)
	Repl. IV.	(1, 8, 9);	(2, 13, 15);	(3, 4, 7);	(5, 11, 14);	(6, 10, 12)
	Repl. V.	(1, 10, 11);	(2, 12, 14);	(3, 5, 6);	(4, 9, 13);	(7, 8, 15)
	Repl. VI.	(1, 12, 13);	(2, 5, 7);	(3, 9, 10);	(4, 11, 15);	(6, 8, 14)
	Repl. VII.	(1, 14, 15);	(2, 4, 6);	(3, 8, 11);	(5, 9, 12);	(7, 10, 13)
(*f*) $k = 4$, $t = 8$,	Repl. I.	(1, 2, 3, 4);	(5, 6, 7, 8)			
$b = 14$, $r = 7$	Repl. II.	(1, 2, 7, 8);	(3, 4, 5, 6)			
	Repl. III.	(1, 3, 6, 8);	(2, 4, 5, 7)			
	Repl. IV.	(1, 4, 6, 7);	(2, 3, 5, 8)			
	Repl. V.	(1, 2, 5, 6);	(3, 4, 7, 8)			
	Repl. VI.	(1, 3, 5, 7);	(2, 4, 6, 8)			
	Repl. VII.	(1, 4, 5, 8);	(2, 3, 6, 7)			
(*g*) $k = 4$, $t = 9$,	(1, 2, 3, 4);	(1, 2, 5, 6);	(1, 2, 7, 8);	(1, 3, 5, 7);		
$b = 18$, $r = 8$	(1, 4, 6, 8);	(1, 3, 6, 9);	(1, 4, 8, 9);	(1, 5, 7, 9);		
	(2, 3, 8, 9);	(2, 4, 5, 9);	(2, 6, 7, 9);	(2, 3, 4, 7);		
	(2, 5, 6, 8);	(3, 5, 8, 9);	(4, 6, 7, 9);	(3, 4, 5, 6);		
	(3, 6, 7, 8);	(4, 5, 7, 8)				
(*h*) $k = 4$, $t = 10$,	(1, 2, 3, 4);	(1, 2, 5, 6);	(1, 3, 7, 8);	(1, 4, 9, 10);		
$b = 15$, $r = 6$	(1, 5, 7, 9);	(1, 6, 8, 10);	(2, 3, 6, 9);	(2, 4, 7, 10)		
	(2, 5, 8, 10);	(2, 7, 8, 9);	(3, 5, 9, 10);			
	(3, 6, 7, 10);	(3, 4, 5, 8);	(4, 5, 6, 7);	(4, 6, 8, 9)		
(*i*) $k = 4$, $t = 16$,	Repl. I.	(1, 2, 3, 4);	(5, 6, 7, 8);	(9, 10, 11, 12);	(13, 14, 15, 16)	
$b = 20$, $r = 5$	Repl. II.	(1, 5, 9, 13);	(2, 6, 10, 14);	(3, 7, 11, 15);	(4, 8, 12, 16)	
	Repl. III.	(1, 6, 11, 16);	(5, 2, 15, 12);	(9, 14, 3, 8);	(13, 10, 7, 4)	
	Repl. IV.	(1, 14, 7, 12);	(13, 2, 11, 8);	(5, 10, 3, 16);	(9, 6, 15, 4)	
	Repl. V.	(1, 10, 15, 8);	(9, 2, 7, 16);	(13, 6, 3, 12);	(5, 14, 11, 4)	

TABLE 11.3 (*Continued*)

Design	Blocks
(*j*) $k = 5$, $t = 9$, $b = 18$, $r = 10$	(1, 2, 3, 5, 9); (1, 2, 3, 7, 8); (1, 2, 4, 6, 8); (1, 2, 5, 6, 8); (1, 2, 6, 7, 9); (1, 3, 4, 5, 6); (1, 3, 4, 5, 7); (1, 3, 6, 7, 9); (1, 4, 5, 8, 9); (1, 4, 7, 8, 9); (2, 3, 4, 6, 9); (2, 3, 4, 7, 8); (2, 3, 5, 8, 9); (2, 4, 5, 6, 7); (2, 4, 5, 7, 9); (3, 4, 6, 8, 9); (3, 5, 6, 7, 8); (5, 6, 7, 8, 9)
(*k*) $k = 5$, $t = 10$, $b = 18$, $r = 9$	(1, 2, 3, 4, 5); (1, 2, 3, 6, 7); (1, 2, 4, 6, 9); (1, 2, 5, 7, 8); (1, 3, 6, 8, 9); (1, 3, 7, 8, 10); (1, 4, 5, 6, 10); (1, 4, 8, 9, 10); (1, 5, 7, 9, 10); (2, 3, 4, 8, 10); (2, 3, 5, 9, 10); (2, 4, 7, 8, 9); (2, 5, 6, 8, 10); (2, 6, 7, 9, 10); (3, 4, 6, 7, 10); (3, 4, 5, 7, 9); (3, 5, 6, 8, 9); (4, 5, 6, 7, 8)
(*l*) $k = 5$, $t = 25$, $b = 30$, $r = 6$	Repl. I. (1, 2, 3, 4, 5); (6, 7, 8, 9, 10); (11, 12, 13, 14, 15); (16, 17, 18, 19, 20); (21, 22, 23, 24, 25) Repl. II. (1, 6, 11, 16, 21); (2, 7, 12, 17, 22); (3, 8, 13, 18, 23); (4, 9, 14, 19, 24) (5, 10, 15, 20, 25) Repl. III. (1, 7, 13, 19, 25); (21, 2, 8, 14, 20); (16, 22, 3, 9, 15); (11, 17, 23, 4, 10); (6, 12, 18, 24, 5) Repl. IV. (1, 12, 23, 9, 20); (16, 2, 13, 24, 10); (6, 17, 3, 14, 25); (21, 7, 18, 4, 15); (11, 22, 8, 19, 5) Repl. V. (1, 17, 8, 24, 15); (11, 2, 18, 9, 25); (21, 12, 3, 19, 10); (6, 22, 13, 4, 20); (16, 7, 23, 14, 5) Repl. VI. (1, 22, 18, 14, 10); (6, 2, 23, 19, 15); (11, 7, 3, 24, 20); (16, 12, 8, 4, 25); (21, 17, 13, 9, 5)
(*m*) $k = 6$, $t = 9$, $b = 12$, $r = 8$	(1, 2, 3, 4, 5, 6); (1, 2, 3, 7, 8, 9); (1, 2, 4, 5, 7, 8) (1, 2, 4, 6, 8, 9); (1, 2, 5, 6, 7, 9); (1, 3, 4, 5, 8, 9) (1, 3, 4, 6, 7, 9); (1, 3, 5, 6, 7, 8); (2, 3, 4, 5, 7, 9) (2, 3, 4, 6, 7, 8); (2, 3, 5, 6, 8, 9); (4, 5, 6, 7, 8, 9)
(*n*) $k = 6$, $t = 10$, $b = 15$, $r = 9$	(1, 2, 4, 5, 8, 9); (5, 6, 7, 8, 9, 10); (2, 4, 5, 6, 9, 10) (1, 2, 4, 6, 7, 8); (3, 4, 7, 8, 9, 10); (2, 3, 4, 6, 8, 10) (1, 2, 6, 7, 9, 10); (1, 3, 5, 6, 8, 9); (1, 2, 3, 8, 9, 10) (2, 3, 4, 5, 7, 9); (1, 4, 5, 7, 8, 10); (1, 2, 3, 5, 7, 10) (2, 3, 5, 6, 7, 8); (1, 3, 4, 5, 6, 10); (1, 3, 4, 6, 7, 9)
(*o*) $k = 6$, $t = 16$, $b = 24$, $r = 9$	(1, 2, 5, 6, 11, 12); (1, 2, 7, 8, 13, 14); (1, 2, 9, 10, 15, 16); (1, 3, 5, 7, 10, 12); (1, 3, 6, 8, 13, 15); (1, 3, 9, 11, 14, 16); (1, 4, 5, 8, 10, 11); (1, 4, 6, 7, 13, 16); (1, 4, 9, 12, 14, 15); (2, 3, 5, 8, 14, 15); (2, 3, 6, 7, 9, 12); (2, 3, 10, 11, 13, 16); (2, 4, 5, 7, 14, 16); (2, 4, 6, 8, 9, 11); (2, 4, 10, 12, 13, 15); (3, 4, 5, 6, 15, 16); (3, 4, 7, 8, 9, 10); (3, 4, 11, 12, 13, 14); (5, 6, 9, 10, 13, 14); (5, 7, 9, 11, 13, 15); (5, 8, 9, 12, 13, 16); (6, 7, 10, 11, 14, 15); (6, 8, 10, 12, 14, 16); (7, 8, 11, 12, 15, 16)

Randomize by permuting blocks (within replicates) and numbers within blocks and by assigning treatments to numbers at random. Designs for the situations marked by *Y* in Table 11.2 are obtained from the Youden square arrangements of Table 11.5.

design with 16 treatments does exist in 5 replicates. Moreover the design is resolvable, i.e., is divided into five separate sections of four blocks each, and each treatment occurs once in each section (see Table 11.3). Thus the experiment can be conveniently set out on 5 tuberculin-sensitized cows, each treatment occurring once on each cow, there being 4 blocks (areas) on each cow within which 16 injections are made. In Wadley's experiment, ten cows were available and two independent replications of the design just mentioned were therefore used.

The advantage of the resolvability in this case is partly the possible gain in precision mentioned above and explained in the next section, and partly that if the treatment effects vary somewhat from cow to cow, an average effect over the cows in the experiment is estimated.

(iii) Analysis of Observations

The general idea involved in the analysis of the results has already been explained in (i); the mean of all the observations on a particular treatment is adjusted for the effect of the blocks in which the treatment did not occur. The estimation of the residual standard deviation is straightforward, but the explanation of the calculations for that is a little beyond the scope of this book and will not be given; see Cochran and Cox (1957, § 11.55). To estimate the treatment effects themselves the procedure of the following example is used.

Example 11.3. Suppose that in the experiment of Table 11.1 the following observations are obtained.

						Block Total
T_4	4.43	T_5	3.16	T_1	1.40	8.99
T_4	5.09	T_2	1.81	T_5	4.54	11.44
T_2	3.91	T_4	6.02	T_1	3.32	13.25
T_5	4.66	T_3	3.09	T_1	3.56	11.31
T_3	3.66	T_4	2.81	T_5	4.66	11.13
T_2	1.60	T_3	2.13	T_1	1.31	5.04
T_3	4.26	T_1	3.86	T_4	5.87	13.99
T_3	2.57	T_5	3.06	T_2	3.45	9.08
T_2	3.31	T_3	5.10	T_4	5.42	13.83
T_5	5.53	T_1	4.46	T_2	3.94	13.93
						111.99

We first calculate the totals of the observations on each block. These are given to the right of the observations. Then the treatment totals are found in the table below; 17.91 is the sum of the 6 observations on treatment T_1.

	Treatment Totals	Sum of all Block Totals in which Treatment Appeared	Adjusted Treatment Total	Adjusted Treatment Mean	Unadjusted Treatment Mean
T_1	17.91	66.51	−12.78	2.88	2.98
T_2	18.02	66.57	−12.51	2.90	3.00
T_3	20.81	64.38	−1.95	3.60	3.47
T_4	29.64	72.63	16.29	4.82	4.94
T_5	25.61	65.88	10.95	4.46	4.27
	111.99	335.97	0.00		

General mean 3.73

Next we find for each treatment the sum of the block totals for all those blocks in which the treatment appeared. Thus, for treatment 1, 66.51 is the sum of the block totals for blocks 1, 3, 4, 6, 7, and 10. (This column adds up to the sum of the previous column times the number of units per block.)

The adjusted treatment total is then calculated as

(no. of units per block) × (treatment total) − (associated sum of block totals);

that is, for treatment 1, this is $3 \times 17.91 - 66.51 = -12.78$.
This column of figures adds up to zero. Finally the adjusted treatment mean is

$$\text{(adjusted treatment total)} \times \frac{\text{(no. of treatments} - 1)}{\text{(total no. of units)} \times \text{(no. of units per block} - 1)}$$
$$+ \text{ (overall mean of original observations)},$$

and these are given in the next to the last column. The final column gives, for comparison, the unadjusted treatment means, obtained by dividing the treatment totals by the number of replications, in this case six. In fact the data analyzed here were constructed artificially in such a way that the true differences from T_1 are $\frac{1}{2}$, 1, $1\frac{1}{2}$, 2 for T_2, T_3, T_4, T_5. In the present case the adjustment has improved the estimates.

The formulas just given can be obtained in two ways. First, the general method of least squares, which is a theoretical technique for obtaining estimates of maximum precision in certain types of situation, may be used to obtain the formulas. Second, suppose that it is accepted as intuitively obvious that the final estimate for a particular treatment must be some linear combination of (*a*) the overall mean, (*b*) the total (or mean) observation on that treatment, and (*c*) the sum of all block totals in which the treatment does (or does not) appear. Then some elementary algebra shows that the formulas above are the only ones to give consistent answers when there are arbitrary block differences, arbitrary treatment differences, and zero random variation within each block.

In the simpler experiments of previous chapters, the standard error of the estimated difference between two treatments is

$$\sqrt{\left(\frac{2}{\text{no. of obs. per treatment}}\right)} \times \text{(residual standard deviation)}. \qquad (1)$$

In the balanced incomplete block design the standard error is larger than (1) due, in effect, to errors in the adjustments applied to correct for block effects. It is usual to express the standard error of the difference in the form

$$\frac{\text{formula (1)}}{\sqrt{\text{(efficiency factor)}}}, \qquad (2)$$

where the efficiency factor is equal to

$$\frac{\text{(no. of treatments)} \times \text{(no. of units per block} - 1)}{\text{(no. of treatments} - 1) \times \text{(no. of units per block)}}. \qquad (3)$$

In the present example this is $(5 \times 2)/(4 \times 3) = 5/6$. The practical interpretation of the efficiency factor will be discussed below.

The analysis is completed by estimating the residual standard deviation, a process not explained here; the answer is 0.75, so that the standard error of the difference between two adjusted treatment means is $\sqrt{(2/6)} \times 0.75 \times \sqrt{(6/5)} = 0.47$. Hence approximate 95 per cent limits of uncertainty for the true difference are the estimated difference plus and minus 0.94.

The general interpretation of the efficiency factor is best seen as follows. The standard error of the estimated difference between two treatments is given by formula (2). If the above system of blocking is abandoned and the same number of units are formed into an ordinary randomized block design, with each treatment once in each block, the standard error is given by (1), where the standard deviation is now measured within large blocks. The point of using the balanced incomplete block design is that one expects to attain an appreciable reduction in the standard deviation; if in fact this reduction is not achieved and the standard deviations are the same in the two designs, the formulas show that

$$\begin{pmatrix}\text{standard error in incomplete}\\ \text{block design}\end{pmatrix} = \frac{1}{\sqrt{(\text{eff.})}} \times \begin{pmatrix}\text{standard error}\\ \text{in randomized}\\ \text{block design}\end{pmatrix},$$

and in the present example the multiplying factor is $\sqrt{1.2} = 1.1$. That is, if we use the incomplete block design and gain no reduction in standard deviation thereby, we get a 10 per cent *increase* in the width of the limits of uncertainty. We must get as least a 10 per cent reduction in standard deviation for the use of the incomplete design to be advantageous.*

Quite generally, the efficiency factor tells us the loss caused by using an incomplete design without reducing the residual standard deviation. For balanced incomplete block designs the efficiency factor is smallest when the number of units per block is small and the number of treatments large; thus with two units per block and a large number of treatments, the efficiency factor is 1/2. The efficiency factor is near one when the number of treatments does not greatly exceed the number of units per block, or when the number of units per block is large.

If the efficiency factor is less than about 0.85 and if the number of blocks exceeds about ten, a more complicated method of analysis may be used that makes any possible loss of precision as compared with a randomized block design small. This method is based on the idea that if the variation between blocks is not too great, the block totals contain a certain amount of information about the treatment effects. The method, which is called the recovery of interblock information, is described in the text-books referred to previously: if it is used, it is an advantage for the design to be resolvable.

* This ignores differences due to a change in the residual degrees of freedom.

11.3 INCOMPLETE DESIGNS FOR THE TWO-WAY ELIMINATION OF ERROR

(i) General

In the preceding section we have considered incomplete designs in which variation between blocks is eliminated, these designs being analogous to randomized block designs. Now we turn to designs, corresponding to a Latin square, in which the units are classified in two ways, analogous to the rows and columns of the Latin square. The need for an incomplete design arises when the admissible number of units in one direction is less than the number of treatments or when the admissible number in both directions is less than the number of treatments. The two cases will be considered separately below.

(ii) Youden Squares

Suppose, for example, that we have an experiment with seven treatments which we would like to lay out in a 7 × 7 Latin square to eliminate the effect of variation between times of day and between days, or between

TABLE 11.4

EXAMPLE OF A YOUDEN SQUARE

T_7	T_1	T_2	T_5
T_5	T_6	T_7	T_3
T_1	T_2	T_3	T_6
T_6	T_7	T_1	T_4
T_4	T_5	T_6	T_2
T_3	T_4	T_5	T_1
T_2	T_3	T_4	T_7

times and between sets of apparatus, etc. Suppose too that it is possible to arrange the units so that there are seven "rows" (days), but that it is not practicable to have more than say four "columns" (times of day). A natural thing is then to arrange the treatments so that each occurs once in each column and so that the rows have the property of the balanced incomplete block design. A design of this sort is called a Youden square.

Table 11.4 shows such an arrangement after randomization. Check that each treatment occurs once in each column and that each pair of treatments occurs together in the same row exactly twice. In fact the seven rows can be regarded as the blocks of the balanced incomplete block design $r = k = 4$, $t = b = 7$, written down in a special order.

The situation for which Youden originally introduced this sort of design (Youden, 1937) gives a more specific example of how the arrangement is used in practice.

Example 11.4. Youden was concerned with a greenhouse experiment on tobacco-mosaic virus. One leaf formed an experimental unit and the observation consisted of the number of lesions produced on the leaf by rubbing with a solution containing virus. Experience showed that the main source of uncontrolled variation was the variation from plant to plant, there also being a steady trend in response down each plant from top to bottom.

This would normally suggest the use of a Latin square to control both types of variation, but when the number of treatments for comparison exceeds the number of leaves per plant, we have just the difficulty discussed above. With four leaves per plant and seven treatments, the arrangement of Table 11.4 would be suitable, each row representing the treatments to one plant and the four columns standing for the leaves in order from the top downwards.

If it was thought that the data provided from one such Youden square would not give sufficient precision, the design would be repeated as many times as seemed desirable, rerandomizing on each occasion.

Example 11.5. An important application of Youden squares has been given by Durbin (1951). In some types of psychological and sociological research, and in other fields too, it is necessary for observers to rank a number of objects in order of preference. If the number of objects to be ranked is large, it will be undesirable to have more than a limited number of objects presented for comparison at a time. A reasonable thing is to divide the objects into blocks in accordance with a balanced incomplete block design and to present the objects in one block for ranking at one time. In the whole experiment each object will be compared with each other object the same number of times in all.

Suppose further that in doing the ranking, the observer is first given one of the objects, then another, and so on until all in the block have been examined. It would then be reasonable to arrange that each object comes in each position the same number of times, and this can be achieved by using a Youden square.

For example if there were seven objects $T_1, \ldots, T_7$ and it was desired to examine only four at a time, design (*b*) of Table 11.5 would give, before randomization,

Order of Presentation	Set 1	Set 2	Set 3	Set 4	Set 5	Set 6	Set 7
1st object	T_1	T_2	T_3	T_4	T_5	T_6	T_7
2nd object	T_3	T_4	T_5	T_6	T_7	T_1	T_2
3rd object	T_4	T_5	T_6	T_7	T_1	T_2	T_3
4th object	T_5	T_6	T_7	T_1	T_2	T_3	T_4

Different sets could be examined by different observers or by the same observer at different times. The observations would consist of noting, say, that in set 1, T_4 was preferred most (rank 1), then T_5 (rank 2), then T_1 (rank 3), last T_3 (rank 4), and so on. Durbin discusses the analysis of such observations, the first step being to calculate the mean rank for each object.

Youden squares can be constructed only in limited cases, in fact only when there is in Table 11.2 a balanced incomplete block design with the number k equal to the number r equal to the required number of columns, and with the number t equal to the number b equal to the required number of treatments. These cases are marked by a Y in the last column of Table 11.2.

Those Youden squares corresponding to the unreduced designs in Table 11.2 with $k = t - 1$ are constructed by omitting the last column of a $t \times t$ Latin square. For example if we write down a 5×5 Latin square, omitting the last column, we get a design with five rows, four columns, and five treatments; the reader should verify that it forms a Youden square by checking that each treatment occurs once in each column and that each pair of treatments occurs together in the same row, the same number of times (three).

Other Youden squares are set out in Table 11.5. A few more requiring more than 150 experimental units per replicate are given by Cochran and Cox (1957, Chapter 13).

In Table 11.5 the designs are specified by writing out in order the treatments occurring in the first row, in the second row and so on. Thus the first design listed would, if written out more fully, appear as

$$\begin{array}{ccc} T_1 & T_2 & T_4 \\ T_2 & T_3 & T_5 \\ T_3 & T_4 & T_6 \\ T_4 & T_5 & T_7 \\ T_5 & T_6 & T_1 \\ T_6 & T_7 & T_2 \end{array}$$

All the designs given in the table can be obtained by the following process of *cyclic substitution* starting from the first row. The process is that, in the example just given, the second row is obtained by adding 1 to the treatment numbers in the first row, and so on. When we come to the next to the last row, the rule gives T_5 T_6 T_8; we adopt the further rule that with seven treatments T_8 means T_1, and so on. In the more lengthy cases of Table 11.5, only the first two rows are explicitly given, the remainder being determined by cyclic substitution as just described.

Randomization follows the method given for Latin squares; that is, rows are randomized, columns are randomized, and then the treatments are numbered at random.

The estimation of treatment effects is done by exactly the same method as for balanced incomplete blocks, the column arrangement being in effect ignored. The precision of the estimated treatment effects is given by formulas (2) and (3), where the residual standard deviation measures the amount of uncontrolled variation remaining when both row and

TABLE 11.5

DETAILS OF SOME YOUDEN SQUARE DESIGNS

(*a*)	3 columns, 7 treatments;	(1, 2, 4); (2, 3, 5); (3, 4, 6); (4, 5, 7); (5, 6, 1); (6, 7, 2); (7, 1, 3).
(*b*)	4 columns, 7 treatments;	(1, 3, 4, 5); (2, 4, 5, 6); (3, 5, 6, 7); (4, 6, 7, 1); (5, 7, 1, 2); (6, 1, 2, 3); (7, 2, 3, 4).
(*c*)	4 columns, 13 treatments;	(1, 2, 4, 10); (2, 3, 5, 11); (3, 4, 6, 12); (4, 5, 7, 13); (5, 6, 8, 1); (6, 7, 9, 2); (7, 8, 10, 3); (8, 9, 11, 4); (9, 10, 12, 5); (10, 11, 13, 6); (11, 12, 1, 7); (12, 13, 2, 8); (13, 1, 3, 9).
(*d*)	5 columns, 11 treatments;	(1, 5, 6, 7, 9); (2, 6, 7, 8, 10); etc.
(*e*)	5 columns, 21 treatments;	(1, 2, 5, 15, 17); (2, 3, 6, 16, 18); etc.
(*f*)	6 columns, 11 treatments;	(1, 2, 3, 7, 9, 10); (2, 3, 4, 8, 10, 11); etc.
(*g*)	6 columns, 16 treatments;	See Cochran and Cox (1957, plan 13.9).
(*h*)	7 columns, 15 treatments;	(1, 2, 3, 5, 6, 9, 11); (2, 3, 4, 6, 7, 10, 12); etc.
(*i*)	8 columns, 15 treatments;	(1, 4, 5, 7, 9, 10, 11, 12); (2, 5, 6, 8, 10, 11, 12, 13); etc.
(*j*)	9 columns, 13 treatments;	(1, 4, 5, 6, 8, 9, 10, 11, 12); (2, 5, 6, 7, 9, 10, 11, 12, 13); etc.

column effects have been removed. The distinction between a balanced incomplete block design and a Youden square is, however, material in the part of the analysis concerned with estimating the residual standard deviation.

To sum up, Youden squares should be considered for use whenever it is desirable to arrange the experimental units in a two-way pattern and the number of units in one direction can be made equal to the number of treatments, while the maximum number of units in the other direction is smaller than the number of treatments.

(iii) Lattice Squares

In a Youden square the number of rows is equal to the number of treatments and it is only in one direction, namely the columns, that the design has an incomplete character. Now it may happen that we desire to eliminate variations in two directions, and in neither can we get enough homogeneous experimental units to include each treatment once. We need a new sort of design to deal with this.

Reasonably simple designs seem to be available only when the number of units per row equals the number of units per column, i.e., if the experiment is "square." If the whole experiment consists of several such squares, such that

(*a*) each treatment occurs once in each square;

(*b*) each pair of treatments occurs together in the same row or column of a square the same number of times in all,

we say we have a set of *balanced lattice squares*.

The simplest example is the set of two 3 × 3 squares shown in Table 11.6(*a*). Written out in full, they are

Square 1			*Square 2*		
T_1	T_2	T_3	T_1	T_6	T_8
T_4	T_5	T_6	T_9	T_2	T_4
T_7	T_8	T_9	T_5	T_7	T_3

The properties defining balanced lattice squares hold, since each treatment occurs once in each square and each pair of treatments occur together in a row or column just once. Thus T_2 and T_7 occur together in the second column of the second square and nowhere else, T_4 and T_6 occur together in the second row of the first square and nowhere else, and so on. In using these squares, every effort should be made to hold the experimental conditions constant within each square, except for allowable row and column differences, but a systematic difference of apparatus or observer between the two squares of the set should not affect the final precision attained.

Balanced lattice squares exist if the number of treatments is the square of a number for which a set of Latin squares with the maximum number of alphabets exists (§ 10.5), i.e., when the number of treatments is 3^2, 4^2, 5^2, the cases we have already had, and 7^2, 8^2, 9^2, 11^2, and so on. The general method of construction is given by Kempthorne (1952, p. 485). The designs should be randomized by permuting rows and columns at random independently for each square of the set.

The analysis of the observations from such an experiment follows the same general principles as for balanced incomplete blocks; details are given by Cochran and Cox (1957, § 12.2) and by Kempthorne (1952, p. 486).

TABLE 11.6

THE SMALLEST BALANCED LATTICE SQUARES

(*a*) Two 3 × 3 Squares (9 Treatments)	Square 1:	(1, 2, 3); (4, 5, 6); (7, 8, 9)	
	Square 2:	(1, 6, 8); (9, 2, 4); (5, 7, 3)	
(*b*) Five 4 × 4 Squares (16 Treatments)	Square 1:	(1, 2, 3, 4);	(5, 6, 7, 8);
		(9, 10, 11, 12);	(13, 14, 15, 16)
	Square 2:	(1, 11, 16, 6);	(12, 2, 5, 15);
		(14, 8, 3, 9);	(7, 13, 10, 4)
	Square 3:	(1, 8, 10, 15);	(2, 7, 9, 16);
		(3, 6, 12, 13);	(4, 5, 11, 14)
	Square 4:	(1, 5, 9, 13);	(6, 2, 14, 10);
		(11, 15, 3, 7);	(16, 12, 8, 4)
	Square 5:	(1, 7, 12, 14);	(8, 2, 13, 11);
		(10, 16, 3, 5);	(15, 9, 6, 4)
(*c*) Three 5 × 5 Squares (25 Treatments)	Square 1:	(1, 2, 3, 4, 5);	(6, 7, 8, 9, 10);
		(11, 12, 13, 14, 15);	
		(16, 17, 18, 19, 20);	
		(21, 22, 23, 24, 25)	
	Square 2:	(1, 9, 12, 20, 23);	(24, 2, 10, 13, 16);
		(17, 25, 3, 6, 14);	
		(15, 18, 21, 4, 7);	
		(8, 11, 19, 22, 5)	
	Square 3:	(1, 14, 22, 10, 18);	
		(19, 2, 15, 23, 6);	
		(7, 20, 3, 11, 24);	
		(25, 8, 16, 4, 12);	
		(13, 21, 9, 17, 5).	

If a second replicate is required of a design with an odd number of treatments, the design listed should be used interchanging rows and columns. If the number of treatments is k^2, the number of squares in the set is $(k + 1)$ when k is even and $\frac{1}{2}(k + 1)$ when k is odd.

The following is an example of the use of lattice squares in a laboratory experiment.

Example 11.6.* This experiment concerned the comparison of the resistance to stem rot of a number of progenies of Broad Red Clover. Seedlings were sown in boxes with space for five "plots" in a box and about 30 seedlings of one progeny to a plot, different plots receiving different progenies.

The number of seedlings on each plot was counted and then the leaves were

* I am grateful to Miss J. Drayner, Plant Breeding Institute, Cambridge, England, for the details of this.

sprinkled with powdered inoculum and the boxes put under humidity cages for the attack to develop. After a suitable time the cages were removed, the boxes allowed to dry out, and the number of seedlings surviving counted. The observation on each plot for analysis was the proportion of seedlings surviving.

It was suspected that there might be systematic differences in the severity of the attack from box to box and also that within any one box, the progenies in the end positions might be subject to a rather less severe attack than those in the center positions. This suggested that a design with two-way elimination of error, between boxes and between positions, would be a good thing.

It was desired to test about 20–30 progenies in one experiment, and this ruled out Latin squares from the start, since a Latin square of this size would not only require 20–30 plots in each box, but also would require 20–30 boxes, an amount of repetition far more than was either practicable or desirable. Youden squares were ruled out for substantially the same reason; thus with 5 units per box and 21 treatments, we might consider the Youden square (*e*) of Table 11.5, but again this would require 21 boxes, more than could be coped with at one time and representing far too large an experiment for the precision desired. Thus a design was required in which *both* the number of rows *and* the number of columns were less than the number of treatments and so lattice squares seemed fitting. With 5 plots per box, we must, with this sort of design, have 5 boxes in a square and 25 treatments (progenies) for comparison. The whole experiment consisted, from Table 11.6, of three such squares that could be dealt with experimentally at different times.

Therefore this was an experiment for comparing quite a large number of progenies, with a small number of repetitions of each progeny, and with balancing of both types of uncontrolled variation. If one set of three squares seemed unlikely to give the desired precision, a further set of three could be added, interchanging rows and columns in accord with the footnote to Table 11.6. This was not thought necessary in the present case.

The snag about the balanced lattice square arrangements is, of course, the severe limitation to square arrangements in which also the number of treatments is the square of the number of rows. Further designs are mentioned briefly in § 11.5, but tend to be rather complicated to analyze.

11.4 FURTHER INCOMPLETE BLOCK DESIGNS

(i) General

In § 11.2 we considered balanced incomplete block designs which were characterized, among other things, by the properties that each pair of treatments is compared with equal precision, that each block contains the same number of experimental units, and that each treatment occurs in all the same number of times. These are the most important incomplete block designs, but the need sometimes arises for others, either because no suitable balanced incomplete block design exists, or because, for example, it is required to make some comparisons more precisely than others.

There is in fact a very wide class of incomplete block designs, called *partially balanced incomplete blocks*, which have some, but not all, of the symmetry of the balanced designs. The following subsections describe briefly some special designs of this type that are sometimes useful; a full description is not attempted and in particular no account is given of methods of analysis, as these tend to be rather complicated. As a general rule, although the experimenter should find it useful to know broadly what possibilities are available, these designs should not be used without statistical help. In particular no gain in precision can be obtained from the designs unless the appropriate methods of analysis are used.

An account of general methods of analysis is given by Rao (1947), although it may often be better to work out methods from first principles. The theory of the general class of designs is outlined by Kempthorne (1952, Chapter 27), where further references are given; there is a very extensive literature on special cases.

(ii) Designs for Differential Precision

It may happen that although a balanced incomplete block design exists that would be suitable for the experiment under contemplation, it is not really the best design for the purpose, because some treatment comparisons are much more important than others. For example, there may be one special treatment, S, and a number of other treatments $T_1, T_2, \ldots$, the main interest attaching to the comparison of $T_1, T_2, \ldots$ individually with S, comparisons of the T's among themselves being of secondary importance. Thus, S may be a control or dummy treatment, the main interest in the early stages of an investigation being to establish which experimental treatments show a difference from the control, rather than in making a detailed comparison of the experimental treatments among themselves.

In the ordinary way, a modified randomized block design would be used in which the special treatment S occurs several times in each block and each other treatment just once. If, however, an incomplete block design is necessary, it may happen that no design with suitably different precisions exists. The procedure is then to have S occur the same number of times, once, twice, or more, within each block and to distribute the other treatments in a balanced incomplete block design over the remaining units. Thus, with S and eight ordinary treatments, and five units per block, we might try S once in each block and have $T_1, \ldots, T_8$ in a balanced incomplete block design with four units per block. A mathematical analysis should always be made before using such a design to assess the precision that will result for the different comparisons.

This idea of adjoining a balanced incomplete block design to an arrangement in which a treatment occurs once or more in each block generalizes in an obvious way when there are two or more "special" treatments each occurring in every block. It is also sometimes possible to build up incomplete block designs in which there are two types of treatment $T_1, T_2, \ldots$, and $S_1, S_2, \ldots$, and it is required to compare each T accurately with each other T and each S accurately with each other S, but comparisons of a T with an S are of subsidiary interest. However such designs, which are of use only in very special cases, are probably best constructed ad hoc.

(iii) More Units per Block than there are Treatments

A somewhat similar modification of a balanced incomplete block design is occasionally useful when the number of units per block in the natural system of blocking is more than, but less than twice, the number of treatments, all comparisons of pairs of treatments being of approximately equal importance. In some situations sufficient units could be omitted from each block to leave a simple randomized block design. If, however, this meant wasting experimental material, the following alternative method would be preferable. First write out an arrangement with each treatment occurring once in each block. This will leave unused a certain number of units in each block and the treatments are distributed over these in a balanced incomplete block arrangement. The whole is then randomized.

(iv) Designs for a Large Number of Treatments

Table 11.2 shows that when the number of treatments is large there are no balanced incomplete block designs with a moderate number of replications and with a smallish number of units per block. For instance, with not more than 10 units per block and with about 20 treatments, there is no design smaller than the one with 21 treatments and five replicates, 105 units; with 25 treatments there is nothing smaller than the design with 6 replicates, 150 units; and if there are more than about 35 treatments, there is no balanced incomplete block design with fewer than 10 replications, and often more are needed.

Now experiments with this number of treatments are usually preliminary trials to select treatments for detailed investigation, and high precision may not be required. There is therefore a need for a systematic class of designs with a fairly small number of replicates of each treatment. The main applications are in plant-breeding work, where it may be required to lay out field trials with as many as several hundred varieties and where it is seldom a good thing to have more than 12–16 plots per block.

Table 11.7 summarizes some of the properties of the wide class of designs called *lattices** that have been introduced for this situation. The first three types in Table 11.7 deal with 9, 16, 25, 49, 64, 81, 100, 121, 144, . . . treatments, cubic lattices give a set of designs for 27, 64, 125, 216, . . . treatments, whereas rectangular lattices are available for 12, 20, 30, 42, 56, 72, 90, 110, . . . treatments, the combined sets of values giving reasonably good coverage of the large numbers of treatments. There are other possible lattice designs but they are probably rarely needed, except in plant-breeding work.

TABLE 11.7

KEY TO THE MAIN TYPES OF LATTICE DESIGN

Name	No. of Treatments	No. of Replications a Multiple of	No. of Blocks a Multiple of
Simple lattice[1]	k^2	2	$2k$
Triple lattice	k^2	3	$3k$
Quadruple lattice[2]	k^2	4	$4k$
Cubic lattice	k^3	3	$3k^2$
Simple rectangular lattice	$k(k + 1)$	2	$2(k + 1)$
Triple rectangular lattice	$k(k + 1)$	3	$3(k + 1)$

Number of units per block is k in all cases.

[1] Sometimes called a square lattice, but must not be confused with a lattice square (§ 11.3).

[2] Does not exist if there is no $k \times k$ Graeco-Latin square (e.g., when $k = 6$).

The reader may obtain from Table 11.7 the type of design most likely to be useful in a given situation, but for full details of the methods of construction and of analysis the thorough accounts in the textbooks by Cochran and Cox (1957, § 10.28) and Kempthorne (1952, Chapter 22) should be consulted. The designs do not compare all pairs of treatments with quite equal precision, but the variation in precision is usually small, except for very small values of k.

(v) Designs with Two Units per Block

Several examples were given in Chapter 3 where a division of the experimental units into blocks of two arose naturally; for instance, the use of pairs of identical twins in animal experiments would frequently enable very precise comparisons to be made. With more than two treatments an incomplete block design is needed and Table 11.2 shows that

* Usually lattice designs are defined quite precisely as arrangements obtained by a particular type of construction. Here, however, we have introduced them in rather a vague way in terms of the situations for which they are most commonly needed.

the only balanced incomplete block designs with two units per block are the unreduced designs formed by taking all possible pairs of treatments. With t treatments this calls for a minimum of $\frac{1}{2}t(t-1)$ blocks and for $(t-1)$ replicates of each treatment. Thus with 10 treatments the

TABLE 11.8

KEYS TO DESIGNS WITH TWO UNITS PER BLOCK

No. of Treatments	No. of Replicates	Companions of Treatment 1	No. of Treatments	No. of Replicates	Companions of Treatment 1
6	4	2, 3, 5, 6	10	4	2, 5, 7, 10
7	4	2, 3, 6, 7	11	8	2, 3, 4, 5, 8, 9, 10, 11
8	6	2, 3, 4, 6, 7, 8	11	6	2, 3, 5, 8, 10, 11
8	5	2, 3, 5, 7, 8	11	4	2, 4, 9, 11
8	4	2, 4, 6, 8	12	10	2, 3, 4, 5, 6, 8, 9, 10, 11, 12
8	3	2, 5, 8	12	9	2, 3, 4, 6, 7, 8, 10, 11, 12
9	6	2, 3, 5, 6, 8, 9	12	8	2, 3, 5, 6, 8, 9, 11, 12
9	4	2, 4, 7, 9	12	7	2, 4, 6, 7, 8, 10, 12
10	8	2, 3, 4, 5, 7, 8, 9, 10	12	6	2, 4, 6, 8, 10, 12
10	7	2, 3, 4, 6, 8, 9, 10	12	5	2, 5, 7, 9, 12
10	6	2, 3, 4, 8, 9, 10	12	4	3, 4, 10, 11
10	5	2, 4, 6, 8, 10	12	3	2, 7, 12

smallest balanced incomplete block design has 45 blocks, each treatment occurring 9 times in all. Although these designs may be satisfactory for the smaller values of t, there is a need for designs for the larger values of t requiring a smaller total number of blocks.

Treatment	Companions
1	2, 3, 6, 7
2	3, 4, 7, 1
3	4, 5, 1, 2
4	5, 6, 2, 3
5	6, 7, 3, 4
6	7, 1, 4, 5
7	1, 2, 5, 6

Zoellner and Kempthorne (1954) have given designs for up to 12 treatments formed as follows. The companions of treatment 1 are found from Table 11.8, the companions of treatment 2 are formed by adding one to these, and so on. Thus the design with seven treatments and four replicates of each, has as the companions of treatment 1, treatments number 2, 3, 6, 7. The companions of treatment 2 are therefore numbers $2 + 1, 3 + 1, 6 + 1$, and $7 + 1$, i.e., numbers 3, 4, 7, and 1, according to the convention that with 7 treatments, number 8 means treatment $8 - 7 = 1$. In this way the key above is formed, and this determines the blocks as:

1 2	2 3	3 5	5 7
1 3	2 4	4 5	6 7
1 6	2 7	4 6	
1 7	3 4	5 6	

Before use the treatments should be randomized within each block and the order of the blocks randomized.

The analysis of the designs is straightforward with the help of some tables provided by Zoellner and Kempthorne. The efficiency factor is approximately 1/2 for all the designs, i.e., the standard error for the estimated difference between two treatments is approximately

$$\sqrt{\left(\frac{2}{\text{no. of replicates}}\right)} \times \frac{1}{\sqrt{(1/2)}} \times \left(\begin{matrix}\text{standard deviation}\\ \text{within pairs}\end{matrix}\right),$$

or $2 \times$ standard deviation within pairs/$\sqrt{}$(no. of replicates). The standard deviation must be reduced by at least a factor $\sqrt{2}$ for the blocking into pairs to be advantageous.

Other designs for blocks of two units have been discussed by Quenouille (1952) and by Youden & Connor (1954) and Clatworthy (1955). Preference studies, such as Example 11.5, also sometimes call for designs with pairs of treatments. For it may be possible to compare only two items, e.g., two foodstuffs, at one trial. Bose (1956) has investigated suitable designs.

(vi) Designs with a Small Number of Replicates

It sometimes happens, particularly in experiments in physics and some types of engineering, that the precision of the observations on individual units is high and that, although an incomplete system of blocking is desirable, the number of replicates of each treatment should be small. Balanced incomplete block designs rarely exist for this. Youden and Connor (1953) have described designs called *chain block designs* that are useful here, each treatment occurring only once or twice in the whole experiment. Their paper should be consulted for details.

(vii) Absence of a Suitable Balanced Design

In Example 11.1 we discussed the process of trial and error that is used to decide on a balanced incomplete block design. It occasionally happens that there are rather precisely defined restrictions on the parameters of the design, and that, although there are balanced incomplete block designs for neighboring values of the parameters, none exists for the values required. In such a case, if it is considered highly desirable to use an incomplete block design, one with as much symmetry as possible should be used in order to get as near as possible to the optimum properties of the balanced incomplete block design.

In simple cases it is possible to construct a design easily by trial and error, but in general the tables of partially balanced incomplete block designs compiled by Bose et al. (1955) should be consulted.

(viii) Summing-Up

In the preceding subsections we have outlined incomplete block designs that are sometimes useful when the number of units per block is restricted and no suitable balanced incomplete block design exists. The situations considered are:

(*a*) where it is required to make some comparisons more precisely than others, § (ii);

(*b*) where the number of units that must be used in each block is greater than the number of treatments and less than twice the number of treatments, § (iii);

(*c*) where the number of treatments is large, § (iv);

(*d*) where there are to be two units per block, § (v);

(*e*) where the number of replicates is to be particularly small, § (vi);

(*f*) where balanced incomplete block designs exist for cases similar to the one under study, but not for the precise combination of values required. § (vii).

In all cases except (*a*), the general idea is to arrange the treatments as symmetrically as possible, with the objects of obtaining maximum precision in the estimation of treatment differences and of simplifying the statistical analysis of the results. Methods of statistical analysis follow the same general pattern in all cases, and are essentially straightforward, but the calculations would prove very time-consuming to anyone not familiar with fairly advanced statistical methods. A rough interpretation of the experimental results can always be made by comparing the mean observations on the different treatments, but this sacrifices any gain in precision arising from the blocking. As stressed in (i) it is pointless to use these designs in setting up the experiment and yet to have no intention of using the proper methods of analysis.

11.5 FURTHER DESIGNS FOR TWO-WAY ELIMINATION OF ERROR

In principle it would be possible to list incomplete designs for the two-way elimination of error that bear the same relation to Youden squares and lattice squares as the designs of the previous section bear to balanced incomplete blocks. This will not be attempted here. Instead we refer briefly to the work of Shrikhande (1951) who gives generalized Youden squares in which

(*a*) each treatment occurs the same number m of times in each row and in which the columns form a balanced incomplete block design. If $m = 1$ these are Youden squares;

(*b*) the number of columns is not a simple multiple of the number of treatments; these designs are more complicated to deal with.

The chief generalization of lattice squares is the set of lattice rectangle designs considered briefly by Kempthorne (1952, p. 503). Mandel (1954) has given interesting two-way designs analogous to the chain-block designs of § 11.4(vi) and has described an application to the testing of car tires.

SUMMARY

Suppose that there are a number of treatments for comparison, the treatments not being arranged in any important factorial form. If possible such an experiment would usually be laid out in randomized blocks or in a Latin square plan, choosing, for example, the blocks so that the units within any one block are as alike as possible.

This requires that each treatment occurs once (or more) in each block, or row, or column. If the number of treatments exceeds the number of units per block in the system of grouping the units that seems likely to achieve greatest reduction in error, the simple randomized block cannot be used. We then need a new set of designs, called *incomplete block* designs, in which each treatment does not occur in each block.

The simplest such designs are the *balanced incomplete blocks* which should be used in preference to other incomplete block designs except when some treatment comparisons are required to have higher precision than others. These designs are such that any pair of treatments occur together in a block the same number of times in all. They exist only in rather limited cases (see Table 11.2) and some trial and error adjustment may be necessary to find the design most appropriate to a given situation.

The estimation of treatment effects in a general incomplete block experiment involves an adjustment of the mean observation on each treatment for the effect of the blocks in which the particular treatment did not appear. These adjustments are simple and relatively precise for balanced incomplete blocks.

The most important incomplete designs analogous to the Latin square, which control variation in two directions simultaneously, are Youden squares and lattice squares. In the first of these the number of columns is equal to the number of treatments, but the number of rows is less than the number of treatments. In lattice squares the number of rows and the number of columns are both equal to the square root of the number of treatments. Both these types of design exist only in rather limited circumstances, but should be considered for use whenever a Latin square would normally be applied but the number of treatments exceeds the number of rows or the number of columns or both.

In addition to the balanced incomplete blocks, Youden squares, and lattice squares, there is a wide variety of so-called partially balanced designs; § 11.4(viii) should be consulted for a brief summary of the main cases.

REFERENCES

Bose, R. C. (1956). Paired comparison designs for testing concordance between judges. *Biometrika*, **43**, 113.

——— W. H. Clatworthy, and S. S. Shrikhande. (1955). *Tables of partially balanced designs with two associate classes*. Institute of Statistics, University of North Carolina.

Clatworthy, W. H. (1955). Partially balanced incomplete block designs with two associate classes and two treatments per block. *J. Res. Nat. Bureau Standards*, **54**, 177.

Cochran, W. G., and G. M. Cox. (1957). *Experimental designs*. 2nd ed. New York: Wiley.

Durbin, J. (1951). Incomplete blocks in ranking experiments. *Brit. J. Statist. Psychol.*, **4**, 85.

Kempthorne, O. (1952). *Design and analysis of experiments*. New York: Wiley.

Mandel, J. (1954). Chain block designs with two-way elimination of heterogeneity. *Biometrics*, **10**, 251.

Mann, H. B. (1949). *Analysis and design of experiments*. New York: Dover.

Pugh, C. (1953). The evaluation of detergent performance in domestic dishwashing. *Applied Statistics*, **2**, 172.

Quenouille, M. H. (1952). *Design and analysis of experiment*. London: Griffin.

Rao, C. R. (1947). General methods of analysis for incomplete block designs. *J. Am. Statist. Assoc.*, **42**, 541.

Shrikhande, S. S. (1951). Designs for two-way elimination of heterogeneity. *Ann. Math. Statist.*, **22**, 235.

Wadley, F. M. (1948). Experimental design in the comparison of allergens on cattle. *Biometrics*, **4**, 100.

Youden, W. J. (1937). Use of incomplete block replications in estimating tobacco-mosaic virus. *Contr. Boyce. Thompson Inst.*, **9**, 41.

——— and W. S. Connor. (1953). The chain block design. *Biometrics*, **9**, 127.

——— (1954). New experimental designs for paired observations. *J. Res. Nat. Bureau Standards*, **53**, 191.

Zoellner, J. A., and O. Kempthorne. (1954). Incomplete block designs with blocks of two plots. *Agric. Expt. Station, Iowa State College*, Research Bulletin 418.

CHAPTER 12

Fractional Replication and Confounding

12.1 INTRODUCTION

In Chapters 6 and 7 we considered the main properties of factorial experiments, i.e., experiments in which each final treatment consists of a combination of a number of more basic treatments, called factors. The experiments described in those chapters were *complete* in that each combination of factor levels occurred the same number of times in all. Also each combination occurred the same number of times in each block, if the experiment was set out in randomized blocks, and the same number of times in each row and column, if the experiment was in a Latin square. The restriction to complete systems meant that the main effects of the separate factors and their interactions could be estimated separately in a simple way.

In the present chapter, two situations are described in which incomplete factorial experiments are useful. In the first, one full replicate involves too many observations for the precision required: for example, in a preliminary investigation it may be desirable to have a rough estimate of the effect of say 6 or 8 factors, yet the smallest complete experiments with these numbers of factors have 64 and 256 units, and this may be many more than would be appropriate. Therefore it is natural to look for designs in which only a suitable selection of the factor combinations is used. Such designs, called *fractional factorials*, are considered in § 12.2.

The second type of situation is analogous to that considered in the previous chapter. We may want to arrange a full, or fractionally replicated, factorial system in blocks with fewer units per block than there are treatments, or in a Latin square with the number of rows and columns less than the number of treatments. The technique by which this is achieved is called *confounding* and is described in § 12.3.

Fractional replication and confounding depend on the idea that certain comparisons among the treatments, the so-called high-order interactions (see Chapter 6), are of relatively little interest. The two methods, especially fractional replication, are of considerable importance in certain fields and have been developed to a high degree of elaboration. What follows is intended solely as an introduction to the general principles involved, nothing like a comprehensive account being attempted. A reader who wishes to apply these methods will need to consult a more comprehensive account of them or to obtain expert advice.

12.2 FRACTIONAL REPLICATION

(i) General Idea

In one full replicate of an ordinary factorial system just enough observations are obtained to estimate separately each main effect and interaction. It is clear then that if the experiment is formed from less than one replicate, it will not be possible to estimate each contrast separately, and a quantity calculated to estimate, say, a particular main effect will in general depend also on the true value of one or more other contrasts, usually interactions. A good design will be one which estimates each main effect, and if possible, each low-order interaction, in such a way as to be entangled only with high-order interactions.

We decide when fractional replication is likely to be useful by finding, by the methods of Chapter 7, the number of experimental units that will give estimates of reasonable precision. For instance, in a two-level experiment, the main effect of a factor is estimated by the difference between two means, each of $\frac{1}{2}N$ observations, where N is the total number of units in the experiment. The standard error of the main effect is thus $2 \times$ residual standard deviation$/\sqrt{N}$.

Suppose, for example, that the standard deviation is thought to be about 10 per cent and that a standard error of about $3\frac{1}{2}$ per cent is considered reasonable. An experiment on about 30 units gives the required precision so that it is natural to plan for an experiment with 32 units, 32 being the nearest power of 2. If there are 5 factors for investigation, a single replicate of the 2^5 system is the thing to use; but suppose that we want to examine more than 5 factors, say 8. A single replicate of the full 2^8 system involves 8 times as many units as are necessary for the precision desired. One possibility is to split the experiment into two halves, each with 32 units, neither half having more than 5 factors. A better method for many purposes, and one that involves 32 not 64 units, is to examine all factors together but not to use the complete set of factor combinations. In fact, it is possible to test 8 factors in 32 units, so as to get information about all main effects and some two-factor interactions.

A slightly different situation arises when the number of units is fixed by practical considerations and when, having satisfied ourselves that sufficient precision will be obtained to make the experiment worth doing, we want to test as many factors as possible with the material available. Here the use of fractional replication is practically obligatory.

(ii) A Simple Case

We begin by examining a very simple case. Suppose that there are three factors A, B, and C and that we consider an experiment in which each occurs at two levels, an upper level and a lower level, i.e., the 2^3 system. The notation of § 6.7 will be used so that, for example, (a_2b_1) denotes the mean of all observations on the treatment a_2b_1 in which A is at the upper level and B at the lower level.

One replicate of the complete factorial system involves only eight experimental units and it would be rather unusual to require an experiment with fewer units. However, suppose that it is required to investigate the 2^3 system using just 4 experimental units. Now the three-factor interaction $A \times B \times C$ divides the treatments into two sets

$$\text{(i)}\quad a_2b_1c_1,\quad a_1b_2c_1,\quad a_1b_1c_2,\quad a_2b_2c_2;$$

$$\text{(ii)}\quad a_2b_2c_1,\quad a_1b_1c_1,\quad a_1b_2c_2,\quad a_2b_1c_2,$$

the set (i) receiving the positive sign in estimating the interaction, the set (ii) the negative sign (see § 6.7), the estimate of the interaction thus being based on a comparison of (i) with (ii). It will not be possible from an experiment with four units to estimate all contrasts among the eight treatments and it is sensible to start by sacrificing completely information about $A \times B \times C$. This will be done if we investigate in the experiment one of the sets (i) and (ii), say the former (although it does not matter which). No estimate of $A \times B \times C$ is now possible.

We shall get four observations $(a_2b_1c_1)$, $(a_1b_2c_1)$, $(a_1b_1c_2)$, $(a_2b_2c_2)$. How shall we estimate the main effects and two-factor interactions? For a main effect, such as C, we compare observations in which C occurs at the upper level with those in which C occurs at the lower level, i.e., we take

$$\tfrac{1}{2}[(a_1b_1c_2) + (a_2b_2c_2) - (a_1b_2c_1) - (a_2b_1c_1)]. \tag{1}$$

This will certainly give a sensible estimate of the main effect of C if the factors A and B are both without effect. Further, the existence of a simple main effect for A or B or both will not affect (1) since the two observations on the lower level of A occur one with a plus and one with a minus sign, etc.

Now consider the estimation of the two-factor interaction $A \times B$. If C were absent we would estimate $A \times B$ by

$$\tfrac{1}{2}[(a_1b_1) + (a_2b_2) - (a_1b_2) - (a_2b_1)] \tag{2}$$

and this suggests that we should try

$$\tfrac{1}{2}[(a_1b_1c_2) + (a_2b_2c_2) - (a_1b_2c_1) - (a_2b_1c_1)], \tag{3}$$

the nearest we can get to (2). This would be an estimate of $A \times B$ if the factor C had no effect. However, (1) and (3) are identical, so that the same quantity appears to estimate both C and $A \times B$. No other estimates of these contrasts are available, so that it is impossible, from the observations on just four units, to distinguish between the main effect C and the two-factor interaction $A \times B$; we say that C and $A \times B$ are *aliases* of one another.

In exactly the same way A and $B \times C$ are aliases and so are B and $C \times A$.

It follows that, even if the uncontrolled variation is negligible, no conclusion can be drawn about the main effect C unless there is some information available about $A \times B$, other than the data of the experiment. If it can be assumed that $A \times B$ is unlikely to be appreciable and if the quantity in formula (3) is larger than could arise by chance, then we can reasonably infer that the apparent main effect of C is real. To reach such a conclusion it is necessary to introduce two assumptions external to the data, one about the interaction, and one about the magnitude of the uncontrolled variation, which cannot be estimated from just the four observations. In the same way we can estimate the main effects of A and B provided that assumptions are made about $B \times C$ and $C \times A$. The three main effects are estimated independently, in the sense that the presence of arbitrary true main effects for two of the factors has no effect on the estimate of the third.

In theory any prior knowledge about, say, $A \times B$ enables something to be inferred about the main effect C. Thus, if $A \times B$ is thought very probably to be in one direction and the estimate from formula (3) comes out to be appreciable and in the opposite direction, the apparent main effect of C is very probably real.

The design with four units that we have been considering is called the one-half replicate of the 2^3 experiment based on the *defining contrast* $A \times B \times C$. The alias of a particular contrast can be obtained by the following formal rule. Multiply the name of the contrast by $A \times B \times C$ by the ordinary rules of algebra and then omit any symbol that is squared. Thus, the alias of A is $A \times (A \times B \times C) = A^2 \times B \times C = B \times C$, omitting A^2 by the general rule. Similarly the alias of $B \times C$ is $(B \times C) \times (A \times B \times C) = A \times B^2 \times C^2 = A$.

It is often useful to set out the treatments and contrasts in a table as in Table 12.1. The sequence of 1's and −1's alongside each main effect and interaction show how each contrast is estimated.

TABLE 12.1

DEFINITION OF CONTRASTS IN THE ONE-HALF REPLICATE OF THE 2^3 SYSTEM

		Treatment			
		$a_2b_1c_1$	$a_1b_2c_1$	$a_1b_1c_2$	$a_2b_2c_2$
Contrast	A	1	−1	−1	1
	B	−1	1	−1	1
	C	−1	−1	1	1
	$B \times C$	1	−1	−1	1
	$C \times A$	−1	1	−1	1
	$A \times B$	−1	−1	1	1

Thus, alongside the symbol C we have 1 when C is at its upper level and −1 when C is at its lower level, corresponding to formula (1). Except for the multiplying factor 1/2, the row of coefficients is equivalent to formula (1). Similarly, the lower section of the table shows how to estimate the two-factor interactions. Note that the coefficients for $A \times B$ can be obtained by multiplying the corresponding coefficients for A and for B. The characteristic alias properties of the design are shown by the lines for $A \times B$ and for C being identical, etc.

These properties generalize directly to more complicated cases provided that the two factors all occur at two levels.

To sum up, we have an experiment from which we can estimate main effects, provided that two-factor interactions may be assumed negligible. Such an experiment might, in rather special circumstances, give useful suggestions as to which factors merit further study, or, say, might suggest remedial action in an industrial process producing too much defective output or lead to the detection of a defective component in a piece of apparatus. But the design's main interest is in indicating what happens in more complicated cases. The essential point to grasp is that we cannot estimate a main effect (or interaction) unless we assume that a certain other contrast is negligible.

(iii) The Two-Level Factorial System

We now consider briefly some more complicated systems, the factors still being all at two levels. The following example illustrates some of the considerations involved in deciding whether to use fractional replication.

Example 12.1.* The experimental set-up consisted of a chemically defined medium on which embryonic chick bones grow. The medium contained among other things 20 amino acids, and preliminary work had shown that 11 of these were necessary for growth, in that the rate of growth, as measured by the wet weight of the tibia, dropped when any one was omitted from the medium.

It was required to design a further experiment in which the interactions among these 11 factors could be explored, the idea being that the discovery of which pairs or groups of factors do or do not interact would give information about the underlying mechanism. A two-level experiment was intended, since the curvatures of the response curves for the separate factors were not of interest at that stage. Now $2^{11} = 2048$, and an experiment with this number of units was quite impracticable and also unnecessary for the precision desired. Further, since there was particular interest in interactions, the factors should be dealt with together in one experiment. Hence fractional replication was clearly called for and tables of these designs (National Bureau of Standards, 1957) were consulted to find the smallest experiment with 11 factors at 2 levels, for which main effects and two-factor interactions could be calculated separately from one another. In fact, there is a suitable design with 128 observations, i.e., a 1/16th replicate.

Table 12.2 shows some of the smaller fractionally replicated designs for factors at 2 levels. The 1/2 replicate designs raise no points essentially different from (ii). As the number of factors increases to five and more, it becomes, however, possible to get designs in which each main effect and two-factor interaction can be estimated separately. Thus, in the 1/2 replicate of the 2^5 experiment, with $A \times B \times C \times D \times E$ as defining contrast, the alias of A is by the general rule $A \times (A \times B \times C \times D \times E) = B \times C \times D \times E$, a high-order interaction, and the alias of $B \times C$ is $A \times D \times E$, etc. The factor combinations occurring in the experiment are the 16 in which 0, 2, or 4 factors occur at the upper level, i.e., are in common with the letters in the defining contrast.

To form a 1/4 replicate a further defining contrast has to be introduced. Thus in the 1/4 replicate of the 2^5, the two defining contrasts are $A \times B \times C$ and $A \times D \times E$. The rules are:

(*a*) the aliases of any contrast are obtained by multiplying into $A \times B \times C$, into $A \times D \times E$ and into $(A \times B \times C) \times (A \times D \times E) = B \times C \times D \times E$. Thus, the aliases of A are $B \times C$, $D \times E$ and $A \times B \times C \times D \times E$, while the aliases of B are $A \times C$, $A \times B \times D \times E$ and $C \times D \times E$.

(*b*) The 8 treatments in the experiment are the factor combinations that have 0 or 2 factors at the upper level in common with both defining contrasts, i.e., $a_1b_1c_1d_1e_1$, $a_1b_2c_2d_1e_1$, $a_1b_1c_1d_2e_2$, $a_1b_2c_2d_2e_2$, $a_2b_2c_1d_2e_1$, $a_2b_1c_2d_2e_1$, $a_2b_2c_1d_1e_2$, $a_2b_1c_2d_1e_2$.

* I am grateful to Dr. J. D. Biggers for the details of this example.

TABLE 12.2

SOME FRACTIONAL FACTORIAL DESIGNS FOR THE 2^n SYSTEM

Fraction	Number of Factors	Number of Experimental Units	Defining Contrasts	Alias of a Main Effect	Alias of a Two-Factor Interaction
1/2	3	4	*ABC*	2 f int.*	Main effect
	4	8	*ABCD*	3 f int.	2 f int.
	5	16	*ABCDE*	4 f int.	3 f int.
	:	:	:	:	:
1/4	5	8	*ABC*, *ADE* (*BCDE*)	One or two 2 f ints.	Main effect or 2 f int.
	6	16	*ABCF*, *ADEF* (*BCDE*)	3 f ints.	One or two 2 f ints.
	7	32	*ABCDE*, *CDFG*, *ABEFG*	3 and 4 f ints.	Mostly 3 f ints. Some 2 f ints.
	8	64	*ABCDE*, *ABFGH* (*CDEFGH*)	4 f ints.	3 f ints.
1/8	6	8	*ADE*, *BCE*, *ACF* (and products)	2 f ints.	Main effects or 2 f ints.
	7	16	*ABCD*, *CDEF*, *ACEG* (and products)	3 f ints.	2 f ints.
	8	32	*ABCD*, *ABEF*, *ACFGH* (and products)	3 f ints. or higher	Some 2 f ints. Some 3 f ints.
	9	64	*ABCD*, *ABEFG*, *ACEHJ* (and products)	3 f ints. or higher	Mostly 3 f ints. Some 2 f ints.
	10	128	*ABCDG*, *ABEFH*, *AGHJK* (and products)	4 f ints.	3 f ints. and higher
1/16	8	16	*ABCD*, *CDEF*, *ACEG*, *EFGH* (and products)	3 f ints.	2 f ints.
	9	32	*ABCD*, *ABEF*, *BCEG*, *ACEHJ* (and products)	3 f ints.	Some 2 f ints. Some 3 f ints.
	10	64	*ADHJ*, *BEGK*, *ABCGH*, *ABFJK* (and products)	3 f ints.	Some 2 f ints. Some 3 f ints.
1/32	10	32	*ABCD*, *CDEF*, *EFGH*, *GHJK*, *ACEGJ* (and products)	3 f ints.	2 f ints.

* This is an abbreviation for two-factor interaction. The plural form, 2 f ints., is used when one contrast has two or more aliases that are two-factor interactions.

The more complicated cases can be dealt with by following the same rules. It is not proposed however, to go here into the justification of the rules and the exact way in which best designs are chosen. Davies (1954) has given a careful account of the designs with particular reference to industrial applications, whereas Brownlee et al. (1948), Daniel (1956), and a publication from the National Bureau of Standards (1957) give extensive lists of available designs. Table 12.3, which is based on their

TABLE 12.3

NUMBERS OF FACTORS THAT CAN BE INVESTIGATED IN FRACTIONAL REPLICATES OF THE 2^n SYSTEM

	Maximum Number of Factors that can be Included		
Number of Experimental Units	Keeping all Main Effects distinct	Keeping all Main Effects distinct from one another and from 2 f Ints.	Keeping all Main Effects and 2 f Ints. distinct
8	7	4	3
16	15	8	5
32	31	13	6
64	63	*	8

* Not known

work, shows the maximum number of factors that can be investigated with a given number of experimental units keeping (*a*) main effects distinct from one another, (*b*) ensuring main effects have as aliases three-factor (or higher) interactions, (*c*) ensuring that main effects and two-factor interactions have as aliases three-factor (or higher) interactions.

In subsection (vi) we consider some of the practical points that have to be watched in using the designs, but first a brief account will be given of fractional replication with factors at more than two levels.

(iv) Factors at More than Two Levels

The methods for factors at two levels can be generalized to deal with experiments in which all factors are at, say, three levels or all at four levels. Unfortunately, this generalization is not altogether satisfactory, and some designs have to be constructed by a more or less ad hoc procedure rather than by the systematic rules.

The most important case is when all factors are at three levels; Table 12.4 lists some of the designs for this situation that can be obtained by the methods of (iii) suitably generalized. Details are given by Davies (1954, Chapter 9) and Kempthorne (1952, Chapter 16).

TABLE 12.4

SOME FRACTIONAL FACTORIAL DESIGNS WITH ALL FACTORS AT THREE LEVELS

Fraction	Number of Factors	Number of Experimental Units	Aliases of Main Effects	Aliases of Two-Factor Interactions
1/3	3	9	2 f ints.	Main effects and 2 f ints.
	4	27	3 f ints.	2 f ints.
	5	81	4 f ints.	3 f ints.
	:	:	:	:
1/9	4	9	2 f ints.	Main effects and 2 f ints.
	6	81	3 f ints.	2 f ints. and 3 f ints.
1/27	6	27	2 f ints.	2 f ints. and 3 f ints.
	7	81	2 f ints.	2 f ints. and 3 f ints.

There are a rather limited number of designs in which main effects and two-factor interactions can be estimated independently of one another. For example with 5 factors there are no arrangements of this type, other than the complete system with 243 observations, in which two-factor interactions can be estimated separately.

An important special case is when the factors are quantitative. If it is desired to fit a second degree response surface (§ 6.9), each factor must occur at three (or more) levels. If there is a design in Table 12.4 of suitable size and allowing main effects and two-factor interactions to be estimated separately, this can be used. But the linear × linear interactions, which enter into the fitting of the second-degree surface, form only part of the two-factor interactions; this makes it very plausible that there should be suitable designs for fitting second-degree response surfaces even when there is no corresponding design in Table 12.4, and this is in fact so. Thus, Box and Wilson (1951), see also Davies (1954), have introduced so-called composite designs for the fitting of response surfaces; see also Box and Hunter (1957).

(v) Latin Squares as Fractional Factorials

It was remarked in the discussion of Latin squares and Graeco-Latin squares, § 10.5, that a certain type of application of these squares could be considered as fractional replication, see especially Example 10.4.

In the first applications of Latin squares that we considered, the experimental units were classified in two ways (corresponding to the rows and columns of the square), and one set of treatments (corresponding to Latin letters) was applied in such a way that the presence of constant differences between rows and between columns did not induce error in the treatment comparisons. This is best not regarded as fractional replication, because the rows and columns of the square correspond to any convenient classification of the units, not to treatments, so that the Latin letters represent the only treatments in the experiment, each treatment thus occurring several times.

If, however, we regard the rows and columns as factors under test, either as treatment factors or as classification factors, the situation is different. A Latin square such as

		Column			
		1	2	3	4
Row	1	*B*	*A*	*D*	*C*
	2	*C*	*D*	*A*	*B*
	3	*D*	*C*	*B*	*A*
	4	*A*	*B*	*C*	*D*

then represents a particular set of 16 out of the $4 \times 4 \times 4 = 64$ possible combinations of the factors, rows, columns, and treatments each at four levels. The defining property of the Latin square then ensures that, say, the mean observation on column 1 minus the mean observation on column 2 is unaffected by constant differences between rows and between treatments. That is, a column main effect, and similarly a row or treatment main effect, can be estimated independently of the main effects of the other factors, provided no interactions are present. Thus, the 4×4 Latin square is a 1/4 replicate of the 4^3 experiment, in which all main effects can be estimated, interactions being assumed absent.

Quite generally a Latin square, or a higher-type square with several alphabets, defines fractional factorial arrangements in which all factors are at the same number of levels, interactions are assumed absent, and all main effects are estimated. The use of the design with the maximum number of alphabets is an extreme form of fractionation; Example 10.4 is a $1/5^4$ replicate of the 5^6 experiment.

(vi) General Discussion

There are a number of general points that need to be thought about when the use of fractional replication is contemplated. First there is, as with other factorial experiments, the choice of the number of levels for each factor. The main situation in which more than two levels are used in this sort of work is probably in fitting a second-degree response surface when the factors are quantitative. With qualitative factors two levels are often sufficient in the type of preliminary work for which factorial experiments with many factors are most appropriate. With quantitative factors it is not reasonable to estimate interactions without also estimating main effect curvatures, if the object is the fitting of a second-degree equation to a quite unknown response surface. Quite often, however, as in Example 12.1, we are interested primarily in the direction and amount of the main effect of a factor and on whether two different factors interact, it being suspected that many pairs of factors should act independently, the corresponding interactions being small. In such cases a two-level design estimating main effects and two-factor interactions is the thing to use. For these reasons the most important designs are (*a*) the two-level ones for which main effects can be estimated free of two-factor, and if possible three-factor, interactions, (*b*) the two-level ones for which all main effects and two-factor interactions can be estimated, and (*c*) three-level designs from which second-degree response surfaces can be fitted.

When the general type of design to use has been settled, some care is usually advisable in identifying the letters A, B, . . . in the key to the design, with the factors that are to be included. For instance, it might happen that, say, the interaction $A \times B$ can be estimated free of main effects and of other two-factor interactions, whereas certain other pairs of two-factor interactions are aliases of one another. It would then be wise to name the factors so that $A \times B$ is an interaction that is considered of particular interest, or particularly likely to be appreciable. In general all special properties of the design under use should, if possible, be exploited.

In the smaller designs estimation of the residual standard deviation is difficult or impossible from the observations themselves, and it is desirable to have an estimate from prior knowledge. In the larger experiments estimates can be found by, for example, assuming that all three-factor and higher-order interactions are negligible. The availability of an estimate of this sort is another special property of each design that needs special consideration in each application.

A final important aspect of the designs is the possibility of extending them if doubtful or ambiguous results are obtained. Thus it might happen that the interaction $A \times B$ has $C \times D$ as one of its aliases and that the sample estimate of $A \times B$ (also of $C \times D$) is appreciable and suggests the presence of some important effect. If the main effects of A and B are appreciable, whereas those of C and D are not, it is somewhat more likely that $A \times B$ is large than that $C \times D$ is. However, it may well happen that the main effects do not permit such an inference and, in any case, further investigation may be desirable. In some cases this is best done by doing a further experiment the same size as the initial one, thus converting, say, a 1/16th replicate into a 1/8th replicate, the combined design having a more favorable alias structure than the initial experiment, and in particular separating out the particular pair or pairs of contrasts whose interpretation is in dispute. Or again, a new experiment may be set up involving, perhaps, factors not included in the first experiment and separating the aliases in question. This problem is discussed to some extent by Davies (1954) and by Daniel (1956), but the possibilities are so rich that special analysis of each particular application by someone with a detailed knowledge of this type of design is really desirable.

To sum up, fractional replication is likely to be a useful tool in situations where many factors are likely to be relevant to the situation being studied and where, in the initial stages of the investigation at least, an economical survey of the effects of many factors, with possibly some ambiguous conclusions, is to be preferred to a detailed examination of a few factors. Under such conditions fractional replication makes practicable investigations that could hardly be undertaken otherwise.

12.3 CONFOUNDING

(i) General

The device of fractional replication just described does, in certain cases, offer the possibility of investigating systems that would otherwise be inaccessible because of their complexity. Confounding, on the other hand, is concerned solely with increasing precision and is the technique for factorial experiments that is analogous to the use of incomplete block designs in nonfactorial experiments.

The general idea of confounding, like that of fractional replication, is that high-order interactions are likely to be of relatively little importance and again, as for fractional replication, designs of considerable complexity can arise. We shall only outline general principles.

(ii) A Special Case

We start by describing a simple special case analogous to that described for fractional replication in § 12.2(ii). Consider a 2^3 experiment, having therefore eight treatments. Several replicates of these eight may be necessary to get the required precision, and if blocks of eight reasonably homogeneous experimental units can be formed, the use of randomized blocks, or possibly 8×8 Latin squares, is natural.

Suppose, however, the natural or most useful grouping is into blocks of four units. If we had a nonfactorial experiment, we should examine first any balanced incomplete block designs with four units per block and eight treatments. With a factorial experiment, however, we have the extra fact that the contrast $A \times B \times C$ is likely to be of less importance than the main effects and two-factor interactions. This suggests that we should form blocks from the two sets into which $A \times B \times C$ divides the eight treatments, as follows:

$$\text{Block 1:} \quad a_2b_1c_1, \quad a_1b_2c_1, \quad a_1b_1c_2, \quad a_2b_2c_2;$$
$$\text{Block 2:} \quad a_1b_1c_1, \quad a_1b_2c_2, \quad a_2b_1c_2, \quad a_2b_2c_1.$$

The whole experiment would consist of several pairs of such blocks appropriately randomized.

In the analysis of the results of such an experiment, the contrasts other than $A \times B \times C$ are estimated in the usual way. Thus, for the two-factor interaction $B \times C$ we take the difference between (*a*) the mean of all observations on treatments $a_1b_1c_1$, $a_2b_1c_1$, $a_1b_2c_2$, $a_2b_2c_2$, and (*b*) the mean of all observations on treatments $a_1b_1c_2, a_2b_1c_2, a_1b_2c_1, a_2b_2c_1$. Each block thus contributes two observations to (*a*) and two to (*b*), so that the existence of constant effects associated with particular blocks has no effect on the estimate of $B \times C$, i.e., the random error of estimation of $B \times C$ is determined by the amount of uncontrolled variation within blocks. Similarly, the main effects and the other two-factor interactions can be estimated by the usual formulas and are free of block effects.

If we try to estimate the three-factor interaction $A \times B \times C$ we find that, as is clear from the way the design has been constructed, pairs of blocks totals (or means) have to be compared, so that the random error of estimation for $A \times B \times C$ is determined by the amount of variation *between* blocks. In forming the experimental units into blocks, the aim is to minimize the variation within blocks, i.e., to maximize the variation between blocks, and therefore we would expect comparisons of block totals to be very imprecise; in fact in practical cases it would usually happen that no useful information about $A \times B \times C$ could be gained from the experiment.

We say in this situation that $A \times B \times C$ is *confounded* with blocks, and call $A \times B \times C$ the *defining contrast* of the lay-out. If the whole experiment consists of pairs of blocks with $A \times B \times C$ confounded in each pair, the confounding is said to be *total*. Unlike the fractional replication based on $A \times B \times C$, this system of confounding is of practical value.

Example 12.2. Bainbridge (1951) has described an experiment on a pilot plant carrying out a gaseous synthesis. The factors under study were temperature, throughput, and concentration, each occurring at two levels. The probable form of the uncontrolled variation suggested that the use of blocks of four units, each dealt with successively in time, would reduce error, and, the three-factor interaction being unlikely to be important, it was confounded in both of the replicates that were run.

A possible form for the design, after randomization of treatments within blocks and treatment sets within replicates, is

Replicate 1	Block 1:	$a_1b_1c_2$,	$a_2b_2c_2$,	$a_2b_1c_1$,	$a_1b_2c_1$
	Block 2:	$a_1b_2c_2$,	$a_1b_1c_1$,	$a_2b_1c_2$,	$a_2b_2c_1$
Replicate 2	Block 1:	$a_1b_1c_1$,	$a_2b_1c_2$,	$a_2b_2c_1$,	$a_1b_2c_2$
	Block 2:	$a_2b_2c_2$,	$a_2b_1c_1$,	$a_1b_1c_2$,	$a_1b_2c_1$

where A, B, and C denote the three factors, and the subscripts 1 and 2 denote the lower and upper level in the usual way.

In an ordinary randomized block experiment it is possible to adjust for a concomitant variable (§ 4.5). This method of getting an additional increase in precision applies almost without change in confounded designs. Bainbridge reports that one important source of uncontrolled variation was gas purity and that this was used as a concomitant variable. Confounding reduced the residual standard deviation by almost 30 per cent and covariance by a further 30 per cent. The two devices together, therefore, almost halved the residual standard deviation and gave an increase in apparent precision equal to that obtained from a fourfold increase in the number of experimental units.

If a number of replicates of an experiment are to be run, there are numerous possibilities for what is called partial confounding. For example, if 4 replicates, 32 units, are to be used, $A \times B \times C$ could be confounded in the first replicate and $B \times C$, $C \times A$, $A \times B$ in the second, third, and fourth replicates respectively. Thus, the third replicate would consist of

$$\begin{array}{ll} \text{Block 1:} & a_1b_1c_1,\quad a_1b_2c_1,\quad a_2b_1c_2,\quad a_2b_2c_2; \\ \text{Block 2:} & a_2b_1c_1,\quad a_2b_2c_1,\quad a_1b_1c_2,\quad a_1b_2c_2, \end{array}$$

having $C \times A$ confounded. In setting out the experiment, the order of treatments is again randomized within each block, the order of blocks is randomized within each pair and finally, the order of the four replicates is randomized. In the analysis of the results from such an experiment, each contrast is estimated by the usual formulas, using, however, only the blocks in which the particular contrast is not confounded. That is, the main effects are estimated from all the observations, and the interactions from observations in three of the four blocks. The interactions are said to be each one-quarter confounded with blocks.

More complicated systems, in which say $A \times B \times C$ is confounded more heavily than the two-factor interactions, are easily constructed.

It is interesting that if an appreciable reduction in variation is achieved by the use of smaller blocks, all contrasts, including those confounded, may be estimated more precisely than they would have been without confounding.

The reader may check that if all contrasts including main effects are equally confounded, requiring seven pairs of blocks, a balanced incomplete block design is obtained. This is to be expected, since by confounding main effects equally with interactions, we are not taking advantage of the factorial structure of the treatments and are dealing with all comparisons on an equal footing.

(iii) Confounding in the 2^n System

We now consider the extension of the method just described to more complex arrangements in which, however, all factors occur at two levels. If the required block size is one-half the total number of factor combinations, the situation is directly analogous to that of (ii). A single contrast, nearly always the highest-order interaction, is used to divide the treatments into two sets to go into different blocks.

For example, with five factors, an arrangement in blocks of $\frac{1}{2} \times 2^5 = 16$ units each can be based on the five-factor interaction $A \times B \times C \times D \times E$. This divides the 32 treatments into two sets determined by the formal multiplication

$$(a_2 - a_1)(b_2 - b_1)(c_2 - c_1)(d_2 - d_1)(e_2 - e_1),$$

the treatments with a positive sign going into one set, those with a negative sign into the other set (see § 6.7). The blocks are, before randomization, therefore:

Block 1: $a_1b_1c_1d_1e_1$, $a_2b_2c_1d_1e_1$, $a_1b_2c_2d_1e_1$, $a_2b_1c_2d_1e_1$,
$a_1b_1c_2d_2e_1$, $a_2b_2c_2d_2e_1$, $a_1b_2c_1d_2e_1$, $a_2b_1c_1d_2e_1$,
$a_1b_1c_1d_2e_2$, $a_2b_2c_1d_2e_2$, $a_1b_2c_2d_2e_2$, $a_2b_1c_2d_2e_2$,
$a_1b_1c_2d_1e_2$, $a_2b_2c_2d_1e_2$, $a_1b_2c_1d_1e_2$, $a_2b_1c_1d_1e_2$.

Block 2: $a_1b_1c_1d_1e_2$, $a_2b_2c_1d_1e_2$, $a_1b_2c_2d_1e_2$, $a_2b_1c_2d_1e_2$,
$a_1b_1c_2d_2e_2$, $a_2b_2c_2d_2e_2$, $a_1b_2c_1d_2e_2$, $a_2b_1c_1d_2e_2$,
$a_1b_1c_1d_2e_1$, $a_2b_2c_1d_2e_1$, $a_1b_2c_2d_2e_1$, $a_2b_1c_2d_2e_1$,
$a_1b_1c_2d_1e_1$, $a_2b_2c_2d_1e_1$, $a_1b_2c_1d_1e_1$, $a_2b_1c_1d_1e_1$.

From one or more replications of this design all contrasts except the five-factor interaction can be estimated free of between-block variation. There would rarely be interest in the five-factor interaction, so that total confounding would be satisfactory.

If the number of units per block is required to be less than one-half of the full number of treatments in the factorial system, more than one contrast has to be confounded. A whole series of designs exists in which the number of units per block is $\frac{1}{4}, \frac{1}{8}, \frac{1}{16}, \ldots$ of the number of treatment combinations, 2^n, in the experiment. We shall not describe the detailed principles for the selection and construction of these designs, but it is worth explaining the following important property. To obtain a block size of $\frac{1}{4} \times 2^n$, two contrasts must be confounded, not one as in the design in (ii). It turns out that when two contrasts are confounded, so is a third, the formal product according to the rule of multiplication of § 12.2. Thus, if $A \times B \times C$ and $A \times D \times E$ are confounded, so also is $(A \times B \times C) \times (A \times D \times E) = A^2 \times B \times C \times D \times E = B \times C \times D \times E$.

To obtain a block size of $\frac{1}{8} \times 2^n$, three independent contrasts must be confounded and when this is done a further four contrasts are confounded, determined by multiplying the first three together in all possible ways. The general consequence of this is that to obtain a block size much smaller than the number of factor combinations, a large number of contrasts which cannot be selected arbitrarily have to be confounded, and it may happen that an arrangement with desired properties does not exist.

Table 12.5 shows useful systems of confounding for the smaller two-level factorial systems. For example, it is possible to confound the 2^6 experiment with eight units per block so that no main effects or two-factor interactions are confounded, but with four units per block three of the 15 two-factor interactions have to be confounded if the confounding of main effects is to be avoided. In an application in which each treatment is used only once, the three two-factor interactions of least interest would, of course, be selected for confounding. With more than one replicate, partial confounding in which different contrasts are confounded in each replicate would be used.

The following is an example of the practical use of a rather complicated system of confounding.

Example 12.3. Campbell and Edwards (1954, 1955) have described a large-scale trial of the effect of semen diluents on conception rate in the artificial insemination of cattle. The experiment was a 2^4 factorial, the sixteen treatments consisting of either citrate or phosphate buffer with all combinations of sulfanilamide, streptomycin, and penicillin. The work was done at four centers and each treatment was replicated twice at each center. This last point need not concern us particularly here.

A central idea of the design from which the need for confounding arose was that each sample of semen, that is each collection from each bull, was split into two parts and different treatments used on the two parts. The aim was that by making comparisons of treatments within samples from the same bull, precision would be increased (see the remarks about paired units in § 3.2).

A design was therefore required for four factors in blocks of two units. Some confounding of main effects cannot be avoided when the block size is so small and in the second of their papers Campbell and Edwards explain the system of balanced partial confounding used. Main effects were estimated in six out of eight replicates and the two-factor interactions in four out of eight replicates.

TABLE 12.5

SOME SYSTEMS OF CONFOUNDING 2^n EXPERIMENTS

Number of Factors	Number of Units per Block	Contrasts Confounded	Number of Two-Factor Interactions Confounded
4	4	*AD*, *ABC*, *BCD*	1
5	8	*ABC*, *ADE*, *BCDE**	0
	4	*AB*, *CD*, *ACE*, and all products*	2
6	16	*ABCD*, *ABEF*, *CDEF*	0
	8	*ACE*, *BDE*, *BCF*, and all products	0
	4	*ABEF*, *CDEF*, *ACF*, *AB*, and all products	3
7	16	*ACEG*, *BDE*, *BCF*, and all products	0
	8	*ABC*, *ADE*, *BDF*, *DEFG*, and all products	0
8	16	*ABC*, *ADE*, *CDF*, *DEFG*, and all products	0
	8	*ABH*, *ADE*, *AFG*, *BDG*, *CH*, and all products	1

Designs in which only one contrast is confounded, i.e., the 2^n system in blocks of 2^{n-1} units, are usually best based on the highest-order interaction. They have not been listed.

* Five replicates of the designs marked by an asterisk can be arranged to give a symmetrical system of partial confounding.

Each pair of treatments was used at one center for about two weeks and the observation was the conception rate for the cows serviced during this period. After two weeks using one pair of treatments at a center, a new pair was used for the next two weeks, and so on until the eight pairs forming the first replicate had been completed. The second replicate was then run in a similar way.

Campbell and Edwards's paper should be consulted for an account of the special problems of analysis connected with this experiment. Their general conclusion about the confounding used was that although a substantial decrease in standard deviation was achieved, the fact that main effects were estimated only in 3/4 of the replicates meant that the actual gain in precision on main effects was slight, whereas there was a loss of information about two- and three-factor interactions. This illustrates an important general conclusion about experiments in which there is partial confounding of the contrasts of interest. If sufficient reduction in standard deviation is not obtained, there may be a decrease in precision, at any rate unless more complex schemes of analysis are used, involving what is called the recovery of interblock information. This snag could not arise in the simpler designs in which the contrasts of main interest are unconfounded in all replicates. Here, any decrease in residual standard deviation leads to an increase in the precision of the estimated effects.

(iv) Factors at More than Two Levels

Satisfactory and fairly simple systems of confounding are available when all factors are at 3 or 4 levels, or less usefully, at 5, 7, . . . levels. For instance, with factors at three levels, there are designs for two or three factors in blocks of three units confounding two-factor interactions, designs for up to four factors in blocks of nine units confounding only three-factor interactions, and so on. Good accounts are given by Yates (1937), Cochran and Cox (1957, Chapter 6), Kempthorne (1952, Chapter 16) and Davies (1954, Chapter 9).

Confounding when some factors are at three levels and others at two tends to be less straightforward, although there are simple designs for the following situations: 3×2^2 in blocks of 6, the number of replicates being a multiple of 3; $3^2 \times 2$ in blocks of 6, the number of replicates being even; 3×2^3 in blocks of 6, the number of replicates being a multiple of 3; $3^3 \times 2$ in blocks of 6. Mixtures of factors at four and at two levels can be dealt with easily. The references given above should be consulted for details.

(v) Confounding in Split Plot Experiments

A further development that is quite often important concerns split plot experiments (§ 7.4), that is, experiments in which information is sacrificed about the main effect of one factor so that greater precision is likely to be attained for the other contrasts. This can be regarded as a special case of confounding in which the defining contrast is a main effect. There is this difference of emphasis, however, that in a split plot experiment we do usually attempt to estimate the main effect in question, whereas in the confounding of interactions we normally regard information about the confounded contrast as lost.

When the subplot treatments are themselves factorial, there are numerous possibilities for confounding additional contrasts in order to reduce the number of subplots within one whole plot. Details will be found in the books referred to in (iv).

(vi) Confounding in Fractionally Replicated Experiments

Confounding and fractional replication, although closely related in their mode of derivation, have quite distinct purposes and are not to be confused with one another. Confounding aims at achieving increased precision by the use of smaller blocks, whereas fractional replication decreases the minimum size of the experiment needed. It is, however, possible to use both devices together, and this is particularly useful in those fractionally replicated experiments in which the total number of experimental units is quite large.

For instance in the application of fractional replication to tissue culture work discussed in Example 12.1 it was not possible to deal with more than 32 experimental units at a time, yet the whole experiment called for 128 units. The experiment must, therefore, be run in four distinct sections at different times. The effect of any systematic variation in the experimental conditions between sections is eliminated if there is a block design with 4 blocks of 32 units, such that the contrasts confounded are 3 or more factor interactions. The National Bureau of Standards tables give a system of confounding with these properties. Had one not existed, it would have been necessary to have randomized the 128 treatments into sets of 32 and to have accepted any resulting increase in error.

Methods for confounding fractional factorials are given in the tables of designs referred to in § 12.2. Suitable systems of confounding by no means always exist, since if any contrast is confounded so are all its aliases, and in highly fractionated designs this means that many contrasts are confounded.

(vii) Double Confounding

All the arrangements discussed so far correspond to randomized blocks, in that the experimental units are grouped in one direction only. Sometimes it is required to set out an experiment with two-way elimination of error, i.e., in a form analogous to a Latin square.

If the maximum possible number of both rows and columns is less than the number of treatments, it will be necessary to confound one set of contrasts with rows and a different set with columns. This procedure is called double confounding and the resulting design, if square, is called a quasi-Latin square. Yates (1937) has discussed these designs and in particular has given arrangements for the following cases: 2^3 experiment in two 4×4 squares; 2^5 experiment in an 8×8 square; 2^6 experiment in an 8×8 square; 3^3 experiment in a 9×9 square; 3^4 experiment in a 9 × 9 square; 3 × 3 × 2 experiment in a 6 × 6 square. See also the general textbooks referred to above.

(viii) Summing up

Confounding is a device for consideration in the following fairly clearly defined circumstances. A factorial experiment, fractionally replicated or not, may be such that a straightforward use of randomized blocks or Latin squares is unlikely to be effective in reducing error, because of the large number of units per block that are needed. In such cases, if the use of small blocks is to be retained, we have seen that some contrasts must be confounded with block differences. Normally we try to arrange that the contrasts confounded are high-order interactions, thereby ensuring that main effects and low-order interactions have a random error arising from the uncontrolled variation within blocks only.

In deciding whether or not confounding should be used, it is sometimes worth examining results of previous experiments on similar material. Statistical methods are available for assessing the gain in precision that would have resulted, or did result, from confounding.

If confounding is to be used, a choice has to be made of the block size, of the contrasts to be confounded, and, in experiments with more than one replicate, as to whether confounding should be partial or total. Usually the permissible block sizes are severely restricted, for example to powers of two, and the appropriate block size will then often be fairly clear. In other cases, a difficult decision may be needed between, on the one hand, having a small block size, with some important contrasts confounded, and, on the other, having larger block size, less confounding of interesting contrasts, but somewhat increased error. In choosing contrasts for confounding, attention is paid to prior knowledge about the system as well as to the practical importance of the different contrasts. For example, in a design in which say one or two two-factor interactions have to be confounded, these should be chosen with some care as being the particular two-factor interactions of least interest or most likely to be negligible.

Total confounding of two-factor, and usually three-factor, interactions should be avoided if possible, i.e., different sets of contrasts should be confounded in different replicates.

The analysis of observations from confounded experiments is straightforward, except in complicated systems of confounding with factors at more than two levels. The general principle is that the formulas of estimation for ordinary factorial experiments are used, each contrast being estimated only from those replicates in which the contrast is not confounded.

SUMMARY

Consider a factorial experiment with a moderate or large number of factors. Even a single replicate will require a large number of experimental units and the precision thereby obtained for the estimates of the contrasts of most importance (for example main effects) may be greater, and possibly very much greater, than is really called for by the nature of the investigation. It is, therefore, sensible to look for arrangements in which only a selection of the possible factor combinations is investigated.

When all factors are at two levels, a series of designs are available in which $\frac{1}{2}$, $\frac{1}{4}$, . . . of the factor combinations are investigated. These are called *fractional factorials*. In them certain contrasts, main effects and if possible two-factor interactions, are estimated independently of one another, but other contrasts get mixed up together. That is, the quantity used to estimate a particular contrast at the same time estimates one or more other contrasts, and it becomes impossible to separate them. The contrasts mixed up in this way are called *aliases* of another. We try to choose a design with sufficient experimental units to give the precision we want, and with a system of aliases such that the effects we are most interested in have as aliases things likely to be negligible.

When all factors are at three levels, $\frac{1}{3}$rd, $\frac{1}{9}$th, . . . replicates can be set up in a similar way. For quantitative factors, it will often be better to use so-called composite designs.

Confounding is a technique for arranging factorial experiments in blocks, with fewer units per block than there are treatments in the experiment, or in square or rectangular designs with fewer rows and columns than there are treatments. Information is sacrificed about certain selected contrasts, usually high-order interactions, in order that the remaining contrasts can be estimated with the higher precision that should result from the use of smaller blocks.

Confounding is thus the device for factorial experiments that is analogous to the use of balanced incomplete block and related designs for experiments in which the treatments have no factorial structure.

REFERENCES

Bainbridge, J. R. (1951). Factorial experiments in pilot plant studies. *Ind. and Eng. Chemistry*, **43,** 1300.

Box, G. E. P., and K. B. Wilson. (1951). On the experimental attainment of optimum conditions. *J. R. Statist. Soc.* B, **13,** 1.

——— and J. S. Hunter. (1957). Multifactor experimental designs. *Ann. Math. Statist.*, **28,** 195.

Brownlee, K. A., B. K. Kelly, and P. K. Loraine. (1948). Fractional replication arrangements for factorial experiments with factors at two levels. *Biometrika*, **35,** 268.

Campbell, R. C., and J. Edwards. (1954). Semen diluents in the artificial insemination of cattle. *Nature*, **173**, 637.

——— and ———. (1955). The effect on conception rates of semen diluents containing citrate or phosphate buffer with all combinations of sulphanilamide, streptomycin, and penicillin. *J. Agric. Sci.*, **46**, 44.

Cochran, W. G., and G. M. Cox. (1957). *Experimental designs*. 2nd ed. New York: Wiley.

Daniel, C. (1956). Fractional replication in industrial research. *Proc. 3rd. Berkeley Symp. on Math. Statist. and Prob.*, **5**, 87. Berkeley: University of California Press.

Davies, O. L. (editor) (1954). *Design and analysis of industrial experiments*. Edinburgh: Oliver and Boyd.

Kempthorne, O. (1952). *Design and analysis of experiments*. New York: Wiley.

National Bureau of Standards (1957). *Fractional factorial experiment designs for factors at two levels*. Washington, D.C.: N.B.S. Applied Mathematics Series.

Yates, F. (1937). *Design and analysis of factorial experiments*. Harpenden, England: Imperial Bureau of Soil Science.

CHAPTER 13

Cross-Over Designs

13.1 INTRODUCTION

If most of the uncontrolled variation in an experiment is due to qualitative variations in the external conditions under which the experimental units respond, the technique of balancing by randomized blocks, Latin squares, and related designs is likely to be effective. Examples are those agricultural field trials in which local variations of fertility predominate, analytical work where variations between observers, sets of apparatus and between days are the largest sources of error, and animal experiments in which systematic differences between animals from different litters account for an appreciable proportion of the total variation.

On the other hand, if most of the variation arises from the peculiarities of individual experimental subjects, balancing into blocks on the basis of obvious spatial, temporal, or similar groupings is unlikely to be satisfactory, at any rate by itself. For example, in many types of experiment on human or animal subjects, very substantial variations may remain even after grouping on obvious features such as age, sex, etc.

One procedure in such cases is to characterize the individual experimental subjects by one or more skilfully chosen concomitant observations, these then being used as a basis either for blocking or for the calculation of adjusted treatment means of the type described in Chapter 4. Another method is artificially to divide the subjects into sections, each section then being regarded as a separate unit, the original subject forming a block. Thus in § 3.2 we mentioned the device of dividing a clover plant into two halves by cutting along the tap root, the experiment being set out in blocks of two units each, using incomplete block methods or confounding when appropriate. Example 11.2, in which different areas on a cow are used as sites for injections, illustrates the same sort of idea; in this case incomplete block designs were used.

No new problems arise if the different sections into which the subject is divided respond independently of one another, in the sense that the observation obtained on one section does not depend on the treatment allotted to other sections. This is certainly true in the first example just mentioned, since after subdivision the two parts of the plant are quite separate; the only special consideration here is as to whether conclusions from divided plants of this sort will apply to ordinary plants.

Sometimes, however, although it is not practicable to divide the original experimental units into independent sections in this way, it is possible to use each subject (plant, animal, etc.) as an experimental unit on several occasions. This will usually eliminate the effect of much of the variation between different individuals. For example, as mentioned in § 2.4, a nutritional experiment comparing the effect of different diets on the milk yield of cows is often best arranged by having each cow fed on a sequence of diets rather than by regarding a cow as an experimental unit and keeping to a fixed diet for each cow. The special problems that this sort of design raises are connected with the possibility that the effect of a treatment applied in one period may extend into subsequent periods and hence that one of the assumptions underlying the preceding discussions, see § 2.4, is untrue. That is, the observation obtained on one experimental unit, i.e., on one individual in one period, may depend in part on the treatment applied to other units, i.e., to the same individual in preceding periods.

Arrangements in which different treatments are applied to the same subject in different periods are called *cross-over* (or change-over) *designs* and in this chapter we consider some of the special problems that they raise.

13.2 EXPERIMENTS WITHOUT CARRY-OVER EFFECTS

It may happen in using a cross-over design that it can reasonably be assumed that the complication mentioned in the previous section does not occur, i.e., that each treatment has no effect in periods subsequent to the one in which it is applied. For example, in the experiment on cows described above it might be decided to separate each experimental period by a period in which a standard treatment is applied, these further periods being sufficiently long for any effect of the earlier treatments to have dissipated. The disadvantage of this is, of course, that, if the total time for which a subject can be under observation is fixed, the number of experimental units that can be formed from each individual is reduced and the precision per individual is lowered.

If the absence of carry-over can be assumed, no really fresh problems of design arise. It would usually be reasonable to expect some time trend and hence to use Latin squares, or related designs, to balance out simultaneously variations between individuals (subjects, animals, etc.) and between times. Thus in an experiment comparing the effect on milk yield of three diets A, B, and C a section of the experiment would consist of a randomized 3×3 Latin square such as

	Experimental Period		
	1	2	3
Cow 1	B	A	C
2	C	B	A
3	A	C	B

The three cows in each group would be chosen to have, so far as possible, lactation curves of similar slopes over the period of the experiment, the whole experiment being built up of a series of such squares. In this type of experiment the number of occasions on which each animal can be used is severely limited so that Youden squares, lattice squares or double confounding may be needed if a considerable number of treatments are to be compared.

A somewhat different situation arises if each subject can be used a large number of times, or even indefinitely. This is so in some psychological experiments, and in certain bioassay procedures, for example in a histamine assay technique described by Schild (1942); see for statistical discussion of these bioassays Finney and Outhwaite (1956). Another example is the industrial experiment of Example 2.6, where a possible carry-over effect of an oiling treatment might, under certain circumstances, be assumed negligible and where a large number of observations can be obtained from a single set of machinery. In these cases sufficient precision may be obtained from one, or at any rate a small number, of individuals.

We then have the problem of arranging say one rather long sequence of treatments in suitable order. This can usually be done satisfactorily by the method of randomized blocks. That is, the periods are divided into fairly short sections and each section used as a randomized block. Thus with four treatments we might have

$$B\ C\ A\ D \mid A\ C\ B\ D \mid D\ A\ B\ C \mid \ldots$$

Incomplete block techniques can be used where appropriate. If the experiment is a small one, and if the time variation over the period of the experiment is likely to be a smooth trend, the special type of design described in § 14.2 may be more appropriate than the method of randomized blocks.

Suppose that the absence of carry-over effects has been ensured by the method, mentioned above, of including intermediate sections of sufficient length in which a standard treatment is applied to the individuals. The experiment then leads to valid estimates of the effect of differences between the treatments in the system investigated, i.e., under conditions in which treatments are being changed frequently. Often, however, our practical interest is in what would happen if treatments were applied continuously to experimental subjects, e.g., in comparing the milk yield of cows fed continuously on diet A with the yield that would have been obtained with continuous feeding on diet B. Thus the danger has to be watched that additional precision may be attained by the cross-over method, only at the cost of distorting the comparisons required.

13.3 CROSS-OVER DESIGNS WITH A LIMITED NUMBER OF PERIODS PER INDIVIDUAL

In the experiments on milk yield discussed above, not more than three or four treatments can usually be applied to each animal, and it is necessary to have a number of animals in the experiment in order to get a satisfactory arrangement. This is quite a common situation and so we consider first designs in which several individuals are used simultaneously.

We must introduce an assumption concerning the effect that a treatment applied in one period may have in subsequent periods. The simplest such assumption is the one mentioned in § 2.4 that the observation obtained on an individual in a particular period is

$$\begin{pmatrix}\text{a quantity depending}\\ \text{only on the individual-}\\ \text{period combination}\\ \text{and independent of}\\ \text{the treatments}\end{pmatrix} + \begin{pmatrix}\text{a quantity}\\ \text{depending on}\\ \text{the treatment}\\ \text{applied in}\\ \text{that period}\end{pmatrix} + \begin{pmatrix}\text{a quantity}\\ \text{depending on}\\ \text{the treatment}\\ \text{applied in the}\\ \text{preceding period}\end{pmatrix}.$$

Thus each treatment is characterized by two quantities, one expressing its *direct effect* in the period in which it is applied, the other giving its *residual effect* in the following period. In this case a natural design is one in which each treatment follows each other treatment the same number of times. At the same time we shall want a Latin square design to ensure that each treatment occurs equally often in each period and on each subject.

Williams (1949) has given suitable arrangements, obtained as follows: Suppose first that the number of treatments n is even. Write down as the first row the numbers

$$1 \quad 2 \quad n \quad 3 \quad n-1 \quad 4 \ldots$$

in which the sequence $1, n, n-1, \ldots$ alternates with the sequence $2, 3, 4, \ldots$. Thus with $n = 6$, the first row is 1 2 6 3 5 4.

The remaining rows of the square are now obtained from the first by successive additions of 1, using the rule that numbers above 6 are to have 6 subtracted from them. The final square for $n = 6$ is thus

$$\begin{array}{cccccc} 1 & 2 & 6 & 3 & 5 & 4 \\ 2 & 3 & 1 & 4 & 6 & 5 \\ 3 & 4 & 2 & 5 & 1 & 6 \\ 4 & 5 & 3 & 6 & 2 & 1 \\ 5 & 6 & 4 & 1 & 3 & 2 \\ 6 & 1 & 5 & 2 & 4 & 3 \end{array}$$

The important property of this square, which is a consequence of the particular choice of the first row, is that not only is it a Latin square, but that also each treatment follows each other treatment just once. Thus treatment five follows treatment three in the first row, treatment six in the second row, and so on.

To use the design, groups of six subjects, likely to show similar time trends and residual effects, are taken. Each group is assigned to one such square and each subject assigned randomly to a row of the square. The six numbers in the row determine the treatments to be applied to the subject in the six periods, i.e., rows correspond to subjects, columns to periods.

To obtain similar designs when the number of treatments is odd, it is necessary to consider pairs of squares simultaneously. For one square the first row is taken to be

$$1 \quad 2 \quad n \quad 3 \quad n-1 \quad 4 \quad n-2 \ldots$$

and for the other square of the pair, the first row is this reversed. Thus with $n = 5$, the first rows are

$$\begin{array}{ccccc} 1 & 2 & 5 & 3 & 4 \end{array}$$

and

$$\begin{array}{ccccc} 4 & 3 & 5 & 2 & 1 \end{array}$$

so that the full squares are

$$\begin{array}{ccccc} 1 & 2 & 5 & 3 & 4 \\ 2 & 3 & 1 & 4 & 5 \\ 3 & 4 & 2 & 5 & 1 \\ 4 & 5 & 3 & 1 & 2 \\ 5 & 1 & 4 & 2 & 3 \end{array} \qquad \begin{array}{ccccc} 4 & 3 & 5 & 2 & 1 \\ 5 & 4 & 1 & 3 & 2 \\ 1 & 5 & 2 & 4 & 3 \\ 2 & 1 & 3 & 5 & 4 \\ 3 & 2 & 4 & 1 & 5 \end{array}$$

These have the property that each treatment follows each other treatment just twice.

To use the squares, make rows correspond to subjects and columns to periods. Thus with five treatments, a multiple of ten subjects is necessary and each subject must be capable of receiving five different treatments. The ten subjects in each pair of squares should be likely to have similar residual effects, and so far as possible the period effects should be constant within each square of five subjects.

The general property of these designs is that each treatment follows each treatment except itself the same number of times. Therefore the mean of all observations on, say, treatment 1 is influenced by the residual effects of all treatments except the first. This implies that the difference between the mean observations on two treatments is not, as it stands, an estimate of the appropriate true direct treatment effect, if different residual effects are present. Thus if treatment 1 has a large positive residual effect and the other treatments do not, the mean observation on treatment 1 is depressed relative to the other treatments. This can be corrected by the calculation of adjustments analogous to those used for balanced incomplete designs.

Example 13.1. Williams (1949) has given the following example. Samples of pulp suspension at varying concentrations were beaten in a Lampén mill to determine the effect of concentration on the properties of the resulting sheets. Observations of the condition of the mill after each beating indicated that certain concentrations of pulp had an effect on the mill which might affect the next beating. Hence a design balanced for residual effects was used. With six treatments, six runs, and six periods per run, the design and observations (burst factors) of Table 13.1(*a*) were obtained: the rows of the 6×6 square given in the text above have been randomized.

The formulas for estimating the treatment and residual effects, and for finding the precision of the estimates, are rather complicated and will not be given here. Full accounts are given in Williams's paper and by Cochran and Cox* (1957, § 4.6*a*).

The effect of the process of adjustment can be judged from Table 13.1(*b*), which gives the unadjusted and adjusted direct and residual effects. For instance the unadjusted mean for T_5 is just the mean of the six observations on this treatment and is 57.75. After adjustment for residual effects, this becomes 57.98. The adjustments to the direct effects are quite small in this example; they would, of course, be greater if large residual effects were present, and they also tend to be greater in smaller squares. In this particular example both direct and residual effects are statistically significant.

If it had not been necessary to apply adjustments, the standard error of the difference in direct effect between two treatments would have been that for the difference of two means each of six observations. In fact, the standard error is slightly greater than this because of random errors in the adjustments that are applied to correct for the presence of residual effects.

Various modifications to the design of Table 13.1 may be worth considering. In some situations the sum of the direct and residual effects

* In the second edition only.

associated with a treatment is of interest for estimating the response that would be obtained if the treatment were applied continuously.* The

TABLE 13.1

A 6 × 6 EXPERIMENT WITH CARRY-OVER EFFECTS

(a) *Plan and Observations*

	Period 1	2	3	4	5	6
Run 1	T_3:56.7	T_6:53.8	T_2:54.4	T_5:54.4	T_4:58.9	T_1:54.5
2	T_5:58.5	T_3:60.2	T_4:61.3	T_6:54.4	T_1:59.1	T_2:59.8
3	T_1:55.7	T_4:60.7	T_5:56.7	T_2:59.9	T_6:56.6	T_3:59.6
4	T_2:57.3	T_1:57.7	T_6:55.2	T_4:58.1	T_3:60.2	T_5:60.2
5	T_6:53.7	T_5:57.1	T_1:59.2	T_3:58.9	T_2:58.9	T_4:59.6
6	T_4:58.1	T_2:55.7	T_3:58.9	T_1:56.6	T_5:59.6	T_6:57.5

(b) *Unadjusted and Adjusted Effects*

	Direct Effects		Residual Effects	
	Unadjusted	Adjusted	Unadjusted	Adjusted
T_1	57.13	57.20	0.92	0.37
T_2	57.67	57.62	−0.48	−0.28
T_3	59.08	59.19	0.24	0.65
T_4	59.45	59.23	−1.62	−1.33
T_5	57.75	57.98	1.22	1.40
T_6	55.20	55.06	−0.26	−0.82

efficiency of estimation of these combined effects is increased by adding a further period at the end for each subject in which the final treatment is repeated. Another method of achieving this is to add a final period in which a uniform control treatment is applied to all subjects. This will be particularly suitable if the residual effects are of intrinsic interest. A final possibility is to add a preliminary period in which the same treatment is applied as in the first experimental period, but in which no observation is made. If the main expense is in making the observation, rather than in applying the treatments, this preliminary period will increase the efficiency with which the treatment effects are estimated.

More complicated designs of this form are needed if the residual effect is suspected to extend for more than one period. It is then natural to look for Latin squares in which each treatment follows each ordered *pair* of

* This assumes that the residual effect when, say, treatment A follows itself is the same as when A is followed by a different treatment, and this may not be true.

other treatments the same number of times. In a further paper Williams (1950) has described such designs; their analysis is rather complicated.

In experiments of the type we are considering, the number of observations that can be made on each subject is severely limited. Hence if a considerable number of treatments is involved, the use of confounding and of balanced incomplete block and related designs is natural. The development of such designs, allowing for the complication of residual treatment effects, has been considered by Patterson (1951).

Finney (1956) has given a careful account of the various types of assumption on which the analysis of this sort of experiment can be based. Patterson (1950) has described a method of analysis that is particularly appropriate when the main difference between subjects is a variation in the slope of their response curves in time.

13.4 DESIGNS WITH A LARGE NUMBER OF OBSERVATIONS PER SUBJECT

Instead of there being a number of subjects with each treatment occurring at most once on each subject, it may happen that there is only one subject, or perhaps two, but that a large number of treatment applications may be made on it. Problems similar to those discussed in § 13.3 will still arise, if there is the possibility of a carry-over of treatment effects from one period to the next.

One example is the textile oiling experiment of Example 2.6, where one set of machinery (subject) is used and where the whole experiment is specified by the sequence of treatments which this subject receives. Another is the bioassay technique (Schild, 1942) mentioned earlier in this chapter, and a third is concerned with the local depression of the milk yield of a cow following injection with insulin, one cow being used and injections of varying amounts and kinds following one another in a long sequence.

One procedure in such cases is to divide the experiment into several separate sections, with a gap between them, to call each section a "subject" and then to apply the methods of § 13.3. Example 13.2 has been set out in this way. Often, however, this is not the best approach, since if the experiment is planned as a single sequence, information about the carry-over effects of treatments is supplied by all observations except the first, so that a single sequence design tends to have higher precision than the design of § 13.3 with the same number of treatment applications.

Finney and Outhwaite (1956) and Sampford (1957) have discussed suitable designs and their papers should be consulted for details.

13.5 SOME OTHER POSSIBILITIES

In the previous sections, we have assumed that any carry-over of treatment effects from one period to another persists for one period, or perhaps two, but is definitely limited in extent. Sometimes other forms of carry-over of treatment effects may seem natural and in such cases an appropriate design has to be found either intuitively or occasionally by theoretical analysis.

For example, it may be suspected that the first treatment that a subject receives has a substantial effect on all its remaining responses, but that there is no other carry-over.

Or, it may be thought that there is a characteristic effect associated with each treatment application, depending in some fairly simple way on the number of times that particular treatment has been applied to the subject before. Pearce (1957) has considered yet another possibility, namely that the units are divided into sets and that each treatment has a direct effect on the unit to which it is applied and an equal carry-over on all units in the same set. In all cases such as these, the general method is to set up a mathematical formula to represent the observations and then, if possible, to construct a design that will allow the quantities in this mathematical expression to be estimated as simply and precisely as possible. Specialist advice may be necessary to do this.

In rotation experiments, treatments are applied in sequence to a subject, e.g., a plot, and there may be a carry-over effect of one treatment into subsequent periods; in these experiments, however, the treatments are applied in definite predetermined sequences and it is the response of the subject to the sequences of treatments that is of interest, rather than the response to the individual treatments. Rotation experiments raise some specialized problems and will not be considered here; see Cochran (1939).

Another possibility is that one may be interested in the result of applying a treatment continuously to a subject for a considerable period. Here, the simplest method is to hold the treatment constant for each subject, and to construct from the observations on each subject (*a*) a measure of the average response of the subject (the mean of all observations on the subject), and (*b*) a measure of the rate of increase or decrease with time of the observation (linear regression coefficient on time). These are then analyzed separately. Stevens (1949) has given an interesting example of an experiment on a perennial crop, coffee, where from yearly observations on each plot he constructed measures of both (*a*) and (*b*), and also a measure of the amplitude of the twoyearly periodic variation in yield. The effect of the treatments on these three aspects of the yield pattern was then analyzed and interpreted separately.

SUMMARY

In situations where a substantial portion of the uncontrolled variation arises from the peculiarities of the individual physical objects (subjects, animals, etc.) that form the experimental units, one method of increasing precision is to use each object as an experimental unit several times. That is we arrange, for example, that each animal receives several treatments, rather than being kept on the same treatment throughout.

Latin squares are a natural design to use in such cases. Sometimes, however, there is the complication that the observation obtained in one period depends on the treatments applied to the object in previous periods, as well as on the current treatment. Special Latin squares are useful when there is such a carry-over effect of treatments.

There are some difficulties connected with these designs. Although a substantial increase in precision may be obtained, the treatment effects when each treatment is applied only for a short period may not be the same as those when each treatment is applied for a long period. If it is the latter that are of interest, the increase in precision will have been attained by answering the wrong question. A second difficulty is that the full analysis of observations, when carry-over effects are present, is a bit complicated.

REFERENCES

Cochran, W. G. (1939). Long-term agricultural experiments. *J. R. Statist. Soc. Suppl.*, **6**, 104.

——— and G. M. Cox. (1957). *Experimental designs*. 2nd ed. New York: Wiley.

Finney, D. J. (1956). Cross-over designs in bioassay. *Proc. Roy. Soc.* B, **145**, 42.

——— and A. D. Outhwaite. (1956). Serially balanced sequences in bioassay. *Proc. Roy. Soc.* B, **145**, 493.

Patterson, H. D. (1950). The analysis of change-over trials. *J. Agric. Sci.*, **40**, 375.

——— (1951). Change-over trials. *J. R. Statist. Soc.* B, **13**, 256.

Pearce, S. C. (1957). Experimenting with organisms as blocks. *Biometrika*, **44**, 141.

Sampford, M. R. (1957). Serially balanced designs. *J. R. Statist. Soc.* B, **19**, 286.

Schild, H. O. (1942). A method of conducting a biological assay on a preparation giving gradual responses illustrated by the estimation of histamine. *J. Physiol.*, **101**, 115.

Stevens, W. L. (1949). Analise estatistica do ensaio de variedades de cafe. *Bragantia*, **9**, 103.

Williams, E. J. (1949). Experimental designs balanced for the estimation of residual effects of treatments. *Austr. J. Sci. Res.* A, **2**, 149.

——— (1950). Experimental designs balanced for pairs of residual effects. *Austr. J. Sci. Res.* A, **3**, 351.

CHAPTER 14

Some Special Problems

14.1 INTRODUCTION

In this chapter we deal briefly with some miscellaneous problems which do not fit conveniently into the previous chapters. In § 14.2 some designs are described for use when the uncontrolled variation is expected to consist predominantly of a smooth trend. Section 14.3 mentions an important class of problems in which a theoretical calculation is possible to find which of a number of systems of observations will lead to estimates of maximum precision. The next section contains an account of designs for finding optimum conditions. The final section deals with the special problems of assays, in particular of bioassays.

14.2 TREND-FREE SYSTEMATIC DESIGNS

Sometimes we have a small number of experimental units and a substantial part of the uncontrolled variation is expected to consist of a smooth trend, for example a trend in space or time. In the ordinary way this situation is dealt with by one of the methods of Chapters 3 and 4, either (*a*) by randomizing the allocation of treatments and adjusting the treatment means for the trend by the method of § 4.3, or (*b*) by using the method of randomized blocks to deal with variations other than the trend, then calculating adjusted treatment means to correct for the trend, or (*c*) by using the method of randomized blocks to control most of the effect of the trend, putting in the first block the units occurring first in space or time, and so on. The first two methods allow the form of the trend to be estimated directly.

If the experiment is a moderate-sized one with, say, more than four replicates of each treatment, these methods will usually be satisfactory. In smaller experiments, however (particularly if the trend is curved), the first two methods tend to give imprecise results. This is because the randomization is quite likely to throw up an arrangement which is markedly unbalanced with respect to the trend. The third method is free from this defect, but is not satisfactory if it is desired to estimate as precisely as possible the form of the trend.

We need therefore a new sort of design for such situations that will enable both the trend and the treatment effects to be estimated simply with maximum precision and will take advantage of our knowledge about the expected form of the uncontrolled variation.

Example 14.1. Consider an experiment to investigate the effect on a textile process of changing the relative humidity and suppose that three relative humidities, 50, 60, and 70 per cent, are to be used. To obtain uniform experimental units a suitable quantity of raw material would be taken and thoroughly mixed and then divided into say nine batches to form nine experimental units. The first batch would be processed at one relative humidity in the first period, the second batch at a different relative humidity in the second period, and so on. Superimposed on any treatment effects and on the random variations remaining, is likely to be a smooth trend due to the ageing of the material. It would often be of interest to estimate this trend explicitly, as well as to set up the experiment so that the trend has little or no influence on the estimates of the treatment effects.

It can be shown that if the treatments are used in the systematic order

$$T_{60} \quad T_{50} \quad T_{70} \quad T_{70} \quad T_{60} \quad T_{50} \quad T_{50} \quad T_{70} \quad T_{60}$$

then

(*a*) any linear trend in time has no effect on the treatment comparisons;

(*b*) any curvature (i.e., second-degree component) in the trend has no effect on the linear part of the treatment effect, that is, on the comparison of 70% relative humidity with 50% relative humidity;

(*c*) there is some mixing of the estimate of the curvature of the trend and of the curvature of the treatment response curve, i.e., of the mean response to 50% and 70% relative humidity minus that for 60% relative humidity. Statistical analysis is necessary to sort these two out.

To see what is meant by the first property imagine that the observations consisted of a pure linear trend, in which values 1, 2, 3, . . . , 9 are obtained. Then the mean observation on T_{50} is $\frac{1}{3}(2 + 6 + 7) = 5$, the mean observation on T_{60} is $\frac{1}{3}(1 + 5 + 9) = 5$, and the mean observation on T_{70} also is 5. That is, this particular linear trend, and in fact any linear trend, leaves the estimates of the treatment differences unaltered.

It can be shown that designs chosen to have these properties of balance with respect to the trend simplify, and maximize the efficiency of, the estimates of both trend and treatment effects.

Cox (1951, 1952) has set out the method for selecting a design and for analyzing it. A few examples are given in Table 14.1. Box and Hay (1953) have described an ingenious and flexible method of dealing with similar experiments in which the treatments correspond to a set of at least two quantitative factors. In such cases there is enough freedom in the choice of factor levels to allow sufficient randomization to be brought into the design. In the simpler situations considered by Cox, such as that of Example 14.1, there is however no randomization, other than possibly in naming the treatments. This would be a defect in a moderate or large sized experiment, in which randomization can be relied on to ensure freedom from systematic error and a correct estimate of the residual error. In a single very small experiment, absence of randomization is much less serious, since we are in any case bound to rely to some extent

TABLE 14.1

SOME DESIGNS BALANCED OR NEARLY BALANCED AGAINST TREND

No. of Treat-ments	Units per Treat-ment	Degree of Trend	Design
2	3	2	$T_1\,T_2\,T_2\,T_1\,T_1\,T_2$
2	4	2	$T_1\,T_2\,T_2\,T_1\,T_2\,T_1\,T_1\,T_2$
3	3	2	(*a*)* $T_2\,T_1\,T_3\,T_3\,T_2\,T_1\,T_1\,T_3\,T_2$
			(*b*)* $T_2\,T_3\,T_1\,T_2\,T_1\,T_3\,T_2\,T_3\,T_1$
3	4	3	$T_1\,T_2\,T_3\,T_3\,T_2\,T_1\,T_1\,T_2\,T_3\,T_3\,T_2\,T_1$
4	3	2	$T_2\,T_3\,T_4\,T_1\,T_1\,T_4\,T_3\,T_2\,T_2\,T_3\,T_4\,T_1$
4	4	3	$T_1\,T_2\,T_3\,T_4\,T_4\,T_3\,T_2\,T_1\,T_1\,T_2\,T_3\,T_4\,T_4\,T_3\,T_2\,T_1$

* The design (*a*) is the one mentioned in Example 14.1 and should be used when the comparison of T_3 with T_1 is of particular interest, as when three equally spaced levels of a quantitative factor are involved. In other cases design (*b*) should be used.

on our prior knowledge of the uncontrolled variation when the number of experimental units is small. Provided that the assumption that the uncontrolled variation is formed from a trend plus random variation is sensible, and that the design adopted is unlikely to correspond to a pattern in the uncontrolled variation, there can be no reasonable objection to the absence of randomization.

Similar principles can be used to pick out systematic Latin squares that are, for example, such that the treatment effects are uninfluenced by particular patterns of diagonal variation across the square. This is useful when the rows and columns of the square correspond to quantitative factors between which there may be a linear × linear component of interaction.

14.3 OPTIMUM ALLOCATION

In this section an outline is given of recent work on a class of problems which, although not all concerned with comparative experiments of the sort we have been discussing up to now, do concern the most efficient distribution of experimental effort. The situations can be described like this: observations may be made on a number of experimental set-ups and the quantities so obtained are known to be expressed statistically in terms of certain unknown parameters. Which set-ups should be observed in order to estimate the parameters with maximum precision?

Example 14.2. In certain experiments on alloys cast from high-purity metals, and in other fields too, the following situation arises. The experimental units are arranged in runs, corresponding to production runs, and within each run they are ordered in time. Thus with each run corresponding to a 20-lb melt, it might be possible to cast, in sequence, four 5-lb ingots. Suppose that the treatments for comparison are different concentrations of the alloying elements and that when an ingot has been cast, the concentration of an alloying element in the remaining melt can be increased by the addition of fresh material but cannot be decreased.

In other words, if we think of just one alloying element, the factor level cannot decrease as we go from unit to unit within a run. This is a severe restriction on the type of design that can be used, since it excludes, for example, any Latin square. For if T_1, T_2, T_3, and T_4 denote successively increasing concentrations of the alloying element, any design which secures that each concentration occurs equally frequently in each run and in each period is bound to involve reversals of order within some runs.

Hence, if it is desired to set up a design in which run and period effects are eliminated from the error of the treatment comparisons, something less direct than a Latin square has to be used. Suppose for simplicity that there is just one alloying element occurring at two levels A_{-1} and A_1. Then with four periods per run, a design must consist of a mixture of sequences of the following five types:

Type				
I	A_{-1}	A_{-1}	A_{-1}	A_{-1}
II	A_{-1}	A_{-1}	A_{-1}	A_1
III	A_{-1}	A_{-1}	A_1	A_1
IV	A_{-1}	A_1	A_1	A_1
V	A_1	A_1	A_1	A_1

The difference between treatments is mixed up both with run and with period differences, unless

(*a*) all the sequences are of Type III, in which case the difference between treatments is completely identified with differences between periods, or

(*b*) one-half the sequences are of Type I and one-half of Type V, when the difference between treatments is completely identified with differences between runs.

Therefore if it is desired to eliminate variations between periods and between runs, a mixture of sequences of the five types must be used.

We can now formulate the basic problem. For any design formed from a mixture of sequences, we can, in theory, adjust the estimated difference between treatments for the lack of balance with runs and periods, and we can also find the standard error of the resulting adjusted estimate. For a given total number of observations, which of the mixtures of sequences minimizes the standard error, i.e., leads to an estimate of the treatment effect of maximum precision? This is a mathematical problem and the solution can be shown to be to have collections of eight production runs, and within each collection one run of Types I and V and two each of the other types, i.e.,

$$\begin{array}{cccc}
A_{-1} & A_{-1} & A_{-1} & A_{-1} \\
A_{-1} & A_{-1} & A_{-1} & A_{1} \\
A_{-1} & A_{-1} & A_{-1} & A_{1} \\
A_{-1} & A_{-1} & A_{1} & A_{1} \\
A_{-1} & A_{-1} & A_{1} & A_{1} \\
A_{-1} & A_{1} & A_{1} & A_{1} \\
A_{-1} & A_{1} & A_{1} & A_{1} \\
A_{1} & A_{1} & A_{1} & A_{1}
\end{array}$$

The mathematical discussion and the extension to designs with several factors have been given by Cox (1954).

The general comment to be made on this example is that if practical considerations severely restrict the arrangement of treatments, it will, if it is important to make the most economical use of the experimental material, be necessary to evaluate theoretically the standard error of the estimated treatment effects for all admissible arrangements and to choose the arrangement which minimizes the standard error. Of course in particular cases it may well be better to adopt a simpler, but less efficient, procedure. Thus in Example 14.2 it would be possible to hold the treatment constant within each run, i.e., to use a whole run as an experimental unit. This would avoid the complication of using a special design, but elimination of the effect of variation from run to run would, of course, no longer be possible.

The next two examples deal with a rather different type of situation where, although there are no alternative treatments under comparison, a theoretical calculation of precision is possible for different experimental set-ups.

Example 14.3. There are a number of situations in which it is desired to estimate the density of particles which are distributed randomly in some medium, for example dust particles in space, bacteria or blood cells in suspension, etc. The direct method is the counting of particles in known small volumes of medium, but this is often tedious. A neat method of avoiding direct counting is based on the following fact. Suppose that the particles are distributed completely randomly through the medium: this requires for example that the proportion of the volume of medium occupied by particles should be negligible, so that there is no "crowding," and that the particles should be uncharged, so that there the particles have no electrostatic effect on one another. Then the mean number of particles in a volume v of original medium is

$$-2.303 \log_{10} \text{ (proportion of a large number of volumes } v \text{ that contain no particles).} \quad (1)$$

Thus if we take a large number of volumes v and observe for each whether or not it contains particles, the density of particles can be estimated. In some cases the determination of whether or not a particular volume contains particles is easy; with bacteria, however, it will be necessary to assume that growth occurs on any plate that contains at least one bacterium.

The problem of design is to decide what volume v of medium should be used in each trial, e.g., what dilution of the original suspension should be taken. If rather a large volume v is taken, nearly all the trial volumes will contain particles, and equation (1) will lead to an estimate with very low precision; the precision will also be low if the trial volume v is too small. It can be shown that maximum precision of the volume v is chosen to contain on the average 1.6 particles (Fisher, 1951, p. 219; Finney, 1952, p. 573).

This result is no use as it stands, since if we knew what volume contains on the average 1.6 particles, we should know the concentration of particles and the estimation that we are considering would be pointless. However it may be shown that very nearly maximum precision is attained if the mean number of particles in the test volume is between 1 and $2\frac{1}{2}$ and that reasonably high precision is retained if the number is between $\frac{1}{2}$ and $3\frac{1}{2}$. Thus if a prior estimate is available correct to within a factor of 2 or 3, the method can be used.

If no such prior value is known, two procedures are available. The one widely used is to make observations at each of a series of volumes, say $v_0, 2v_0, 4v_0, 8v_0, \ldots$ (series of volumes progressing by factors of 4 or 10 are also sometimes used), chosen to be sure of covering the optimum value. The disadvantage of this is that it will usually happen that the observations at several of the volumes give little information about the density under estimate. The second procedure is to do a small preliminary trial with a range of volumes in order to estimate the single volume at which the main series of observations should be made. Obviously this cannot be done if the whole experiment must be laid out at one time.

Example 14.4. Andrews and Chernoff (1955) have discussed the following problem connected with the estimation of the virulence of a strain of bacteria. There are available 30 test animals and 10 ml of material containing this strain of bacteria in suspension. It is thought that the concentration of bacterial organisms in this suspension is about four organisms per ml and that the probability that a dose of one organism applied to a test animal leads to a response is about 1/5. This latter probability is to be estimated more accurately.

Part of the suspension must be used for a plate count to estimate the concentration and the remainder allocated among the test animals to determine virulence.

To deal with this problem it is again necessary to set up a statistical model to represent the observations. Briefly this is that there is a certain unknown chance α that one organism administered to a test animal will lead to a positive response, whereas if several organisms are administered these act independently and no response is obtained if and only if no response would have been obtained with each organism separately. Further it is assumed that the numbers of organisms in a certain nominal dose have a particular random distribution called the Poisson distribution. From these assumptions it can be shown that the probability that a nominal dose of D organisms produces a negative response is $e^{-\alpha D}$, where e is the base of natural logarithms.

Now if the concentration of the suspension of organisms were known and if there were an unlimited supply of test animals and organisms, we should have a situation mathematically the same as that of Example 14.3 and the optimum dose would be $1.6/\alpha$, where the best initial estimate, 0.2, would be used, for α, indicating an optimum dose of 8 organisms. We have, however, the additional features described above, and these appreciably complicate the mathematical solution. Andrews and Chernoff show that optimum procedure is approximately the following:

(*a*) Every test animal receives the same dose.

(*b*) The fraction of the suspension given to the test animals is approximately the smaller of $1/(1 + \sqrt{\alpha})$ and $1.6s/(\alpha\lambda)$, where α is the best initial estimate of the probability that a single organism will lead to a positive response, s is the number of test animals, and λ is the best initial estimate of the total number of organisms in the available suspension. The initial estimates of α and λ do not have to be very precise.

For example, with the values of 0.2 and 40 for α and λ, and with $s = 30$, $1/(1 + \sqrt{\alpha}) \simeq 0.69$ and $1.6s/(\alpha\lambda) \simeq 6$, so that 69% of the suspension should be divided into 30 equal portions for application to the test animals and 31% of the suspension used for a plate count. The estimation of α from the resulting observations is straightforward.

Equivalent problems occur in estimating the unknown parameters in theoretical relationships. For example, in a diffusion problem we might be able to measure the concentration of diffusing solute at different distances x into the diffusing medium and after different times t from the start of the diffusion process. Theory gives a relation between concentration and x, t involving as unknown parameters the diffusion coefficient and a constant defining boundary conditions. The problem is to decide at what values of x and t to observe concentration in order to get estimates of the unknown parameters of maximum precision.

It is characteristic of these problems that a statistical model has to be assumed to represent the situation and that the conclusions about the optimum arrangement would be wrong if the model were seriously wrong. For instance, in Example 14.4 the idea that all animals should receive the same nominal dose would certainly not be acceptable if there were a serious doubt about the formula $e^{-\alpha D}$ for the chance that a nominal dose D produces a negative response. We can sometimes cover this possibility, however, by including additional parameters in the model to represent departure from the form first assumed. A second characteristic feature is that, except in especially fortunate cases, initial estimates have to be available for one or more of the quantities in the experiment. In the example just discussed such estimates are needed for the quantities α and λ. It is a matter for investigation in each case how precise these initial estimates need to be; if values of sufficient precision are not available, a small part of the experimental material may be used to obtain rough estimates and from these a suitable method of using the remaining material then determined. Elfving (1952) and Chernoff (1953) have given general mathematical discussions of these problems, assuming that any necessary prior estimates are available. The details of special cases are liable to be complicated.

14.4 SEARCH FOR OPTIMUM CONDITIONS

In the designs discussed in previous chapters it has been assumed that the object is the estimation of the differences between alternative treatments. Quite often, however, particularly in technological experiments, the ultimate object is the selection of the treatment or set of treatments which are in some sense the best. It may, however, still be necessary to estimate all relevant treatment effects, both in order to get added understanding of the system being investigated and also because the criterion determining the optimum conditions may be rather imprecisely defined. Thus if there are several treatments which differ only slightly with respect to, say, total cost per unit yield, we may decide to use the process which is best by some other standard, e.g., estimated long-run reliability (assuming that no allowance for this has been included in the calculation of cost). To be able to do this satisfactorily, it will be necessary to estimate, not just which is the best treatment according to the different criteria, but also the amounts by which other treatments depart from the optimum. Even in such cases, however, the need does arise for designs that will select a group of treatments, or range of experimental conditions, for fuller investigation.

Therefore it is of interest to examine procedures where the emphasis is on the estimation of optimum conditions, rather than on the estimation of treatment differences. First suppose that there is one quantitative factor v that can be varied. To determine the value of v for which a response y is maximized (or minimized) it will usually be best to proceed in two stages. In the first, a rough determination is made of the region within which the maximum lies. This can be done either by setting out an experiment with, say, about six equally spaced levels covering the range of interest or by proceeding in steps. In one form of the latter method, observations are taken at two levels v_0 and $v_0 + \Delta$; if the first level gives higher response than the second, observations are continued at a level $v_0 - \Delta$; if the second gives the higher response observations are taken at $v_0 + \Delta$ whereas if both levels give about the same response, further observations are taken at both $v_0 - \Delta$ and $v_0 + \Delta$. This procedure is continued, the aim being at each step to shift the treatment into a region in which higher response is obtained. There are clearly many ways in which such a procedure can be formalized, but it is probably best to leave scope for special considerations, such as the ease with which the treatment may be changed, the amount of random variation relative to the slope of the response curve, and so on.

When the general form of the response curve has been established, the second stage of the experiment consists of a three-level (or possibly four-level) experiment centered on the suspected position of the maximum, the lower and upper levels being chosen as far apart as possible, subject to the response curve being reasonably parabolic in the region covered by the three levels. A second-degree equation (§ 6.8) is then fitted to the results of the second stage and any relevant results from the first stage, and the maximum of the fitted equation determined by plotting or by differentiation. A more detailed discussion is given by Hotelling (1941).

It is assumed here that the response curve is approximately of a special mathematical form within the range investigated in the second stage. Kiefer and Wolfowitz (1952) have discussed a very interesting procedure which requires only weak assumptions about the response curve; however in most practical cases it seems likely that more precise estimates of the position of the maximum can be obtained by judicious use of the parabolic approximation.

Box and Wilson (1951), see also Davies (1954), have suggested a procedure for use when there are several quantitative factors, $v_1, v_2, \ldots,$ that can be varied independently. Their idea is first to determine, by a two-level factorial experiment, the direction near the starting point in which the response surface rises most steeply. For example in Fig. 14.1,

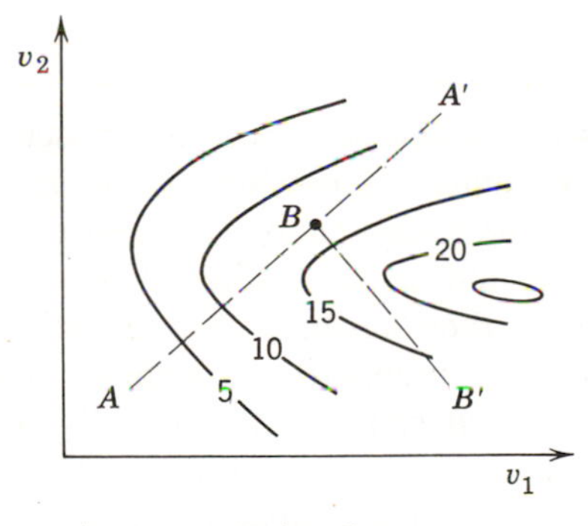

Fig. 14.1. Paths of steepest ascent.

if it is decided to start near A, an experiment will show that changes in the direction AA' produce maximum increments in yield. Since this line is at about 45° to the axes representing the two factors, equal changes in v_1 and in v_2 should be made, and this is done in the next stage, the optimum position B along the line AA' being determined. A further two-level experiment centered at B determines the line BB' of steepest ascent from B and so on. When the optimum has been determined approximately, a three-level experiment is used to investigate the shape of the response surface near the optimum. A difficulty is that the direction of the line of steepest ascent depends on the units in which v_1 and v_2 are measured. A full account, with examples, of this method is given by Box and Wilson and by Davies.

When the different treatments are qualitative, the problems of design are different, although stage-by-stage investigation is again likely to be useful. Apart from the particular problems of selection in genetics (Cochran, 1951), little work seems to have been done on experimental designs for such situations, although there have been a number of theoretical investigations of so-called decision rules for selecting best treatments.

14.5 ASSAYS

An assay is a system of observations intended to give a number measuring a particular property of some experimental material, for example the potency of a drug, the strength of an insecticide, the mean fiber diameter of a consignment of wool, and so on. All such measurement involves in the last analysis comparison with some sort of standard. We can, however, for practical purposes distinguish between assays in which (*a*) a standard material similar to the experimental material is used and the observations on it compared with the observations on the experimental material and (*b*) no such standard material is used directly. We have already touched on this distinction in the discussion of Example 1.4.

For instance, in early attempts at standardizing insulin, potency was measured directly by the amount necessary to produce a certain response in mice, giving "animal units" of potency. This is the second type of assay and was found to be unsatisfactory because of variations between different batches of mice and between different times. In particular the comparison of different workers' results was unreliable. The introduction of international standard preparations against which experimental preparations could be tested enabled this difficulty to be avoided; in each trial, some mice receive the standard, others the experimental drug. The set-up then has the form of a comparative experiment of the type we have been discussing in this book, the object being to assess the unknown by direct comparison with the standard. Variations between batches of mice affect unknown and standard almost equally, and reproducible measures of potency are thereby obtained.

As another example, consider the measurement of the mean fiber diameter of wool. One quick method is to form, in a controlled way, a plug of fibers and to measure the rate of flow of air through the plug when a fixed pressure drop is applied across it. From the rate of flow, the pressure drop, and the mass of wool in the plug, a quantity Q can be calculated that is closely correlated with the mean fiber diameter. In practice a value for estimated mean fiber diameter is derived from Q by the use of a calibration curve, obtained by testing plugs made of fibers of known mean diameter, determined by a more laborious optical method. This is an example of the second type of assay with no direct use of standards. That is, the standards, the fibers of known mean diameter, are used to construct a calibration curve, but this curve is assumed to remain fixed and the standards are not introduced directly in each individual determination. The calibration curve has in fact been shown to be quite constant and this method is satisfactory.

If this were not so and the relation between Q and mean fiber diameter tended to shift in time, the first type of assay could be used. A series of standard lots of wool would be taken each of known mean fiber diameter, the series covering a range of fiber diameters. To test a new batch, a group of two or three standards would be selected whose diameters are likely to straddle the diameter of the new batch. The experimental quantity Q would be determined for all, and the diameter of the experimental batch then estimated by an obvious graphical or statistical technique.

These examples are typical of a wide range of procedures used both in the physical and the biological sciences. Physical applications tend on the whole to be like the second example, where a fixed relation can be assumed between the quantity measured experimentally and the quantity it is required to estimate, or where the observation obtained can be used directly to measure the property of interest. The use of standards is restricted to initial calibration and occasional recalibrations. Biological applications tend on the whole to be like the first example, where it is desirable to introduce a standard explicitly into the determination. A very thorough and authoritative account of the statistical problems of design and analysis connected with this sort of assay has been given by Finney (1952), and only a few salient points will be mentioned here.

The assay of insulin mentioned above is *direct* in that the observation for each experimental unit is the quantity of drug, unknown or standard, necessary to produce a certain response. No new problems of design are involved in such an assay, the methods for the comparison of two or more alternative treatments being used. Most bioassays, however, are *indirect*, in that for one reason or another it is necessary to apply a predetermined dose to each experimental unit and to observe the resulting response, quantitative or qualitative, not to increase gradually the dose until a fixed response is obtained. This raises some new problems.

The simplest design for indirect assays is the so-called symmetrical four-point assay. Two concentrations of the standard are used, one, say, λ times the other, and two concentrations of the unknown, the one also λ times the other. The concentrations should be adjusted so that the observations on corresponding doses of standard and unknown are expected to be about the same; prior knowledge is, of course, necessary to do this. The ratio λ between the two concentrations should be chosen to be as large as possible, subject to the requirement that the relation between observation and log dose should be linear over the whole range of concentrations of both drugs in the assay. (Considerable prior investigation of the dose-response is assumed to have been done in setting up the method in the first place.)

An experiment is now arranged to compare these four treatments, if possible increasing precision by some of the techniques discussed in previous chapters. This experiment will in effect be a 2×2 factorial experiment, one factor being the type of drug, the other the level of dose, high or low. From the results the relative potencies of unknown and standard can be estimated quite easily.

Notice that we are interested not in the treatment effects in their own right, but in using the treatment effects to estimate a special type of relation existing between the treatments. The assumptions on which the assay is based are two. First the relation between response and log-dosage must be linear over the range used; this cannot be checked from the data. Secondly the response curves for the two preparations must be parallel if the hypothesis that the test preparation is equivalent to an unknown dilution of the standard is to be maintained. This can be checked from the data.

If little reliable information is available about the shape of the response curve, a six-point assay will be suitable, in which each preparation occurs at three levels, equally spaced on the log-dosage scale. From the results, tests of both linearity and parallelism are possible; this corresponds to an ordinary factorial experiment with a quantitative factor, in which three levels are necessary to get an estimate of the curvature of the response curve. More than three levels will not normally be advisable in routine work.

Since the final experiment is set out as a comparative trial, all the methods discussed in earlier chapters, incomplete block techniques, confounding, cross-over designs, etc. are from time to time found useful in designing assays. Finney's book should be consulted for details and examples, as well as for careful discussion of the various special complications that arise.

SUMMARY

A number of unrelated special topics have been discussed briefly:

(*a*) Trend-free systematic designs are available that are sometimes suitable for small experiments in which a smooth trend is superimposed on the treatment effects;

(*b*) the optimum allocation of observations can sometimes be determined when a theoretical calculation of precision is possible for each of a number of possible set-ups, any of which may be observed;

(*c*) special methods are available for determining optimum conditions, i.e., the experimental conditions under which a suitable quantity is maximized;

(*d*) a few of the problems connected with assays, particularly bioassays, have been outlined.

REFERENCES

Andrews, F. C., and H. Chernoff. (1955). A large-sample bioassay design with random doses and uncertain concentration. *Biometrika*, **42**, 307.

Box, G. E. P., and W. A. Hay. (1953). A statistical design for the efficient removal of trends occurring in a comparative experiment, with an application in biological assay. *Biometrics*, **9**, 304.

——— and K. B. Wilson. (1951). On the experimental attainment of optimum conditions. *J. R. Statist. Soc.* B, **13**, 1.

Chernoff, H. (1953). Locally optimal designs for estimating parameters. *Ann. Math. Statist.*, **24**, 586.

Cochran, W. G. (1951). Improvement by selection. *Proc. 2nd. Berkeley Symp. on Math. Statist. and Prob.*, 449. Berkeley: University of California Press.

Cox, D. R. (1951). Some systematic experimental designs. *Biometrika*, **38**, 312.

——— (1952). Some recent work on systematic experimental designs. *J. R. Statist. Soc.* B, **14**, 211.

——— (1954). The design of an experiment in which certain treatment arrangements are inadmissible. *Biometrika*, **41**, 287.

Davies, O. L. (editor) (1954). *Design and analysis of industrial experiments.* Edinburgh: Oliver and Boyd.

Elfving, G. (1952). Optimum allocation in linear regression theory. *Ann. Math. Statist.*, **23**, 255.

Finney, D. J. (1952). *Statistical method in biological assay.* London: Griffin.

Fisher, R. A. (1951). *Design of experiments.* 6th ed. Edinburgh: Oliver and Boyd.

Hotelling, H. (1941). Experimental determination of the maximum of a function. *Ann. Math. Statist.*, **12**, 20.

Kiefer, J., and J. Wolfowitz. (1952). Stochastic estimation of the maximum of a regression function. *Ann. Math. Statist.*, **23**, 462.

General Bibliography

Nearly all the general ideas described in this book are contained in or stem from Sir Ronald Fisher's pioneer work, the first five chapters of his book (Fisher, 1935) and an early paper (Fisher, 1926) forming excellent introductions to the subject. Other introductory accounts are those of Finney (1955) and Wilson (1952), the latter giving a very readable introduction to a number of aspects of scientific research.

There are several books which, although nonmathematical, give a reasonably detailed account both of the construction of designs and of the full methods for statistical analysis. Cochran and Cox (1957) have given numerous detailed plans as well as worked numerical examples of the statistical analysis of biological experiments. In a book by a group of workers at the I.C.I. (Davies, 1954) there is a very thorough account of the design and analysis of industrial experiments, including modern developments. Quenouille (1953), as well as giving more general material, has discussed long-term experiments and various complications of analysis that can arise. Federer (1955) has in particular given an exhaustive account of the analysis of the more complicated designs. Yates's book (Yates, 1937) should be consulted for a well-exemplified account of the design and analysis of factorial experiments.

The reader who knows something of the theory of statistics should read the book by Kempthorne (1952) for a detailed exposition of the theory of the main types of design. Mann (1949) has dealt particularly with the elegant mathematics underlying the construction of incomplete block designs and systems of confounding.

The more mathematical research on new designs is abstracted in *Mathematical Reviews*; other current work of interest will be found in *Biometrics* and in *Applied Statistics*.

Cochran, W. G., and G. M. Cox. (1957). *Experimental designs.* 2nd ed. New York: Wiley.

Davies, O. L. (editor) (1954). *Design and analysis of industrial experiments.* Edinburgh: Oliver and Boyd.

Federer, W. T. (1955). *Experimental design.* New York: Macmillan.

Finney, D. J. (1955). *Experimental design and its statistical basis.* London: Cambridge University Press.

Fisher, R. A. (1926). The arrangement of field experiments. *J. Min. of Agric.*, **33**, 503. Reprinted in *Contributions to mathematical statistics.* New York: Wiley, 1950.

Fisher, R. A (1935). *Design of experiments*. Edinburgh: Oliver and Boyd (and subsequent editions).

Kempthorne, O. (1952). *Design and analysis of experiments*. New York: Wiley.

Mann, H. B. (1949). *Analysis and design of experiments*. New York: Dover.

Quenouille, M. H. (1953). *Design and analysis of experiment*. London: Griffin.

Wilson, E. B. (1952). *Introduction to scientific research*. New York: McGraw-Hill.

Yates, F. (1937). *Design and analysis of factorial experiments*. Harpenden, England: Imperial Bureau of Soil Science.

Appendix

TABLES OF RANDOM PERMUTATIONS AND RANDOM DIGITS

Tables A.1 and A.2 give random permutations and Table A.3 gives random digits. Examples of the use of the tables will be found in § 5.2.

TABLE A.1

Random Permutations of 9

5 5 6 7 1	4 3 3 7 3	8 7 4 6 3	9 7 4 9 4	9 2 2 8 8	2 7 9 3 5	8 3 1 9 4
4 1 2 8 2	7 1 1 2 9	9 5 7 8 2	8 9 3 6 6	1 7 7 2 4	4 8 5 7 3	3 7 4 5 6
9 3 3 2 9	8 8 8 4 5	2 4 6 1 6	3 6 7 7 8	7 4 4 7 1	7 3 2 8 6	6 1 2 2 2
7 9 7 4 3	5 5 2 9 2	1 6 5 3 5	7 8 5 1 9	5 1 9 1 3	6 5 1 4 9	2 9 8 7 8
1 6 9 6 5	6 9 4 3 6	4 3 9 2 9	5 1 8 2 3	8 3 3 3 2	8 9 6 1 2	4 5 7 6 9
6 4 4 3 6	2 4 6 8 1	7 9 3 4 1	6 2 6 4 2	2 9 8 5 9	9 2 4 2 8	9 6 9 8 1
8 7 8 1 7	1 2 5 6 8	3 1 2 9 8	4 4 1 8 7	6 5 1 6 7	5 4 3 5 1	1 4 3 1 7
3 2 1 9 4	3 6 7 5 7	6 8 8 7 7	2 5 9 5 1	3 8 5 4 6	3 6 7 9 4	5 2 5 4 5
2 8 5 5 8	9 7 9 1 4	5 2 1 5 4	1 3 2 3 5	4 6 6 9 5	1 1 8 6 7	7 8 6 3 3
7 4 6 1 5	9 2 2 2 9	2 8 1 7 3	2 4 2 1 9	2 4 8 3 1	2 6 5 4 8	8 4 9 4 2
9 3 8 3 2	1 1 1 9 8	9 4 9 5 4	8 8 8 8 6	7 7 5 4 6	5 3 2 7 6	9 3 8 2 1
1 6 3 4 7	6 5 8 4 5	6 1 7 1 9	5 2 5 6 3	8 5 7 5 5	6 9 9 8 1	3 6 7 9 7
6 8 2 8 4	4 8 7 8 6	5 7 5 4 5	9 6 7 5 8	5 9 9 7 7	8 5 3 3 5	6 9 4 6 9
4 1 4 7 8	2 3 9 3 4	4 2 2 3 6	4 7 4 2 5	6 3 3 6 9	1 7 8 5 4	4 5 2 1 4
2 9 1 9 3	7 9 6 6 2	1 6 4 6 1	7 9 9 7 4	1 8 4 1 8	9 2 7 9 3	1 8 3 5 5
5 5 5 5 1	3 7 4 7 7	8 5 8 9 2	1 5 1 3 2	9 6 2 8 4	3 8 1 1 9	5 7 1 3 3
8 2 9 2 9	8 6 5 5 3	7 9 6 8 8	3 1 6 9 7	4 1 6 9 3	4 4 6 6 2	7 2 6 8 8
3 7 7 6 6	5 4 3 1 1	3 3 3 2 7	6 3 3 4 1	3 2 1 2 2	7 1 4 2 7	2 1 5 7 6
9 7 7 5 5	9 9 9 3 8	9 8 6 1 7	5 8 6 1 2	1 9 8 3 3	3 1 7 7 3	7 6 6 5 5
3 8 1 7 2	6 2 7 1 6	4 1 3 4 2	3 6 2 4 3	2 6 1 2 8	8 8 6 2 7	8 9 7 4 7
4 3 4 2 7	7 3 1 7 2	1 5 4 8 6	6 2 1 6 1	7 8 5 1 7	5 9 1 3 6	3 1 2 3 1
5 9 2 8 3	3 7 5 8 9	2 9 1 7 1	2 3 8 3 4	3 5 9 9 9	7 2 3 4 1	5 7 1 7 8
1 6 5 1 1	5 6 4 4 1	7 3 7 2 3	4 7 3 8 8	9 3 2 5 6	6 6 9 5 9	9 8 9 1 2
6 2 8 3 6	8 4 6 2 5	5 2 2 6 8	9 1 7 5 6	4 7 4 6 4	1 7 4 6 4	1 2 8 8 6
2 4 9 6 4	1 8 3 5 4	3 6 5 9 4	8 5 9 7 9	8 1 6 8 1	4 5 5 9 5	2 4 5 9 4
8 5 6 9 9	2 5 2 6 7	8 7 8 3 9	1 9 4 2 5	6 4 7 4 5	2 3 2 8 2	6 3 3 2 3
7 1 3 4 8	4 1 8 9 3	6 4 9 5 5	7 4 5 9 7	5 2 3 7 2	9 4 8 1 8	4 5 4 6 9
7 4 9 8 7	9 7 1 7 1	9 2 3 8 7	7 8 5 3 5	5 1 6 4 9	7 8 6 1 8	2 9 7 3 4
5 6 1 1 2	6 4 6 1 4	5 9 1 2 8	2 4 6 8 7	7 3 7 6 1	5 1 7 4 1	9 3 4 7 7
4 9 3 5 6	1 1 8 4 8	3 5 4 9 3	3 6 1 2 3	2 6 8 7 7	4 5 3 8 5	8 5 9 5 1
3 3 2 2 8	5 2 3 2 2	7 3 8 6 9	4 1 8 6 1	1 9 2 3 6	3 9 5 7 7	1 2 8 1 2
2 1 4 9 4	4 6 2 8 3	2 7 6 5 1	5 7 3 1 2	9 8 4 1 3	6 3 1 2 9	6 1 5 8 8
9 7 5 4 5	3 9 7 9 9	1 4 2 3 4	6 9 7 4 4	3 2 5 2 2	8 4 2 6 3	5 6 3 6 3
6 2 6 3 9	8 8 5 5 5	8 6 7 7 2	9 3 4 5 8	8 7 9 9 4	9 2 4 9 4	4 8 1 2 9
8 5 8 7 1	2 3 9 3 7	4 1 5 1 5	8 5 9 7 6	4 5 3 5 8	1 6 8 5 2	3 4 6 4 5
1 8 7 6 3	7 5 4 6 6	6 8 9 4 6	1 2 2 9 9	6 4 1 8 5	2 7 9 3 6	7 7 2 9 6
8 4 6 8 6	2 1 9 9 7	2 2 1 8 9	5 1 9 2 4	5 2 6 2 8	1 6 8 8 3	8 1 9 4 1
9 9 4 5 8	4 4 8 7 8	8 7 5 9 7	3 6 4 7 7	3 8 5 3 6	4 4 6 7 7	6 6 8 7 8
6 6 3 1 1	6 8 3 1 9	7 5 7 5 5	6 5 1 8 5	2 4 3 8 2	5 1 4 3 6	4 9 7 8 6
7 3 7 7 2	7 3 6 2 2	3 8 9 4 6	4 7 2 6 9	7 9 7 4 1	3 8 2 6 5	3 5 3 1 4
2 8 9 3 4	1 5 5 5 1	5 4 3 6 4	7 8 7 5 3	9 5 8 6 5	8 2 7 9 2	5 3 4 3 5
3 7 2 6 9	8 6 4 6 3	4 1 8 2 1	1 9 6 4 8	4 7 2 1 3	6 3 5 5 1	2 2 6 9 9
5 1 8 4 5	9 9 1 8 4	1 9 4 3 2	8 2 8 9 6	6 3 4 9 9	2 7 1 2 4	9 8 2 6 2
4 5 5 2 7	3 2 7 3 6	9 3 2 1 8	9 3 5 1 2	1 6 9 7 7	9 5 9 1 8	7 7 1 5 7
1 2 1 9 3	5 7 2 4 5	6 6 6 7 3	2 4 3 3 1	8 1 1 5 4	7 9 3 4 9	1 4 5 2 3

Taken, by permission of the authors and publisher, from § 15.5 of *Experimental designs* by W. G. Cochran and G. M. Cox, John Wiley and Sons, New York, 1957.

TABLE A.1. RANDOM PERMUTATIONS OF 9 (*Continued*)

86224	54583	96522	49829	86656	35665	11391
22339	81837	65138	36916	37435	93252	58135
75695	42661	74496	13375	94888	54187	22882
54453	75224	59984	58594	61277	11549	44579
68182	96145	47869	74282	25544	48813	97668
19911	39419	33375	65141	48911	29938	86257
91848	17976	21743	82737	79169	66426	65724
33566	28752	82617	27463	53792	87791	33413
47777	63398	18251	91658	12323	72374	79946

65348	95382	16446	84991	43379	93459	32561
23759	61754	84591	37185	71913	69815	64492
74982	47219	67287	55662	88437	25731	55325
52296	72595	32618	62439	65621	34563	93244
47517	29868	59925	93228	29248	42642	79756
89433	36127	28869	48774	12594	76127	28673
11621	84641	91132	21857	37862	58296	47137
38164	13933	75354	19316	56155	87384	11918
96875	58476	43773	76543	94786	11978	86889

74242	12262	86522	61847	12183	97745	46147
85987	83923	11659	77293	54478	34117	65816
96459	31375	74288	99458	29617	46336	98378
17763	66748	97167	15785	65246	62598	14592
22875	59159	55946	23364	76534	25674	71954
59121	48516	69394	34929	91921	79261	33229
63616	77894	38413	56511	88892	53452	57785
41598	25437	43835	82636	47769	88923	22461
38334	94681	22771	48172	33355	11889	89633

97597	32358	11761	22663	73827	41767	89437
25775	26672	27813	48859	81989	79388	92192
72339	51744	53276	73232	38665	83826	56281
44852	49519	82455	99945	14346	95693	78849
53988	15296	64594	55577	96291	68442	37768
66146	93127	48942	67128	47773	54939	11615
39464	74935	99628	31784	62554	16174	44926
88221	67863	76187	86491	29418	37551	23374
11613	88481	35339	14316	55132	22215	65553

25665	42832	48626	42413	98197	47496	13414
81157	65718	77278	31168	37265	85672	41959
49239	36666	34347	58235	81671	32824	32873
78481	94493	56835	19396	62832	53169	54765
54572	23945	99482	93549	15344	64948	25341
32396	18177	65169	75624	76788	11583	67136
16923	57329	23711	67771	53556	98331	76698
63718	89284	11993	24957	24929	26217	98527
97844	71551	82554	86882	49413	79755	89282

Taken, by permission of the authors and publisher, from § 15.5 of *Experimental designs* by W. G. Cochran and G. M. Cox, John Wiley and Sons, New York, 1957.

TABLE A.2

RANDOM PERMUTATIONS OF 16

7	12	15	15	1	2	7	16	10	2	14	15	7	13	13	10	6	1	8	10
13	3	8	16	7	10	11	10	13	5	11	7	13	16	7	7	5	13	2	14
3	1	4	5	14	13	3	14	9	13	13	2	9	15	6	2	8	4	5	8
11	8	16	14	15	6	2	6	2	16	8	5	12	3	9	13	4	3	10	4
14	9	1	6	3	9	14	13	8	6	5	8	14	7	3	15	13	11	4	7
2	16	10	13	5	5	13	2	11	7	3	12	5	14	12	16	2	2	9	15
4	6	13	7	2	15	1	9	1	4	7	10	6	9	11	9	7	6	16	11
6	14	6	10	4	14	4	15	3	3	4	16	2	6	5	1	12	10	6	9
10	15	2	1	13	12	16	3	4	8	10	1	15	5	14	12	14	12	3	2
12	10	7	12	9	11	9	8	12	14	15	4	11	8	16	8	9	14	14	1
15	7	5	2	10	7	8	12	6	15	6	13	16	12	15	4	11	8	12	6
16	2	11	8	8	8	15	5	16	1	1	9	8	1	8	14	16	5	13	5
9	13	14	3	6	4	10	11	5	12	9	3	10	4	4	3	10	9	1	3
8	11	9	4	11	3	12	7	7	10	12	14	3	10	1	6	15	16	15	12
1	5	12	11	16	16	5	4	14	9	16	11	1	2	10	5	1	15	7	13
5	4	3	9	12	1	6	1	15	11	2	6	4	11	2	11	3	7	11	16

11	8	16	5	5	13	1	13	2	16	14	12	9	8	7	5	13	3	13	3
2	2	8	8	14	16	4	3	8	11	10	14	15	1	2	11	4	5	15	9
6	13	2	13	6	5	9	15	11	10	12	6	16	15	16	9	10	12	16	15
14	12	4	16	16	11	14	10	5	12	3	3	12	14	15	13	6	4	1	16
8	6	3	9	4	10	6	4	16	2	2	9	8	16	4	6	5	15	7	8
9	15	12	10	3	2	12	6	1	15	4	13	7	7	9	12	14	8	8	11
3	10	11	12	13	12	5	11	7	8	9	5	14	11	10	1	3	13	3	5
16	1	13	14	8	14	15	5	3	7	11	15	6	12	5	7	11	1	14	4
1	14	14	2	9	15	16	14	6	14	7	8	3	13	11	8	7	7	12	7
4	4	6	4	12	3	11	8	15	9	8	1	13	6	3	3	15	9	9	12
15	5	1	11	10	6	3	7	10	5	5	11	10	10	12	15	16	14	5	2
5	3	5	6	7	7	13	2	14	3	16	4	5	5	13	4	9	16	2	6
12	7	15	15	15	9	8	12	12	13	15	10	1	4	6	16	2	6	11	1
10	11	10	3	2	4	2	1	4	6	6	7	11	9	14	10	8	11	4	13
7	9	7	7	11	1	7	16	13	1	13	2	4	2	1	2	12	2	10	14
13	16	9	1	1	8	10	9	9	4	1	16	2	3	8	14	1	10	6	10

1	6	7	4	8	6	5	2	8	15	4	6	6	1	4	5	7	13	2	10
9	15	11	3	11	15	9	10	1	3	8	2	15	7	9	8	16	1	14	3
10	16	4	5	12	9	16	11	7	1	7	16	11	8	3	3	12	2	3	4
4	14	1	9	5	5	4	13	6	8	15	5	12	5	7	16	5	11	8	1
7	3	13	14	15	2	1	14	16	5	14	9	2	16	1	12	6	14	4	13
16	11	2	1	14	16	6	9	3	4	16	14	3	15	11	11	3	9	12	5
3	10	16	16	13	7	13	1	11	14	9	10	16	2	10	2	10	7	10	16
11	13	9	13	4	13	8	3	5	13	10	12	5	12	5	14	13	16	5	6
15	2	3	12	9	12	2	4	13	10	3	13	14	4	2	1	14	8	6	12
14	1	14	6	10	1	3	12	4	2	2	4	13	3	16	9	9	3	7	14
13	12	5	11	3	11	15	8	2	7	11	7	8	14	6	4	4	4	15	11
12	5	10	7	2	14	7	15	14	16	13	1	9	10	12	10	11	10	9	8
8	9	8	10	6	4	11	7	10	11	6	8	4	9	8	15	8	6	11	9
2	7	6	2	1	8	10	6	15	12	1	11	7	11	13	6	1	15	13	15
6	4	15	8	16	10	14	16	9	6	12	3	10	6	14	7	2	12	16	7
5	8	12	15	7	3	12	5	12	9	5	15	1	13	15	13	15	5	1	2

13	4	10	4	16	13	16	13	5	3	6	14	1	16	8	7	2	3	3	12
5	14	4	6	8	2	15	1	13	14	16	4	15	4	3	12	12	1	4	7
2	2	2	15	14	16	9	12	16	6	10	15	14	9	10	1	14	8	8	16
7	12	15	8	12	3	5	14	7	12	5	13	16	1	7	5	11	2	9	3
6	9	7	14	9	14	10	11	15	11	12	1	12	12	14	16	3	11	11	8
14	5	16	7	10	8	11	8	14	13	7	11	6	3	11	4	4	6	6	9
15	11	8	9	7	12	8	7	1	15	9	3	3	7	13	11	10	4	5	1
11	6	6	1	4	1	3	16	12	5	4	9	13	13	6	8	15	9	1	14
4	10	3	16	2	11	7	9	6	9	1	8	4	11	5	2	16	10	12	4
1	8	1	13	1	15	4	4	11	4	2	16	5	8	1	9	5	12	16	6
9	7	14	2	6	4	14	10	9	8	15	10	7	10	9	10	6	14	10	11
12	1	9	10	15	5	2	15	10	2	14	2	8	2	4	13	8	5	15	5
3	3	12	11	5	9	6	6	3	10	13	12	9	6	2	15	7	15	7	13
10	15	11	5	13	7	12	5	2	7	11	5	10	15	12	3	1	13	13	10
8	13	13	3	3	10	13	2	4	1	8	6	11	14	15	6	9	16	2	2
16	16	5	12	11	6	1	3	8	16	3	7	2	5	16	14	13	7	14	15

Taken, by permission of the authors and publisher, from § 15.5 of *Experimental designs* by W. G. Cochran and G. M. Cox, John Wiley and Sons, New York, 1957.

TABLE A.2. Random Permutations of 16 (*Continued*)

9	16	15	12	2	11	4	16	11	10	2	5	5	14	11	2	14	13	16	6
11	3	2	6	15	13	10	1	4	13	11	8	16	16	4	3	5	15	5	15
14	14	8	16	11	15	5	14	14	11	1	14	15	15	13	5	7	11	11	16
4	13	1	3	5	7	6	2	16	1	14	9	14	3	3	1	6	16	6	10
6	6	10	7	13	10	16	7	2	12	6	12	6	13	8	9	15	9	1	11
2	10	14	9	12	3	3	10	5	6	5	16	12	10	15	10	11	4	9	8
5	15	11	14	10	4	14	13	6	4	12	4	11	5	10	14	16	5	7	9
16	5	13	10	3	9	12	6	3	7	3	7	3	11	14	7	3	14	4	12
8	12	7	11	7	8	13	15	13	9	4	3	8	1	12	6	9	8	15	14
1	8	3	2	1	5	15	9	9	3	10	11	13	8	5	13	12	3	3	5
13	9	9	1	6	2	11	3	8	8	15	1	7	9	7	8	8	6	2	3
15	1	5	5	9	6	9	4	10	5	8	13	10	7	9	15	2	10	8	4
7	4	12	13	16	1	2	11	12	2	16	15	2	4	2	11	1	7	13	1
10	2	4	15	4	16	1	12	7	15	9	10	9	12	16	4	13	2	10	13
3	7	6	8	8	14	7	5	1	14	13	2	4	2	1	16	4	1	12	7
12	11	16	4	14	12	8	8	15	16	7	6	1	6	6	12	10	12	14	2

12	6	13	4	5	7	2	1	9	2	5	1	15	2	14	13	13	11	2	13
6	11	4	15	12	12	6	15	6	15	6	3	12	5	15	11	16	9	8	1
13	5	1	6	7	6	13	5	7	8	15	6	4	15	1	14	5	14	10	4
11	1	11	7	8	15	8	4	12	13	16	9	3	10	7	2	12	3	9	8
3	7	3	14	15	4	12	11	4	10	8	12	1	4	16	6	2	2	16	7
10	12	15	11	4	13	5	10	3	14	11	2	9	11	2	9	9	12	12	11
15	9	16	16	9	2	16	2	15	6	7	15	8	1	8	12	4	13	6	9
14	15	2	13	3	16	10	14	13	9	10	7	14	9	6	5	6	4	11	12
1	2	12	9	1	8	15	3	8	11	2	5	10	3	3	10	10	7	13	10
5	10	5	3	13	9	9	13	10	1	3	8	7	8	9	4	15	15	7	15
7	14	9	2	11	14	11	6	14	12	9	10	16	12	13	3	7	5	4	14
9	8	10	1	6	3	3	8	5	5	14	16	2	7	12	16	14	10	15	5
2	3	7	5	10	1	1	12	2	7	1	4	6	16	10	8	8	1	5	16
16	13	14	10	2	5	7	16	1	16	13	11	11	6	5	1	11	16	3	3
4	4	6	8	14	10	14	7	11	3	4	13	13	13	11	15	3	6	14	6
8	16	8	12	16	11	4	9	16	4	12	14	5	14	4	7	1	8	1	2

3	14	11	8	9	14	14	2	13	1	8	4	15	16	7	6	15	13	13	13
12	9	6	9	8	10	12	13	14	5	11	10	10	12	9	10	5	16	6	3
11	11	7	1	11	13	11	4	2	7	16	5	8	3	11	12	6	12	5	11
1	16	9	3	1	7	8	15	5	4	3	7	16	8	12	15	7	5	9	4
13	3	1	2	13	5	4	9	7	6	5	15	4	6	4	1	10	6	1	14
7	12	10	10	5	15	5	8	16	2	12	3	5	13	14	13	13	2	3	7
10	15	15	4	14	1	16	16	12	11	9	16	1	2	10	11	8	7	16	8
15	7	4	14	7	4	7	10	6	10	1	1	2	11	3	16	2	4	2	1
9	5	2	7	3	3	13	14	15	15	6	12	9	15	15	9	16	15	15	10
8	6	16	5	15	8	2	12	1	3	10	8	3	14	13	2	1	10	8	12
2	10	5	11	4	9	3	6	11	12	15	9	7	5	2	8	14	1	4	5
5	4	3	15	2	2	15	11	10	14	7	14	14	7	6	3	11	11	10	2
4	1	12	12	16	6	1	3	4	16	13	11	11	4	1	7	12	3	7	9
6	13	14	6	12	16	9	1	8	8	4	13	12	10	5	5	4	9	12	16
16	8	8	13	10	11	10	5	9	13	14	2	6	9	8	14	9	14	11	15
14	2	13	16	6	12	6	7	3	9	2	6	13	1	16	4	3	8	14	6

1	2	14	12	4	4	3	6	12	7	11	11	9	13	13	7	4	10	16	9
9	3	10	13	3	5	5	13	15	9	14	13	14	9	9	4	8	4	15	2
13	6	15	10	11	3	15	12	4	5	5	4	3	6	4	5	12	14	14	3
8	5	5	15	8	9	8	8	2	3	1	12	8	3	11	2	9	16	10	12
11	12	9	14	16	11	4	15	1	4	3	15	5	15	7	11	16	15	7	1
10	4	13	6	1	13	12	9	8	6	7	8	15	7	3	8	13	9	8	10
16	11	11	16	7	15	9	5	7	2	6	10	16	10	6	1	3	6	1	13
5	7	4	3	2	1	14	2	10	13	16	1	6	4	15	6	15	12	11	16
3	15	12	2	14	8	11	16	14	16	9	7	13	8	2	16	2	11	2	15
7	9	7	9	13	6	2	4	13	14	15	6	10	11	8	12	10	3	3	8
2	10	8	8	15	14	6	3	5	1	4	3	7	2	14	15	14	2	6	4
15	8	3	1	6	2	10	7	3	10	10	2	4	1	5	3	7	13	13	14
6	13	2	5	5	7	13	10	16	15	12	16	1	14	16	14	6	1	9	7
12	16	16	4	10	10	7	1	9	8	2	5	12	16	12	9	11	7	4	5
4	1	1	7	9	12	16	11	6	11	8	14	2	5	1	10	1	5	12	6
14	14	6	11	12	16	1	14	11	12	13	9	11	12	10	13	5	8	5	11

Taken, by permission of the authors and publisher, from § 15.5 of *Experimental designs* by W. G. Cochran and G. M. Cox, John Wiley and Sons, New York, 1957.

TABLE A.3

RANDOM DIGITS

12 67	73 29	44 54	12 73	97 48	79 91	20 20	17 31	83 20	85 66
06 24	89 57	11 27	43 03	14 29	84 52	86 13	51 70	65 88	60 88
29 15	84 77	17 86	64 87	06 55	36 44	92 58	64 91	94 48	64 65
49 56	97 93	91 59	41 21	98 03	70 95	31 99	74 45	67 94	47 79
50 77	60 28	58 75	70 96	70 07	60 66	05 95	58 39	20 25	96 89
00 31	32 48	23 12	31 08	51 06	23 44	26 43	56 34	78 65	50 80
01 67	45 57	55 98	93 69	07 81	62 35	22 03	89 22	54 94	83 31
24 00	48 34	15 45	34 50	02 37	43 57	36 13	76 71	95 40	34 10
77 52	60 27	64 16	06 83	38 73	51 32	62 85	24 58	54 29	64 56
36 29	93 93	10 00	51 34	81 26	13 53	26 29	16 94	19 01	40 45
94 82	03 96	49 78	32 61	17 78	70 12	91 69	99 62	75 16	50 69
23 12	21 19	67 27	86 47	43 25	25 05	76 17	50 55	70 32	83 36
77 58	90 38	66 53	45 85	13 93	00 65	30 59	39 44	86 75	90 73
92 37	51 97	83 78	12 70	41 42	01 72	10 48	88 95	05 24	44 21
28 93	48 44	13 02	49 32	07 95	26 47	67 70	72 71	08 47	26 18
09 68	01 98	80 27	49 78	56 67	49 22	13 66	61 33	53 18	36 03
61 73	92 33	89 48	20 42	32 33	79 37	68 88	44 59	35 17	97 61
82 35	37 33	53 42	52 04	16 54	08 25	48 89	57 87	59 89	96 76
39 20	77 72	55 19	66 58	57 91	38 43	67 97	52 66	45 29	74 67
51 90	71 05	82 38	37 40	94 52	24 09	35 44	37 33	35 20	65 89
97 49	53 79	17 25	02 65	77 70	88 45	53 51	63 30	89 66	42 03
73 18	91 38	25 82	29 71	56 89	86 74	68 58	75 36	93 13	33 31
17 79	34 97	25 89	01 17	67 92	62 25	54 70	52 88	28 05	61 17
97 27	26 86	17 67	59 56	95 07	49 05	70 06	70 35	21 35	26 18
56 06	63 00	07 40	65 87	09 49	70 34	67 02	33 39	04 40	01 51

Taken, by permission of University College, London, and the authors, from *Tables of random sampling numbers*, Tracts for computers, No. XXIV, by M. G. Kendall and B. Babington Smith, Cambridge University Press, 1939.

TABLE A.3. RANDOM DIGITS (*Continued*)

43 83	39 24	50 74	10 05	38 11	25 80	44 14	98 31	87 41	02 74
63 19	91 27	08 59	02 28	47 13	05 53	02 28	81 96	46 90	95 52
23 87	60 31	98 97	76 57	82 47	64 87	50 45	73 54	26 47	62 10
07 04	47 34	36 03	87 67	03 28	72 19	98 99	32 98	78 76	85 40
98 61	67 62	09 89	73 50	06 81	29 09	43 43	30 21	32 69	82 19
36 86	50 21	42 18	20 55	00 90	01 96	42 12	68 18	45 93	52 99
70 64	92 95	09 09	79 63	09 29	69 99	98 26	19 83	94 88	95 37
41 71	91 61	31 86	38 01	71 79	44 75	67 69	35 31	69 47	81 64
23 48	32 36	88 50	29 07	27 32	21 28	73 41	77 39	00 78	92 65
13 32	99 81	00 28	87 13	00 86	56 16	81 20	63 29	37 45	08 91
70 55	85 27	24 96	91 83	89 17	89 98	51 31	17 29	05 77	62 95
12 50	84 01	63 40	74 86	88 90	63 76	97 74	08 70	88 88	98 96
97 00	24 63	47 63	47 66	21 79	28 66	67 24	33 20	01 52	09 59
16 99	63 29	67 89	14 55	70 31	45 56	05 71	84 30	48 32	90 94
57 95	93 54	30 74	11 18	31 26	75 39	81 28	63 34	31 23	77 67
01 32	91 11	23 65	44 58	69 77	58 86	35 20	92 12	48 15	56 67
00 30	26 68	89 38	13 99	47 38	06 82	49 47	40 33	23 72	01 50
48 15	27 13	97 70	18 48	14 28	26 30	74 16	13 07	36 21	94 84
58 86	65 76	67 05	99 53	33 56	92 61	63 98	55 39	15 77	61 67
75 07	14 81	41 16	12 21	79 82	16 42	70 43	73 33	78 22	63 25
86 19	97 09	64 04	21 26	65 11	20 32	82 38	52 94	79 21	85 07
66 17	52 10	35 14	21 89	54 32	61 49	63 06	36 25	63 84	78 24
56 70	95 77	25 19	21 15	29 88	57 75	51 19	31 06	48 50	09 65
14 43	67 32	81 78	19 72	32 70	34 86	11 90	37 02	54 39	45 87
04 17	91 71	96 90	85 68	32 35	77 20	71 43	55 95	28 90	51 69

Taken, by permission of University College, London, and the authors, from *Tables of random sampling numbers*, Tracts for computers, No. XXIV, by M. G. Kendall and B. Babington Smith, Cambridge University Press, 1939.

Author Index

Subject Index